AF675358

Pretreatment and Bioconversion of Crop Residues

Pretreatment and Bioconversion of Crop Residues

Editor

Carlos Martín

MDPI • Basel • Beijing • Wuhan • Barcelona • Belgrade • Manchester • Tokyo • Cluj • Tianjin

Editor
Carlos Martín
Department of Chemistry
Umeå University
Umeå
Sweden

Editorial Office
MDPI
St. Alban-Anlage 66
4052 Basel, Switzerland

This is a reprint of articles from the Special Issue published online in the open access journal *Agronomy* (ISSN 2073-4395) (available at: www.mdpi.com/journal/agronomy/special_issues/Pretreatment_Bioconversion_Crop).

For citation purposes, cite each article independently as indicated on the article page online and as indicated below:

LastName, A.A.; LastName, B.B.; LastName, C.C. Article Title. *Journal Name* **Year**, *Volume Number*, Page Range.

ISBN 978-3-0365-1410-9 (Hbk)
ISBN 978-3-0365-1409-3 (PDF)

© 2021 by the authors. Articles in this book are Open Access and distributed under the Creative Commons Attribution (CC BY) license, which allows users to download, copy and build upon published articles, as long as the author and publisher are properly credited, which ensures maximum dissemination and a wider impact of our publications.
The book as a whole is distributed by MDPI under the terms and conditions of the Creative Commons license CC BY-NC-ND.

Contents

About the Editor

Carlos Martín

Carlos Martín is Full Professor of Biomass and Bioresources Utilization at Inland Norway University of Applied Sciences (INN University), and Researcher within Bio4Energy research environment at Umeå University, Sweden. He received a MSc degree (1989) from the Leningrad Forest Technical Academy, USSR, as well as a PhD degree (2002) from the University of Matanzas, Cuba, in a sandwich model with Lund University, Sweden. After postdoctoral stays at Risø DTU National Laboratory for Sustainable Energy, Denmark, and Karlstad University, Sweden, he started his independent career in 2004 at the University of Matanzas. As an Alexander von Humboldt Foundation fellow, he worked for three years at Th ünen Institute of Wood Research, Germany. Since 2013, he has been working at Umeå University, and in 2021 he started at INN University. His general research interests are focused on studies on biorefineries, biomass pretreatment and cellulosic ethanol.

Preface to "Pretreatment and Bioconversion of Crop Residues"

Crop residues are exceptionally important as an alternative to raw materials for producing energy carriers, chemicals, and new materials in a post-petroleum scenario. Next-generation bioconversion processes in future biorefineries will make possible to convert biomass components into most of the products that are produced today from fossil feedstocks, as well as to not-yet-developed products that will become basic commodities for the future society. Crop residues are a realistic feedstock option for biorefineries considering their large availability, low cost, and renewable nature.

A basic process step in biorefineries is the pretreatment, which is directed to separate biomass resources into their main components and to activate cellulose towards enzymatic saccharification. Biomass constituents are then converted into different kinds of value-added products following a concept that is analogous to the way that today's petroleum refineries produce fuels and products from petroleum.

Pretreatment research is today a hot issue, attracting attention not only from the academia, but also from the industrial sector, which looks forward to implementing the most promising methods in commercial-scale biorefineries. Pretreatment effectiveness is feedstock-dependent, and new research is required to develop efficient methods for different materials. This book covers some of the latest advances in research in developing efficient pretreatment methods and bioconversion approaches to be applied to crop residues of different nature.

The materials published in this book were written by experts in the pretreatment and bioconversion fields. I am confident that that reading the book will serve as a source of inspiration to scientists and entrepreneurs interested in making next-generation biorefineries of crop residues a commercial reality.

Carlos Martín
Editor

agronomy

Editorial

Pretreatment of Crop Residues for Bioconversion

Carlos Martín

Department of Chemistry, Umeå University, SE-901 87 Umeå, Sweden; carlos.martin@umu.se;
Tel.: +46-90-786-6099

Decreasing the dependence on fossil resources as raw materials for the production of fuels, platform chemicals, and commodities is an imperative requirement of today's industry and society in order to alleviate the threats related to climate change. Processing lignocellulosic biomass in biorefineries provides an alternative route for producing fuels and most of the chemicals that today are produced from fossil resources. Crop residues are a realistic option of lignocellulosic feedstock for biorefineries considering their large availability, low cost, and renewable nature. The generation of crop residues is increasing as a result of the expansion of the agricultural production necessary to support the increase in the global population. The harvest of cereals [1], other food crops [2], and non-food agricultural products [3] generates large amounts of residues, whose proper management poses huge challenges. Although the environmental impact of disposing crop residues by burning has not been as mediatic as that of industrial and vehicular emissions, its contribution to air pollution is an issue of major significance in many countries [4].

The bioconversion of crop residues via sugar-platform processes, in which polysaccharides are hydrolyzed to sugars for further conversion through microbial, enzymatic, or chemical processing, is a highly promising option for producing fuels and chemicals required for the sustainable development of human society. The enzymatic saccharification of cellulose is a selective approach for deconstructing biomass, but it requires a pretreatment for removing or weakening the barriers causing the inherent recalcitrance of lignocellulosic feedstocks [5]. In spite of the intensive research in the area, pretreatment of crop residues is still an open question. Pretreatment effectiveness is feedstock-dependent [6–8], and further research is required to develop efficient methods enabling the commercial operation of bioconversion processes based on crop residues of different nature.

check for updates

Citation: Martín, C. Pretreatment of Crop Residues for Bioconversion. *Agronomy* **2021**, *11*, 924. https://doi.org/10.3390/agronomy11050924

Received: 12 April 2021
Accepted: 6 May 2021
Published: 8 May 2021

Publisher's Note: MDPI stays neutral with regard to jurisdictional claims in published maps and institutional affiliations.

Copyright: © 2021 by the author. Licensee MDPI, Basel, Switzerland. This article is an open access article distributed under the terms and conditions of the Creative Commons Attribution (CC BY) license (https://creativecommons.org/licenses/by/4.0/).

This Issue

In this Special Issue, fourteen original research papers and one review covering some of the latest advances in pretreatment and bioconversion of crop residues are presented. Research results dealing with wheat straw, corn stover, sweet sorghum bagasse, hazelnut shells, oil palm empty fruit bunch, olive tree pruning biomass, soybean husks, oat husks, sugar beet pulp, rice straw, wheat bran, barley straw, as well as biomass from trees used in intercropping systems, and other residues of crop harvest and processing are discussed. Pretreatment methods such as auto-catalyzed and acid-catalyzed hydrothermal processing, steaming, alkaline methods, and different organosolv approaches are reported. Bioconversion with enzymes and microbes for producing fermentable sugars, xylitol, and biomethane are also included.

Wheat (*Triticum aestivum* L.) straw is a crop residue of high relevance for bioconversion, and hydrothermal pretreatment is one of the methods of higher interest for agricultural residues. In a comprehensive study backed with advanced analytical techniques, Ilanidis et al. clarified correlations between pretreatment conditions and critical aspects of biochemical conversion of wheat straw, including susceptibility to enzymatic saccharification, by-product formation, and inhibitory effects on enzymatic hydrolysis and ethanolic fermentation [9]. It was found that auto-catalyzed hydrothermal pretreatment is a better approach than sulfuric acid-catalyzed hydrothermal pretreatment to achieve high sugar yields from wheat straw while minimizing by-product formation.

Pretreatment of corn (*Zea mays* L.) stover, another well-studied crop residue, was also included in this Special Issue. Krafft et al. pretreated corn stover using a combination of steam refining and alkaline extraction of lignin, and they evaluated how different operational conditions affect enzymatic hydrolysis [10]. The proposed approach was found to be suitable for corn stover. According to the authors, steam refining at severities below 3.5 combined with alkaline extraction deserves being tested for other agricultural residues as well.

Organosolv pretreatment as a fractionation strategy for biorefinery applications is presented in two articles. In the first one, Mondylaksita et al. assess the feasibility of low acid-catalyzed organosolv pretreatment using ethanol as a solvent for fractionating oil palm (*Eleis guineensis* J.) empty fruit bunch (OPEFB) [11]. A high delignification degree (90%) with consequent recovery of high-purity lignin (71% purity) and highly-digestible glucan (94% enzymatic digestibility) were achieved using low sulfuric acid concentration (0.07%). The study demonstrated the possibility of deconstructing OPEFB into high-quality fractions using remarkably lower acid concentrations than the typically reported ones. In the second article, Domínguez et al. report formosolv pretreatment for fractionating biomass of Paulownia (*Paulownia elongata* × *fortunei*), which is a fast-growing tree used in intercropping agricultural systems [12]. Under optimal conditions, formosolv treatment resulted in the solubilization of 78.5% of the initial lignin and in a cellulose-enriched pulp (80%). The study included also a complete characterization allowing us to verify lignin structural changes resulting from the formosolv process.

Processing hazelnut (*Corylus avellana* L.), one of the major nut crops commercialized worldwide, generates large amounts of waste shells. Two articles in this Special Issue present novel studies on the pretreatment of hazelnut shells for biorefining them to value-added products. Rivas et al. report on the hydrothermal pretreatment of hazelnut shells at different temperatures, with further refining of the resulting liquor by membrane processing to yield substituted oligosaccharides (OS) with the purity required for food use [13]. The antioxidant activity found in the produced OS reveals their potential as bioactive components of functional food formulations, cosmetics, or pharmaceuticals, and the presented data provide the basis for assessing large-scale manufacture. In another study, López et al. evaluate the biorefining of hazelnut shells by a combination of autohydrolysis for removal hemicelluloses and delignification by either alkaline treatment or different organosolv approaches [14]. By reaching an effective fractionation of the main components, the investigated biorefinery schemes provide an innovative way for the integral valorization of hazelnut shells. Consecutive stages of autohydrolysis and acid-catalyzed organosolv resulted in 47.9% lignin removal, yielding solids of increased cellulose content (55.4%) and low content of hemicelluloses. The enzymatic hydrolysis of cellulose resulted in 74.2% conversion.

The production of cellulose nanofibers as a valorization alternative for olive tree (*Olea europaea* L.) pruning biomass (OTPB), a crop residue of relevance in Mediterranean countries, is presented by Sánchez-Gutiérrez et al. [15]. In that study, mechanical pretreatment and TEMPO-mediated oxidation were applied to OTPB, and the influence of residual lignin content on pretreatment effectiveness was evaluated. The characterization in terms of chemical composition, morphology, thermal stability, and crystallinity revealed the high potential of the produced nanofibers for application in different sectors.

A techno-economic study revealing the challenges of producing biodiesel by transesterification of castor oil with ethanol produced from lignocellulosic residues of the same plant (*Ricinus communis* L.) is reported by Rahimi et al. [16]. The oil extraction residues are subjected to alkaline pretreatment, enzymatic hydrolysis, and fermentation, and the resulting ethanol is combined with oil for biodiesel production. Although the results showed that biodiesel production is more profitable when fossil-based methanol is used, using ethanol produced from castor plant residues is a highly relevant approach because of its environmental benefits. The study on the castor biorefinery is a good reference for other non-edible oilseeds, whose potential for ethanol production has previously been reported [17,18].

Bioconversion of the large variety of crop residues available in different latitudes requires efficient cocktails of hydrolytic enzymes. This Special Issue includes four papers on assessment of in-house developed enzymes. Karp et al. report an evaluation of crude preparations of *Penicillium verruculosum* cellulases and xylanases in saccharification of acid- and alkali-pretreated sugarcane (*Saccharum officinarum* L.) bagasse, soybean (*Glycine max* L.) husks, and OPEFB [19]. Soybean husks, regardless of pretreatment method, and alkali-pretreated sugarcane bagasse were efficiently saccharified, while OPEFB resulted in low sugar yields. *P. verruculosum* B1 cellulase/xylanase preparation was the best enzyme choice for bagasse, while a combination of B1 preparation with the crude obtained by recombinant expression of *Penicillium canescens* xylanase in the *P. verruculosum* B1 host strain gave the highest hydrolysis yield for soybean husks. In another work using the *P. verruculosum* cellulase complex, Osipov et al. evaluated the hydrolysis of 69 samples of pretreated and non-pretreated agricultural and agro-industrial residues, such as wheat straw, sunflower peels, sugarcane bagasse, sugarbeet pulp, oat husks, soybean husk, and corn stover [20]. The effectivity of *P. verruculosum* cellulases for hydrolyzing cellulose-containing materials to sugars was shown.

Another two articles covering in-house developed enzymes for the bioconversion of crop residues deal with novel termite metagenome-derived cellulases and hemicellulases. In the first one, Mkabayi et al. optimize the synergistic action of two termite feruloyl esterases and an endoxylanase from *Thermomyces lanuginosus*; then, they applied them to hydrothermally pretreated and acid-pretreated corn cobs, resulting in a high production of xylo-oligosaccharides (XOS) and hydroxycinnamic acids [21]. In the second paper, Mafa et al. compare the efficacy of two alkaline pretreatment approaches, with either lime or NaOH, for sweet sorghum (*Sorghum bicolor* (L.) Moench) bagasse and corn cobs, and they assess the hydrolysis of the pretreated biomass using a holocellulolytic cocktail of termite metagenome-derived enzymes [22]. The study concluded that alkaline pretreatment was more effective for corn cobs than for sweet sorghum bagasse, and it revealed the multifunctionality of the enzyme system, which makes it suitable for the hydrolysis of hemicelluloses and amorphous cellulose. The authors propose using their holocellulolytic enzyme cocktail for the hydrolysis of crop residues in the biorefinery industry.

The bioconversion of pretreated wheat bran and rice (*Oryza sativa* L.) straw to xylitol is reported by Bedö et al. [23]. After investigating the effects of xylose concentration and aeration on xylitol production by *Candida boidinii* in a semi-defined xylose medium, models predicting the conditions leading to maximum yield and volumetric productivity were developed. The adequacy of the models was tested in fermentations of hemicellulosic hydrolysates obtained by acid pretreatment of wheat bran and rice straw. The models were successfully verified for wheat bran hydrolysate, which was found to be an outstanding substrate for xylitol production by *C. boidinii*, while the use of rice straw hydrolysate requires further research. In another article on bioconversion, Morales-Polo et al. report the anaerobic digestion of barley (*Hordeum vulgare* L.) crop residues for energy uses [24]. Substrate characterization and analysis of the anaerobic digestion process through all phases are discussed, and the potential for biogas and methane generation is determined. Using Spain as a case study, the article estimates the energy potential of biogas from barley crop residues and the expected reduction of CO_2 emissions from using this renewable energy source to replace fossil sources.

A review paper on consolidated bioprocessing (CBP) of crop residues for the production of biofuels completes this Special Issue. In the review, Olguin-Maciel et al. present an in-depth assessment of CBP, which involves not only technical issues of this innovative approach but also societal and economic aspects [25]. The role of CBP in the bioconversion of crop residues and other lignocellulosic materials to ethanol and other products in a single process and in a sustainable way is highlighted.

The guest editor cordially thanks all the authors who contributed to this Special Issue. In their contributions, the authors focus from different perspectives on the study of the pretreatment of crop residues for bioconversion. The papers compressing this Special Issue show the level of advancement achieved in this research area. However, the remaining challenges are still significant. Therefore, the current efforts for the development of technologies enabling massive implementation of commercial-scale biorefineries based on crop residues are expected to continue for the foreseeable future.

Conflicts of Interest: The author declares no conflict of interest.

References

1. Jeevan Kumar, S.P.; Sampath Kumar, N.S.; Chintagunta, A.D. Bioethanol Production from Cereal Crops and Lignocelluloses Rich Agro-Residues: Prospects and Challenges. *SN Appl. Sci.* **2020**, *2*, 1673. [CrossRef]
2. Carvalho, J.L.N.; Nogueirol, R.C.; Menandro, L.M.S.; de Bordonal, R.O.; Borges, C.D.; Cantarella, H.; Franco, H.C.J. Agronomic and Environmental Implications of Sugarcane Straw Removal: A Major Review. *Gcb Bioenergy* **2017**, *9*, 1181–1195. [CrossRef]
3. Martín, C.; Fernández, T.; García, R.; Carrillo, E.; Marcet, M.; Galbe, M.; Jönsson, L.J. Preparation of Hydrolysates from Tobacco Stalks and Ethanolic Fermentation by *Saccharomyces cerevisiae*. *World J. Microbiol. Biotechnol.* **2002**, *18*, 857–862. [CrossRef]
4. Kaushal, L.A.; Prashar, A. Agricultural Crop Residue Burning and Its Environmental Impacts and Potential Causes–Case of Northwest India. *J. Environ. Plan. Manag.* **2021**, *64*, 464–484. [CrossRef]
5. Jönsson, L.J.; Martín, C. Pretreatment of Lignocellulose: Formation of Inhibitory by-Products and Strategies for Minimizing Their Effects. *Bioresour. Technol.* **2016**, *199*, 103–112. [CrossRef]
6. Nitsos, C.; Matsakas, L.; Triantafyllidis, K.; Rova, U.; Christakopoulos, P. Investigation of Different Pretreatment Methods of Mediterranean-Type Ecosystem Agricultural Residues: Characterisation of Pretreatment Products, High-Solids Enzymatic Hydrolysis and Bioethanol Production. *Biofuels* **2018**, *9*, 545–558. [CrossRef]
7. López, Y.; Gullón, B.; Puls, J.; Parajó, J.C.; Martín, C. Dilute Acid Pretreatment of Starch-Containing Rice Hulls for Ethanol Production: 11th EWLP, Hamburg, Germany, August 16–19, 2010. *Holzforschung* **2011**, *65*, 467–473. [CrossRef]
8. Martín, C.; Peinemann, J.C.; Wei, M.; Stagge, S.; Xiong, S.; Jönsson, L.J. Dilute-Sulfuric Acid Pretreatment of de-Starched Cassava Stems for Enhancing the Enzymatic Convertibility and Total Glucan Recovery. *Ind. Crop. Prod.* **2019**, *132*, 301–310. [CrossRef]
9. Ilanidis, D.; Stagge, S.; Jönsson, L.J.; Martín, C. Hydrothermal Pretreatment of Wheat Straw: Effects of Temperature and Acidity on Byproduct Formation and Inhibition of Enzymatic Hydrolysis and Ethanolic Fermentation. *Agronomy* **2021**, *11*, 487. [CrossRef]
10. Krafft, M.J.; Bendler, M.; Schreiber, A.; Saake, B. Steam Refining with Subsequent Alkaline Lignin Extraction as an Alternative Pretreatment Method to Enhance the Enzymatic Digestibility of Corn Stover. *Agronomy* **2020**, *10*, 811. [CrossRef]
11. Mondylaksita, K.; Ferreira, J.A.; Millati, R.; Budhijanto, W.; Niklasson, C.; Taherzadeh, M.J. Recovery of High Purity Lignin and Digestible Cellulose from Oil Palm Empty Fruit Bunch Using Low Acid-Catalyzed Organosolv Pretreatment. *Agronomy* **2020**, *10*, 674. [CrossRef]
12. Domínguez, E.; del Río, P.G.; Romaní, A.; Garrote, G.; Gullón, P.; de Vega, A. Formosolv Pretreatment to Fractionate Paulownia Wood Following a Biorefinery Approach: Isolation and Characterization of the Lignin Fraction. *Agronomy* **2020**, *10*, 1205. [CrossRef]
13. Rivas, S.; Moure, A.; Parajó, J.C. Pretreatment of Hazelnut Shells as a Key Strategy for the Solubilization and Valorization of Hemicelluloses into Bioactive Compounds. *Agronomy* **2020**, *10*, 760. [CrossRef]
14. López, L.; Rivas, S.; Moure, A.; Vila, C.; Parajó, J.C. Development of Pretreatment Strategies for the Fractionation of Hazelnut Shells in the Scope of Biorefinery. *Agronomy* **2020**, *10*, 1568. [CrossRef]
15. Sánchez-Gutiérrez, M.; Espinosa, E.; Bascón-Villegas, I.; Pérez-Rodríguez, F.; Carrasco, E.; Rodríguez, A. Production of Cellulose Nanofibers from Olive Tree Harvest—A Residue with Wide Applications. *Agronomy* **2020**, *10*, 696. [CrossRef]
16. Rahimi, V.; Shafiei, M.; Karimi, K. Techno-Economic Study of Castor Oil Crop Biorefinery: Production of Biodiesel without Fossil-Based Methanol and Lignoethanol Improved by Alkali Pretreatment. *Agronomy* **2020**, *10*, 1538. [CrossRef]
17. Hernández, E.; García, A.; López, M.; Puls, J.; Parajó, J.C.; Martín, C. Dilute Sulphuric Acid Pretreatment and Enzymatic Hydrolysis of *Moringa oleifera* Empty Pods. *Ind. Crop. Prod.* **2013**, *44*, 227–231. [CrossRef]
18. Martín, C.; García, A.; Schreiber, A.; Puls, J.; Saake, B. Combination of Water Extraction with Dilute-Sulphuric Acid Pretreatment for Enhancing the Enzymatic Hydrolysis of *Jatropha curcas* Shells. *Ind. Crop. Prod.* **2015**, *64*, 233–241. [CrossRef]
19. Karp, S.G.; Osipov, D.O.; Semenova, M.V.; Rozhkova, A.M.; Zorov, I.N.; Sinitsyna, O.A.; Soccol, C.R.; Sinitsyn, A.P. Effect of Novel Penicillium Verruculosum Enzyme Preparations on the Saccharification of Acid- and Alkali-Pretreated Agro-Industrial Residues. *Agronomy* **2020**, *10*, 1348. [CrossRef]
20. Osipov, D.O.; Dotsenko, G.S.; Sinitsyna, O.A.; Kondratieva, E.G.; Zorov, I.N.; Shashkov, I.A.; Satrutdinov, A.D.; Sinitsyn, A.P. Comparative Study of the Convertibility of Agricultural Residues and Other Cellulose-Containing Materials in Hydrolysis with Penicillium Verruculosum Cellulase Complex. *Agronomy* **2020**, *10*, 1712. [CrossRef]
21. Mkabayi, L.; Malgas, S.; Wilhelmi, B.S.; Pletschke, B.I. Evaluating Feruloyl Esterase—Xylanase Synergism for Hydroxycinnamic Acid and Xylo-Oligosaccharide Production from Untreated, Hydrothermally Pre-Treated and Dilute-Acid Pre-Treated Corn Cobs. *Agronomy* **2020**, *10*, 688. [CrossRef]

22. Mafa, M.; Malgas, S.; Bhattacharya, A.; Rashamuse, K.; Pletschke, B.I. The Effects of Alkaline Pretreatment on Agricultural Biomasses (Corn Cob and Sweet Sorghum Bagasse) and Their Hydrolysis by a Termite-Derived Enzyme Cocktail. *Agronomy* **2020**, *10*, 1211. [CrossRef]
23. Bedő, S.; Fehér, A.; Khunnonkwao, P.; Jantama, K.; Fehér, C. Optimized Bioconversion of Xylose Derived from Pre-Treated Crop Residues into Xylitol by Using Candida Boidinii. *Agronomy* **2021**, *11*, 79. [CrossRef]
24. Morales-Polo, C.; del Cledera-Castro, M.M.; Revuelta-Aramburu, M.; Hueso-Kortekaas, K. Bioconversion Process of Barley Crop Residues into Biogas—Energetic-Environmental Potential in Spain. *Agronomy* **2021**, *11*, 640. [CrossRef]
25. Olguin-Maciel, E.; Singh, A.; Chable-Villacis, R.; Tapia-Tussell, R.; Ruiz, H.A. Consolidated Bioprocessing, an Innovative Strategy towards Sustainability for Biofuels Production from Crop Residues: An Overview. *Agronomy* **2020**, *10*, 1834. [CrossRef]

agronomy

MDPI

Article

Hydrothermal Pretreatment of Wheat Straw: Effects of Temperature and Acidity on Byproduct Formation and Inhibition of Enzymatic Hydrolysis and Ethanolic Fermentation

Dimitrios Ilanidis, Stefan Stagge, Leif J. Jönsson and Carlos Martín *

Department of Chemistry, Umeå University, SE-901 87 Umeå, Sweden; dimitrios.ilanidis@umu.se (D.I.); stefan.stagge@umu.se (S.S.); leif.jonsson@umu.se (L.J.J.)
* Correspondence: carlos.martin@umu.se; Tel.: +46-90-786-6099

Abstract: Biochemical conversion of wheat straw was investigated using hydrothermal pretreatment, enzymatic saccharification, and microbial fermentation. Pretreatment conditions that were compared included autocatalyzed hydrothermal pretreatment at 160, 175, 190, and 205 °C and sulfuric-acid-catalyzed hydrothermal pretreatment at 160 and 190 °C. The effects of using different pretreatment conditions were investigated with regard to (i) chemical composition and enzymatic digestibility of pretreated solids, (ii) carbohydrate composition of pretreatment liquids, (iii) inhibitory byproducts in pretreatment liquids, (iv) furfural in condensates, and (v) fermentability using yeast. The methods used included two-step analytical acid hydrolysis combined with high-performance anion-exchange chromatography (HPAEC), HPLC, ultra-high performance liquid chromatography-electrospray ionization-triple quadrupole-mass spectrometry (UHPLC-ESI-QqQ-MS), and pyrolysis-gas chromatography/mass spectrometry (Py-GC/MS). Lignin recoveries in the range of 108–119% for autocatalyzed hydrothermal pretreatment at 205 °C and sulfuric-acid-catalyzed hydrothermal pretreatment were attributed to pseudolignin formation. Xylose concentration in the pretreatment liquid increased with temperature up to 190 °C and then decreased. Enzymatic digestibility was correlated with the removal of hemicelluloses, which was almost quantitative for the autocatalyzed hydrothermal pretreatment at 205 °C. Except for the pretreatment liquid from the autocatalyzed hydrothermal pretreatment at 205 °C, the inhibitory effects on *Saccharomyces cerevisiae* yeast were low. The highest combined yield of glucose and xylose was achieved for autocatalyzed hydrothermal pretreatment at 190 °C and the subsequent enzymatic saccharification that resulted in approximately 480 kg/ton (dry weight) raw wheat straw.

Keywords: wheat straw; hydrothermal pretreatment; bioconversion inhibitors; enzymatic hydrolysis; ethanolic fermentation

check for updates

Citation: Ilanidis, D.; Stagge, S.; Jönsson, L.J.; Martín, C. Hydrothermal Pretreatment of Wheat Straw: Effects of Temperature and Acidity on Byproduct Formation and Inhibition of Enzymatic Hydrolysis and Ethanolic Fermentation. *Agronomy* **2021**, *11*, 487. https://doi.org/10.3390/agronomy11030487

Academic Editors: Tony Vancov and Peter Langridge

Received: 19 January 2021
Accepted: 1 March 2021
Published: 5 March 2021

Publisher's Note: MDPI stays neutral with regard to jurisdictional claims in published maps and institutional affiliations.

Copyright: © 2021 by the authors. Licensee MDPI, Basel, Switzerland. This article is an open access article distributed under the terms and conditions of the Creative Commons Attribution (CC BY) license (https://creativecommons.org/licenses/by/4.0/).

1. Introduction

Growing global energy demand and the need to replace fossil fuels with renewable fuels are the major challenges of modern society [1]. The current production of liquid biofuels, such as bioethanol, is dominated by biochemical conversion routes mainly based on food-related feedstocks, such as corn starch and sugarcane sugar. Lignocellulosic materials can serve as additional resources for the biofuel sector [1,2]. Agricultural and agroindustrial residues, e.g., wheat straw, corn stover, and sugarcane bagasse, are lignocellulosic materials of interest in many countries due to their availability [3,4]. Co-utilization of lignocellulosic residues with starch or sugar can further boost bioethanol production through integration into 1.5 G processes [5].

Wheat (*Triticum aestivum* L.) is the world's most widely cultivated crop [6]. Based on the data on wheat production worldwide (FAOSTAT) and on a residue/crop ratio of 1.3 [7], around 980 million tons of wheat straw are estimated to be available on a yearly basis. Dry wheat straw consists mainly of cellulose (30–40%), hemicelluloses (20–30%), and lignin

(10–20%) [8]. Theoretically, bioconversion processes can yield around 270 L of ethanol per ton of dry wheat straw [9]. Biochemical conversion of agricultural residues, such as corn stover and wheat straw, has been significantly improved [7,10], but further research is still required for designing highly competitive and sustainable processes for the production of biofuels and other bio-based commodities.

The production of biofuels from lignocellulosic biomass comprises several steps, such as pretreatment, enzymatic saccharification, and microbial fermentation [11,12]. For achieving effective enzymatic saccharification of cellulose and, consequently, a higher release of fermentable sugars, a pretreatment step is indispensable. By disrupting the lignocellulosic matrix, pretreatment exposes cellulose and makes it more reactive towards cellulases [13]. Among different existing pretreatment methods, hydrothermal processing is an attractive option for agricultural residues, and it has good potential for industrial implementation [14]. Hydrothermal pretreatment (HTP) can be catalyzed by either hydronium ions generated by water autoionization or externally added acid species [15]. In HTP, moist lignocellulosic biomass is heated to around 200 °C for a certain period of time. Under those conditions, degradation of hemicelluloses and relocation of lignin can occur, and that leads to better accessibility of cellulose, which facilitates enzymatic hydrolysis [15,16].

While aiming at producing pretreated biomass with highly digestible cellulose, most pretreatment methods unavoidably lead to the formation of byproducts, resulting mainly from partial degradation of polysaccharides and lignin [14]. The formed byproducts can negatively affect the efficiency of enzymatic saccharification and microbial fermentation [14,17]. Certain groups of substances that inhibit microorganisms, such as aliphatic carboxylic acids, furan aldehydes, and phenolic compounds, have been extensively studied [18], while the importance of other groups of inhibitors has only recently started to emerge. Recent studies [19,20] have shown the significance of small aliphatic aldehydes as inhibitors of microbes used for biochemical conversion of biomass. Furthermore, the presence of benzoquinones in pretreatment liquids of different materials and their inhibitory effects on *Saccharomyces cerevisiae* have recently been discovered [21]. With regard to the inhibition of enzymes, aromatic substances such as phenolics have been found to play a role [14]. Pseudolignin, another byproduct, consists of thermal degradation products of carbohydrates. Pseudolignin remains insoluble and is accounted for as Klason lignin in the analytical two-step treatment with sulfuric acid (TSSA). The formation of pseudolignin can affect the enzymatic saccharification process [22,23]. In previous studies on hydrothermal pretreatment of wheat straw, the inhibition problem has generally been limited to substances such as furfural, HMF (5-hydroxymethylfurfural), acetic acid, formic acid, and certain phenolic compounds [8,9,24], while newly discovered inhibitors, such as aliphatic aldehydes and benzoquinones, have not been considered.

There are still several issues about hydrothermal pretreatment of wheat straw that require further research and innovation efforts. Issues that have so far not received enough attention include (i) how autocatalyzed (A-HTP) and sulfuric acid-catalyzed (SA-HTP) hydrothermal pretreatments affect the formation of byproducts, including pseudolignin and newly discovered inhibitors, (ii) how the pretreatment liquids inhibit the enzymatic hydrolysis of cellulose, and (iii) how xylan and lignin affect the digestibility of pretreated solids. In the current study, hydrothermal pretreatment of wheat straw under different temperatures between 160 and 205 °C, using two catalytic approaches, auto-catalysis and catalysis with sulfuric acid, was investigated. The investigation covered the evaluation of the effects of pretreatment conditions on (i) release of sugars and bioconversion inhibitors, (ii) chemical composition and enzymatic digestibility of pretreated solids, and (iii) inhibitory effects of pretreatment liquids on enzymatic saccharification and yeast fermentation. The results were backed by advanced analytical techniques, such as pyrolysis-gas chromatography/mass spectrometry (Py-GC/MS), high-performance anion-exchange chromatography (HPAEC), and ultra-high performance liquid chromatography-electrospray ionization-triple quadrupole-mass spectrometry (UHPLC-ESI-QqQ-MS).

2. Materials and Methods

2.1. Materials

Wheat straw was provided by RISE Processum AB (Örnsköldsvik, Sweden). The wheat straw was previously collected by the Swedish University of Agricultural Sciences (SLU) from Mälardalen, Sweden, in the autumn of 2015. The moisture content of the biomass was determined using an HG63 moisture analyzer (Mettler-Toledo, Greifensee, Switzerland).

2.2. Pretreatment of Wheat Straw

The severity factor (SF; Equation (1)) of different pretreatment conditions was determined as the logarithm of the reaction ordinate, as proposed by Overend and Chornet [25] (Equation (2)).

$$SF = Log\ Ro \tag{1}$$

$$Ro = t\ \times \exp\left(\frac{Tr - 100}{14.75}\right) \tag{2}$$

where t is the holding time of pretreatment in minutes, and Tr is the pretreatment temperature in °C.

For each pretreatment batch, 37.5 g of wheat straw (on dry weight (DW) basis) was mixed with tap water (for A-HTP) or aqueous sulfuric acid (for SA-HTP) at a 7:1 liquid-to-solid ratio in a 1-L reactor (Parr 4520, Moline, IL, USA). The loading of sulfuric acid (96%) was 0.5 g per 100 g wheat straw (DW). The reaction time was 15 min. For A-HTP, the temperature settings were 160, 175, 190, and 205 °C, which correspond to SF values of 2.9, 3.4, 3.8, and 4.3, respectively. For SA-HTP, the temperature settings were 160 and 190 °C, which correspond to SF values of 2.9 and 3.8, respectively. The pretreatment conditions were chosen within the ranges typically reported for wheat straw [8–10]. At the cooling stage of pretreatment, when the temperature was approximately 90 °C, a sample was carefully taken from the gas phase on the top of the reaction suspension. The sample was received under a water column in a 15-mL Falcon tube, and it was analyzed immediately (Section 2.6). After pretreatment, the solid and liquid phases were separated by vacuum filtration, and the solid phase was washed with 1 L water. A portion of the filter cake was stored frozen for analytical enzymatic saccharification assay, and the rest was air-dried for around one week; the yield of pretreated solids was determined gravimetrically based on DW. The weight of the portion used for enzymatic saccharification was included in the calculation of the DW of pretreated solids (DW of PSs).

The yield of pretreated solids (*PSs*) was calculated as follows:

$$Yield\ of\ PSs\ (\%) = \frac{DW\ of\ PSs\ (g)}{Initial\ DW\ of\ biomass\ (g)} \times 100 \tag{3}$$

2.3. Compositional Analysis of the Raw Material and Pretreated Solids

The content of extractives in raw wheat straw and pretreated solids was determined by extracting aliquots of each material with 200 mL ethanol in a Soxhlet system (Extraction System B-811, Büchi, Flawil, Switzerland). Structural carbohydrates and lignin were determined using TSSA. TSSA was performed essentially according to an NREL protocol [26] but using HPAEC for quantification of monosaccharides. HPAEC was performed using a Dionex ICS-5000 (Sunnyvale, CA, USA) instrument with pulsed amperometric detection (PAD), a separation column (4 × 250 mm), and a guard column (4 × 50 mm) (both CarboPac PA1, Dionex, Sunnyvale, CA, USA). Prior to the analysis, the samples were diluted with ultra-pure water and filtered through 0.2 μm Millex-GN nylon membranes (Merck Millipore Ltd., Cork, Ireland). Acid-insoluble (Klason) lignin was determined gravimetrically and acid-soluble lignin (ASL) was determined spectrophotometrically at λ 240 nm. Ash content was determined using an NREL protocol [27]. All analyses were performed in triplicates. Values are stated as mass fractions of initial masses of samples (DW).

The recoveries of glucan and xylan were calculated using the following equations:

$$Glucan\ recovery\ (\%) = \frac{G_{PS} \times Yield\ of\ PS}{G_{RM}} \times 100 \quad (4)$$

$$Xylan\ recovery\ (\%) = \frac{X_{PS} \times Yield\ of\ PS}{X_{RM}} \times 100 \quad (5)$$

Here, G_{PS} and X_{PS} are the mass fractions of glucan and xylan, respectively, in pretreated solids (PS), and G_{RM} and X_{RM} are their mass fractions in the raw material.

Total lignin recovery was calculated using the following equation:

$$Lignin\ recovery\ (\%) = \frac{(KL + ASL)_{PS} \times Yield\ of\ PS}{(KL + ASL)_{RM}} \times 100 \quad (6)$$

Here, *KL* is the mass fraction of Klason lignin, and *ASL* is the mass fraction of acid-soluble lignin in pretreated solids (*PSs*) and raw material (*RM*).

2.4. Pyrolysis-Gas Chromatography/Mass Spectrometry (Py-GC/MS)

Py-GC/MS analysis was performed at the Biopolymer Analytical Facility of the KBC Chemical–Biological Center (Umeå, Sweden). Prior to Py-GC/MS analysis, the pretreated and raw samples were freeze-dried and ball-milled (mixer mill MM400, Retsch, Haan, Germany). Then, 50 µg biomass powder was applied to a pyrolyzer equipped with an autosampler (PY-2020iD and AS-1020E, Frontier Lab, Fukushima, Japan) connected to a GC/MS machine (7890A/5975C, Agilent Technologies AB, Kista, Sweden). The pyrolysate was separated and analyzed, as described by Gerber et al. [28]. Areas were determined for peaks assigned as carbohydrates (C), guaiacyl (G), syringyl (S), *p*-hydroxyphenyl (H), generic phenolic constituents (P), known spectra unknown identification (U), and unknown spectra (0). Values are stated as percentages of total peak areas (C + G + S + H + P + U + 0). The fraction of lignin (L) corresponds to the sum of the fractions of G, S, H, and P.

2.5. Analytical Enzymatic Saccharification of Pretreated Solids

The enzymatic digestibility of pretreated solids and the inhibitory effect of the pretreatment liquids on the cellulolytic enzymes were evaluated following an analytical enzymatic saccharification procedure [15]. In one set of assays, aliquots of 250 mg (DW) of pretreated solids or raw wheat straw were suspended in 50 mM citrate buffer (pH 5.2) in 15-mL Falcon tubes, with a total reaction mixture of 5 mL. In another set of assays, the pretreated solids were instead suspended in pretreatment liquids, whose pH had previously been adjusted to pH 5.2. No citrate buffer was included in the assays with pretreatment liquids as carboxylic acids present in the pretreatment liquid make the addition of buffer unnecessary. In all assays, cellulolytic enzymes (Cellic CTec2 enzyme blend, Sigma-Aldrich Chemie GmbH, Steinheim, Germany) were added so that the loading was 100 CMCase units/g biomass. The reaction mixtures were incubated at 50 °C and 180 rpm for 72 h in an Ecotron orbital incubator (INFORS HT, Bottmingen, Switzerland). After 72 h, the tubes were centrifuged, and the supernatants were collected for further analysis using HPLC. Enzymatic digestibility was determined using the following equation:

$$Enzymatic\ digestibility\ (\%) = \frac{G_H}{G_{APS}} \times 100 \quad (7)$$

Here, G_H is the amount of hydrolyzed glucan based on the glucose concentration in the enzymatic hydrolysate, and G_{APS} is the amount of glucan in pretreated solids.

The degree of inhibition of enzymatic saccharification of glucan by the pretreatment liquids was determined using Equation (8).

$$Degree\ of\ inhibition\ (\%) = \frac{\left(ED_{buf} - ED_{PL}\right)}{ED_{buf}} \times 100 \quad (8)$$

Here, $EDbuf$ is the enzymatic digestibility of glucan (%) resulting from enzymatic hydrolysis of pretreated solids suspended in sodium citrate buffer, and $EDPL$ is the enzymatic digestibility of glucan (%) resulting from enzymatic hydrolysis of pretreated solids suspended in pretreatment liquid.

2.6. Analysis of Sugars and Inhibitors in Liquid Samples

Analysis of monosaccharides (glucose, xylose, arabinose, galactose, and mannose) was performed using HPAEC (as described in Section 2.3). Acid posthydrolysis of pretreatment liquids with 4% (w/w) sulfuric acid was performed in order to hydrolyze oligosaccharides to allow their quantification.

Determination of furfural and HMF was carried out using an HPLC system (Dionex UltiMate 300; ThermoFisher, Waltham, MA, USA) with a diode-array detector and a 3 × 50 mm, 1.8-μm Zorbax RRHT SB-C 18 column. The temperature was set to 40 °C.

Determination of formaldehyde, acetaldehyde, *p*-benzoquinone, vanillin, coniferyl aldehyde, syringaldehyde, *p*-hydroxybenzaldehyde, and acetovanillone was performed using UHPLC-ESI-QqQ-MS after derivatization with 2,4-dinitrophenylhydrazine (DNPH). A 1290 Infinity system (Agilent Technologies, Santa Clara, CA, USA) coupled to a 6490 triple-quadrupole mass spectrometer (QqQ-MS) was used. The MS parameters were set as follows: gas temperature 290 °C, gas flow 20 L/min, nebulizer 30 psi, sheath gas temperature 400 °C, sheath gas flow 12 L/min, capillary voltage—3000 V, and nozzle voltage—2000 V. A 2.1 × 150 mm XTerra MS C18 column was used. Eluent A consisted of aqueous 0.1% (v/v) formic acid, and Eluent B was a 75:25 (v/v) mixture of acetonitrile and 2-propanol with 0.1% formic acid. Elution was performed using a gradient profile containing the following fractions of Eluent B: 0.0–4.5 min 30–40%, 4.5–9.0 min 40–50%, 9.0–11.0 min 50–70%, 11.0–11.01 min 70–95%, 11.01–15.0 min 95%, 15.0–15.01 min 95–30%, 15.01–18.0 min 30%, and, at the end, two min post-time with 30% for further re-equilibration. Data evaluation was done with MassHunter Quant software. The calibration and derivatization processes were based on previous studies [19,21].

The determination of aliphatic carboxylic acids (formic acid, acetic acid, and levulinic acid) was performed by MoRe Research Örnsköldsvik AB, Sweden. Acid determination was done using HPAEC.

Total aromatic content (TAC) in pretreatment liquids was determined by measuring the absorbance at 280 nm using a UV1800 spectrophotometer (Shimadzu, Kyoto, Japan). Total carboxylic acid content (TCAC) was determined by titration from pH 2.8 to pH 7.0 using an aqueous solution of sodium hydroxide (200 mM).

The determination of total phenolic compounds was carried out by using the Folin–Ciocalteu method [29], with vanillin as the calibration standard. A SpectraMax i3x (Molecular Devices, LLC, San Jose, CA, USA) multimode microplate reader was used for reading the absorbance at 760 nm of the color generated after the incubation of reaction mixtures containing the sample and Folin–Ciocalteu reagent at 23 °C for two hours. Reactions were performed in triplicates.

2.7. Fermentability of Pretreatment Liquids

Fermentability tests with freeze-dried *S. cerevisiae* yeast (Ethanol Red, Fermentis, Marcq en Baroeul, France) were performed using microtiter plates (maximum well volume 330 μL; Nunc, Roskilde, Denmark). Freeze-dried yeast was rehydrated by suspending it in 5 times its weight of sterile tap water for 30 min at 35 °C, and it was then inoculated at an initial loading of 0.2 g (DW)/L. Prior to fermentation, the pH of the pretreatment liquids was adjusted to 5.5 using a 1-M solution of NaOH. The final volume in each well was 300 µL. The mixture contained either 120 µL pretreatment liquid (diluted with deionized water to 40% of the initial concentration) or 234 µL pretreatment liquid (100% of the initial concentration), 6 µL of a nutrient solution (consisting of 150 g/L yeast extract, 75 g/L $(NH_4)_2HPO_4$, 3.75 g/L $MgSO_4{\cdot}7\ H_2O$, and 238.2 g/L $NaH_2PO_4{\cdot}H_2O$), and deionized water. Glucose was added to all the media until a final concentration of 20 g/L. Reference fermentations without any pretreatment liquid were performed, while control samples with only culture medium (no yeast) were also processed in order to confirm that there was no cross-contamination. The microtiter plates were incubated in an orbital shaker at 150 rpm and 30 °C.

Yeast growth was monitored by measurements of optical density (OD) at 600 nm using the SpectraMax i3x multimode microplate reader after 0, 12, 24, 36, and 48 h. Growth medium without yeast was used as blank. All experiments were performed in triplicates, and mean values were used in the evaluation. The growth rate in each medium was calculated with the following equation:

$$Growth\ rate = \frac{Cell\ density\ (OD_{600})}{Incubation\ time\ (\mathrm{h})} \tag{9}$$

The relative growth rate (RGR) of cultures with a medium of pretreatment liquid was calculated by comparing the growth rate of those cultures to the average growth rate of the reference cultures with sugar-based medium (Equation (10)).

$$RGR = \frac{Growth\ rate\ of\ cultures\ with\ PL}{Growth\ rate\ of\ reference\ cultures} \tag{10}$$

2.8. Statistical Processing of the Results

Two-way analysis of variance (ANOVA) and Tukey's posthoc test were applied to evaluate the statistical difference between the results of enzymatic saccharification experiments corresponding to different pretreatment conditions and different liquid media (either sodium citrate buffer or pretreatment liquids) used in the assays. Levene's test for homogeneity of variance and Shapiro's test for normal distribution of residuals were used for verifying that the data met the model assumptions. Statistical tests were run using R package stat [30].

3. Results

3.1. Chemical Composition of Raw and Pretreated Solids

The wheat straw used in this study contained 39.8% glucan, 24.3% hemicellulosic carbohydrates, 22.8 ± 0.1% lignin, 5.1 ± 0.3% ash, and 4.7 ± 0.1% ethanol extractives (Table 1). That composition is in agreement with previous reports on wheat straw, showing similar values [31].

The pretreatment led to different degrees of reduction of the gravimetric yield of the resulting solid material with respect to the starting amount of wheat straw (Figure 1). Depending on the pretreatment conditions, the yields of pretreated solids ranged between 64.3% and 87.6%, and a steady decrease was observed with increasing temperature and increasing severity factor. The yield of pretreated solids resulting from the most severe autocatalyzed pretreatment (the one performed at 205 °C) was 27% lower than that of the least severe one (at 160 °C). The yields were lower for SA-HTP than for A-HTP, but the differences were relatively small. The most noticeable difference was found for 190 °C (69.4% for SA-HTP and 73.0% for A-HTP), while the values for 160 °C (85.5% for SA-HTP and 87.6% for A-HTP) were closer. It is noteworthy that the differences in yield between the pretreatment approaches (A-HTP and SA-HTP) at a given temperature were smaller than the differences in yield between two consecutive temperatures in the A-HTP series, for instance, 190 °C (73.0%) and 205 °C (64.3%) or 175 °C (80.3%) and 190 °C (73.0%). Evidently, a temperature increase of 15 °C had a stronger effect than the addition of sulfuric acid.

Table 1. Chemical composition of raw wheat straw and pretreated solids, mass fractions in % (dry weight) [a].

Experimental Conditions [b]	Glucan [b]	Xylan	Arabinan	Galactan	Mannan	Klason Lignin	ASL	Ash	Extractives
Raw	39.8 (<0.1)	19.7 (<0.1)	2.2 (<0.1)	1.3 (<0.1)	1.1 (0.1)	20.1 (0.1)	2.7 (0.1)	5.1 (0.3)	4.7 (0.1)
A-160	42.5 (<0.1)	19.5 (<0.1)	ND	ND	ND	21.1 (1.8)	2.8 (0.3)	4.7 (0.1)	3.9 (0.1)
A-175	47.1 (2.0)	18.5 (<0.1)	ND	ND	ND	22.2 (1.9)	2.9 (0.1)	4.9 (<0.1)	4.5 (0.1)
A-190	52.4 (0.2)	11.1 (<0.1)	ND	ND	ND	31.1 (3.3)	2.7 (0.1)	5.9 (0.3)	5.9 (0.1)
A-205	57.3 (<0.1)	2.7 (0.1)	ND	ND	ND	39.3 (0.1)	2.8 (0.1)	6.6 (0.1)	5.6 (0.1)
SA-160	43.5 (<0.1)	19.7 (0.5)	ND	ND	ND	28.8 (2.0)	2.7 (0.1)	5.5 (<0.1)	4.1 (0.1)
SA-190	54.8 (<0.1)	8.6 (<0.1)	ND	ND	ND	33.3 (0.1)	2.3 (0.2)	5.5 (<0.1)	7.5 (0.1)

[a] Average values of solid fractions resulting from hydrothermal pretreatment of wheat straw (mass fractions in % of dry weight). Numbers in parenthesis indicate standard deviation. [b] The letters in the codification of the experimental conditions stand for the pretreatment approach (A for A-HTP or SA for SA-HTP), and the numerals (160, 175, 190, and 205) stand for temperatures. ND, not detected.

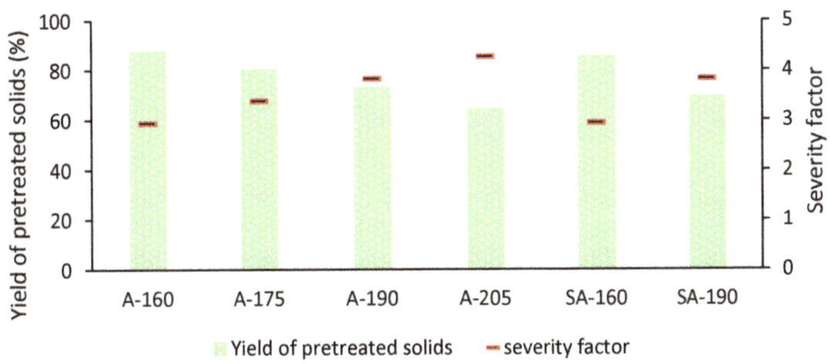

Figure 1. Gravimetric yield of pretreated solids after autocatalyzed hydrothermal pretreatment (A-HTP) at 160 °C (A-160), 175 °C (A-175), 190 °C (A-190), and 205 °C (A-205), and sulfuric acid-catalyzed (SA-HTP) hydrothermal pretreatment at 160 °C (SA-160) and 190 °C (SA-190).

The glucan content was higher in the pretreated solids than in raw wheat straw, and, for both pretreatment approaches, it increased with the increase in temperature (Table 1). For A-HTP, the glucan content increased from 42.5% at 160 °C to 57.3% at 205 °C, while for SA-HTP, it increased from 43.5% at 160 °C to 54.8% at 190 °C. For the same pretreatment temperature, A-HTP and SA-HTP did not exhibit any major difference in glucan content. The same applies for glucan recovery, which was comparable for both pretreatment approaches, and displayed values above 92% for all experimental conditions (Figure 2).

The xylan content decreased with increasing temperature (Table 1). Independently of the catalytic approach, the xylan content of solids pretreated at 160 °C was comparable to that of raw wheat straw. After pretreatment at 205 °C, the xylan content was below 3 ± 0.1%. The xylan recovery for pretreatments at 160 °C corresponded to approx. 87 ± 0.1%, whereas the recovery at 205 °C was below 9% (Figure 2). At 160 °C, the xylan recovery was not affected by acid addition, whereas at 190 °C, the value for A-HTP (41.1%) differed substantially from that of SA-HTP (30.3%). Other hemicellulosic carbohydrates were readily solubilized already at 160 °C as no arabinose, galactose, or mannose were detected in the analytical acid hydrolysates of the pretreated solids (Table 1).

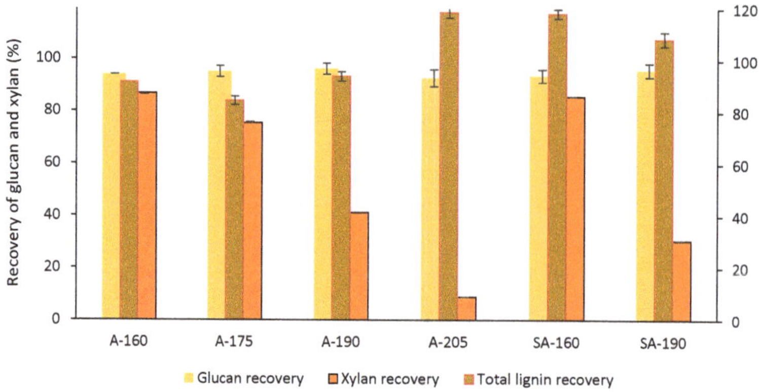

Figure 2. Recovery of glucan, xylan, and total lignin (Klason lignin plus acid-soluble lignin (ASL)) in the pretreated solids. The codification of the experimental conditions is the same as in Table 1.

A gradual increase of Klason lignin content in the pretreated solids was observed with increasing temperature. For experiments performed at the same temperature, the Klason lignin content was higher for SA-HTP than for A-HTP (Table 1). The content of acid-soluble lignin (ASL) did not display a clear trend regarding pretreatment temperature, but it was lower for SA-HTP than for A-HTP.

For the autocatalyzed pretreatment at 205 °C and for both acid-catalyzed pretreatments, the recovery of total lignin was higher than the theoretically possible value (Figure 2), indicating the formation of pseudolignin. To clarify this issue, Py-GC/MS analysis was performed. For the A-HTP experiments at 190 and 205 °C and for the SA-HTP experiments, Py-GC/MS revealed lower lignin content than the fraction of total lignin determined by the TSSA procedure (Klason lignin + acid-soluble lignin; Table 2). Higher Δ_{Lignin} factor values (Table 2) than for raw wheat straw indicate pseudolignin formation. The increase in Δ_{Lignin} factor value was especially apparent after pretreatment at 205 °C. At the same temperature, the Δ_{Lignin} factor was higher for SA-HTP than for A-HTP, indicating that increasing acidity stimulated pseudolignin formation.

Table 2. Pyrolysis-gas chromatography/mass spectrometry (Py-GC/MS) analysis of raw wheat straw and pretreated solids.

Exp. Conditions	Carbohydrates (%) [a]	Lignin (%) [a]	Δ_{Lignin} [b]	G:S:H Ratio [c]
Raw	70.4 (0.5)	27.8 (0.6)	−5.0	44:50:5
A-160	73.4 (1.0)	24.7 (1.1)	−0.8	46:46:7
A-175	70.5 (0.6)	27.7 (0.5)	−2.6	44:49:7
A-190	66.5 (0.4)	31.6 (0.4)	2.2	45:48:7
A-205	68.6 (0.6)	28.4 (0.6)	13.7	48:46:5
SA-160	71.1 (0.4)	27.1 (0.5)	4.4	45:47:9
SA-190	65.0 (0.6)	32.7 (0.4)	2.9	45:48:7

[a] Mean values from triplicates. Standard deviations are shown in parentheses. [b] The Δ_{Lignin} factor was calculated by subtracting the peak area fraction assigned to lignin in Py-GC/MS analysis (this table) from the mass fraction of total lignin (Klason lignin and ASL; Table 1). [c] Relative proportion of lignin guaiacyl (G), syringyl (S), and *p*-hydroxyphenyl (H) units. Same codification of experimental conditions as in Table 1.

Negative Δ_{Lignin} factor values, particularly for raw wheat straw (Table 2), can be explained by the fact that the wheat straw contains relatively large fractions of ash (5.1%) and extractives (4.7%) (Table 1). While these substances are included in initial mass values used for TSSA calculations (Table 1), they (especially ash/inorganic substances) might not be covered by total peak areas used for Py-GC/MS calculations (Table 2). This explanation is supported by the fact that for raw wheat straw, it was not only the TSSA value for total lignin (22.8%) that was lower than the Py-GC/MS counterpart (27.8%); the TSSA value for the sum of carbohydrates (64.1%) was also lower than the Py-GC/MS counterpart (70.4%).

Py-GC/MS analysis of the G:S:H ratio (Table 2) indicated that pretreatment caused a relative increase of G (guaiacyl) units, a relative decrease of S (syringyl) units, and a relative increase of H (*p*-hydroxyphenyl) units. The relative decrease of S units might be associated with the cleavage of β-*O*-4 linkages in lignin, whereas corresponding increases for G and H units might be indirect effects.

The ash content was not much affected by the pretreatment but was slightly higher for the highest pretreatment temperature (A-HTP at 205 °C; Table 1). The increase in ash content at high temperatures can be an indirect effect of the decrease in hemicellulosic carbohydrates.

The mass fraction of extractives, which was 4.7% in raw wheat straw, decreased slightly for pretreatments at low temperature (160/175 °C; Table 1). At higher temperatures (190/205 °C), there was an increase. The decrease at lower temperatures can be explained by the solubilization of original extractives during pretreatment, whereas the increase at higher temperatures can be explained by the formation of new extractives due to the fragmentation of carbohydrates and lignin.

3.2. Carbohydrates in the Pretreatment Liquids

Carbohydrates in the pretreatment liquids are reported as monosaccharides and oligosaccharides (Table 3), with disaccharides included among the latter. Xylose, arabinose, and glucose were the main monosaccharides identified, whereas there were very little or no galactose and mannose. The concentrations of xylose, glucose, galactose, and mannose increased with increasing pretreatment temperatures. However, for the A-HTP series, the arabinose concentration was highest at intermediate temperatures (175/190 °C).

Table 3. Concentrations of monosaccharides and oligosaccharides (OS) (g/L) and pH in the pretreatment liquids [a].

	A-160	A-175	A-190	A-205	SA-160	SA-190
Glucose	0.1 (<0.1)	0.1 (<0.1)	0.1 (<0.1)	0.7 (<0.1)	0.2 (<0.1)	0.2 (0.1)
Gluco-OS	1.8 (0.1)	2.4 (0.1)	2.7 (0.1)	2.0 (0.1)	2.4 (0.1)	2.7 (0.1)
Xylose	0.1 (<0.1)	0.2 (<0.1)	1.6 (0.1)	4.4 (<0.1)	0.1 (<0.1)	2.5 (<0.1)
Xylo-OS	1.4 (0.1)	7.5 (0.3)	15.0 (0.3)	0.9 (0.1)	2.7 (0.1)	14.1 (0.3)
Arabinose	0.4 (<0.1)	1.0 (0.1)	0.9 (<0.1)	0.2 (<0.1)	0.6 (0.1)	1.0 (0.1)
Arabino-OS	0.8 (<0.1)	1.1 (0.1)	0.6 (<0.1)	<0.1 (<0.1)	0.8 (0.1)	0.6 (0.1)
Galactose	ND	0.1 (<0.1)	0.2 (<0.1)	0.3 (<0.1)	0.1 (<0.1)	0.3 (<0.1)
Galacto-OS	0.5 (<0.1)	0.7 (<0.1)	0.6 (<0.1)	0.4 (<0.1)	0.4 (<0.1)	0.4 (<0.1)
Mannose	ND	ND	0.1 (<0.1)	0.3 (0.1)	ND	0.1 (<0.1)
Manno-OS	0.3 (<0.1)	0.5 (<0.1)	0.6 (<0.1)	0.3 (0.1)	0.4 (<0.1)	0.8 (<0.1)
pH	5.0	4.5	4.0	3.5	4.5	3.8

[a] Mean values from triplicates; ND, not detected. Standard deviations shown in parentheses. Same codification of experimental conditions as in Table 1.

For monosaccharides derived from oligosaccharides/disaccharides, the highest concentrations were 15.0 ± 0.3 g/L for Xylo-OS (xylose derived from oligosaccharides/disaccharides) and 2.7 ± 0.1 g/L for Gluco-OS (glucose derived from oligosaccharides/disaccharides) (Table 3). The maximum concentrations of the others (Arabino-OS, Galacto-OS, and Manno-OS) were around 1 g/L or lower. For A-HTP, Xylo-OS, Gluco-OS, and Manno-OS reached the highest concentrations at 190 °C, whereas the highest concentrations of Arabino-OS and Galacto-OS were instead found at 175 °C. This indicates that oligosaccharides containing arabinose and galactose units were more heat-labile. This is supported by data from the SA-HTP series showing higher concentrations of Xylo-OS, Gluco-OS, and Manno-OS for SA-190 than for SA-160, whereas Arabino-OS and Galacto-OS do not follow that pattern. In most cases, the concentrations of monosaccharides derived from oligosaccharides/disaccharides were higher than the concentrations of the corresponding monosaccharides in the same pretreatment liquid (Table 3).

The pH of the A-HTP liquids decreased with pretreatment temperature, from 5.0 for 160 °C to 3.5 for 205 °C (Table 3). As further discussed below, the decrease of pH with increasing pretreatment temperature results from the formation of carboxylic acids. For the SA-HTP series, the pH was only slightly (0.2–0.5 pH units) lower than for the corresponding pretreatment liquids in the A-HTP series.

3.3. Effects of Pretreatment Conditions on Byproduct Formation

For both the A-HTP and SA-HTP series, the furan aldehyde concentrations always increased with increasing pretreatment temperature (Table 4). The furfural concentrations ($\leq$77.1 $\pm$ 2.4 mM) in the pretreatment liquids were always higher than the corresponding HMF concentrations ($\leq$2.4 mM). The very low values after pretreatment at 160 °C are difficult to compare, but furfural values after pretreatment at 190 °C show clearly higher concentration for SA-HTP (24.1 $\pm$ 0.5 mM) than for A-HTP (14.4 mM). As acid conditions promote hydrolysis of hemicelluloses to pentose sugars, the precursors of furfural, this is an expected result. The condensate of A-205 was found to contain furfural that had been volatilized during the pretreatment.

Table 4. Concentration of bioconversion inhibitors in the pretreatment liquids [a].

	A-160	A-175	A-190	A-205	SA-160	SA-190
Furfural [b]	1.2 (<0.1)	2.0 (<0.1)	14.4 (0.5)	77.1 (2.4)	0.7 (<0.1)	24.1 (0.5)
Furfural [b,e]	<0.1 (<0.1)	<0.1 (<0.1)	<0.1 (<0.1)	1.9 (<0.1)	<0.1 (<0.1)	<0.1 (<0.1)
HMF [b]	0.1 (<0.1)	0.1 (<0.1)	0.3 (<0.1)	2.4 (<0.1)	0.1 (<0.1)	0.6 (<0.1)
Acetic acid [b]	15.0 (0.1)	22.5 (0.1)	36.1 (0.1)	66.4 (0.4)	11.7 (0.1)	33.8 (0.1)
Formic acid [b]	1.0 (0.1)	2.4 (0.1)	8.1 (0.1)	24.1 (0.1)	0.6 (0.1)	7.9 (0.1)
Levulinic acid [b]	<0.1 (<0.1)	<0.1 (<0.1)	0.2 (<0.1)	0.4 (<0.1)	<0.1 (<0.1)	0.2 (<0.1)
TCAC	31 (1.5)	41 (1.5)	52 (1.2)	96 (2.5)	32 (1.0)	54 (2.2)
Formaldehyde [b]	ND	ND	ND	ND	0.1 (0.1)	ND
Vanillin [c]	89 (7)	124 (7)	190 (7)	255 (7)	75 (7)	214 (7)
Syringaldehyde [c]	26 (4)	41 (4)	75 (4)	108 (4)	25 (4)	84 (4)
p-Hydroxybenzaldehyde [c]	26 (2)	27 (3)	34 (3)	47 (3)	22 (3)	41 (3)
Coniferyl aldehyde [c]	29 (3)	34 (3)	50 (3)	38 (3)	30 (3)	59 (3)
p-Coumaraldehyde [c]	3 (1)	4 (1)	5 (1)	4 (1)	3 (1)	5 (1)
Acetovanillone [c]	5 (2)	8 (2)	15 (3)	9 (2)	5 (2)	12 (2)
p-Benzoquinone	ND	ND	ND	ND	ND	ND
Total phenolics [d]	1.0 (0.1)	2.0 (0.1)	3.0 (0.1)	7.2 (0.5)	1.6 (0.1)	4.1 (0.2)
TAC	0.2 (<0.1)	0.2 (<0.1)	0.6 (<0.1)	1.5 (<0.1)	0.2 (<0.1)	0.6 (<0.1)

[a] Mean values from triplicates. The standard deviations are shown in parentheses. Same codification of experimental conditions as in Table 1. [b–d] Values given in mM ([b]), µM ([c]), or g/L ([d]). [e] Furfural concentration in the condensate of the gas phase. Total aromatic content (TAC) was determined as UV absorption at 280 nm, with a dilution factor of 500. Total carboxylic acid content (TCAC; mM) was determined by titration. ND, not detected.

The main aliphatic carboxylic acid in the pretreatment liquids was acetic acid (11.7–66.4 mM). Pretreatment liquids also contained formic acid ($\leq$24.1 $\pm$ 0.1 mM) and levulinic acid ($\leq$0.4 mM). The concentrations of acids increased with the pretreatment temperature, and there were only minor differences between the A-HTP and SA-HTP series. As expected, TCAC values were higher than the sum of concentrations of acetic acid, formic acid, and levulinic acids and followed the same trend.

The formaldehyde concentrations were mostly below the detection level. A reason for this can be that during hydrothermal pretreatment, formaldehyde is probably formed mainly from lignin, and wheat straw has rather low lignin content compared to, for example, softwood [19,20].

The concentration of total phenolics ranged from 1.0–7.2 g/L (Table 4). The SA-HTP series exhibited slightly higher values than corresponding pretreatment liquids in the A-HTP series. The concentrations of total phenolics and individual phenolics with a one-carbon side chain (i.e., vanillin, syringaldehyde, and *p*-hydroxybenzaldehyde) increased with increasing pretreatment temperature (Table 4). Phenolics with two- or three-carbon side chains, such as acetovanillone, coniferyl aldehyde, and *p*-coumaraldehyde, increased with the pretreatment temperature up to 190 °C, but then decreased for A-205. This suggests that phenolics with longer side chains are more susceptible to degradation at higher temperatures, whereas phenolics with one-carbon side chains are more stable.

p-Benzoquinone was not detected in any of the pretreatment liquids (Table 4). Studies of yeast have shown that relevant toxic concentrations of *p*-benzoquinone are several orders of magnitude lower than for other inhibitors in Table 4, as concentrations in the micromolar range have clearly inhibitory effects [21]. The methodology used would have also revealed concentrations in that range.

The TAC measurement covers a wide range of aromatics, including phenylic compounds, such as phenolic and nonphenolic aromatics, and heteroaromatics, such as furan aldehydes. The TAC values increased with the temperature (Table 4). At corresponding temperatures, the TAC values for the SA-HTP series were identical to those of the A-HTP series.

3.4. *Enzymatic Saccharification of Pretreated Wheat Straw*

Since the susceptibility to enzymatic saccharification is a key issue in the bioconversion of lignocellulosic biomass into bio-based products, the impact of pretreatment conditions on enzymatic saccharification was carefully examined using reaction mixtures with both buffer and pretreatment liquids. Reaction mixtures with buffer provide a clean analytical view of the susceptibility of pretreated solids to enzymatic saccharification. Reaction mixtures with pretreatment liquid provide a more complex and, perhaps, more industrially relevant view, where both the susceptibility of the solid phase and the influence of potential enzyme inhibitors in the pretreatment liquid are taken into account.

All pretreatment conditions resulted in a clear enhancement of saccharification of glucan, as revealed by the increase of enzymatic digestibility from 18% for raw wheat straw to 45–98% for pretreated solids (Figure 3). For A-HTP solids suspended in sodium citrate buffer, the enzymatic digestibility increased with the pretreatment temperature (from 46 ± 0.5% for A-160 to 98 ± 0.8% for A-205). For SA-HTP, the enzymatic digestibility values were comparable with those from A-HTP at corresponding temperatures. Xylan contained in the pretreated solids was also hydrolyzed during saccharification trials (data not shown).

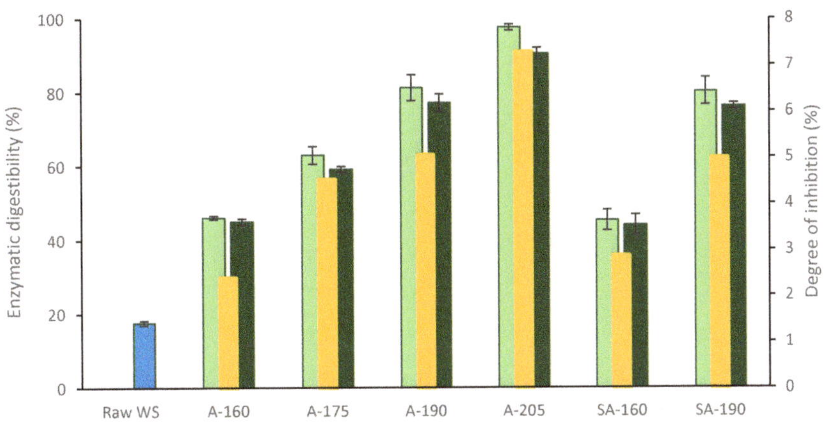

Figure 3. Enzymatic digestibility of glucan in cut raw wheat straw (WS) in buffer (blue bar) and pretreated solids suspended in either sodium citrate buffer (light green bars) or pretreatment liquid (dark green bars). Yellow bars represent the degree of inhibition (DI) of enzymatic saccharification by the pretreatment liquids. Same codification of experimental conditions as in Table 1.

In the experiments with the pretreated solids suspended in pretreatment liquids, the enzymatic digestibility of glucan was lower than in the series with the buffer (Figure 3). The enzymatic digestibility ranged from 44 ± 0.8% for SA-160 to 90 ± 1.6% for A-205. Based on the differences between the hydrolysis series, the degree of inhibition (DI) by the pretreatment liquids was calculated. For A-HTP, the DI was 2.4 ± 0.1% for A-160, and it gradually increased with the temperature, reaching up to 7.3 ± 0.3% for A-205. At 160 °C, the DI was slightly higher for SA-HTP than for A-HTP, but at 190 °C, no discernible difference was found.

The statistical significance of the differences between enzymatic digestibility values achieved in saccharification trials of pretreated solids from different pretreatment conditions ($p < 0.001$), suspended in different liquid media ($p < 0.001$), and their interactions ($p = 0.0155$) was confirmed by two-way analysis of variance (Table 5) and Tukey's posthoc test. Levene's test for homogeneity of variance ($F = 1.210$, $p = 0.333$), and Shapiro's test for normal distribution of residuals ($W = 0.957$, $p = 0.173$) confirmed that the data met the model assumptions.

Table 5. Two-way ANOVA for enzymatic digestibility in saccharification trials with pretreatment solids suspended in either sodium citrate buffer or pretreatment liquids for all the pretreatment conditions.

Source of Variation	Degree of Freedom	Sum of Squares	Mean Squares	*F*-Value	*p*-Value
Pretreatment conditions (P)	5	11990	2398.1	1148.47	<0.001
Liquid medium (L)	1	109	108.5	51.97	<0.001
P:L interactions	5	37	7.4	3.54	0.0155
Residuals	24	50	2.1		

Potential correlations between the enzymatic convertibility of glucan and the contents of xylan and lignin in the pretreated solids were investigated (Figure 4). For both the A-HTP and SA-HTP series, enzymatic digestibility displayed an obvious negative correlation (R^2 0.92) with xylan content (Figure 4a). For lignin content, there was a slightly positive trend (Figure 4b) that can be attributed to indirect effects (hemicellulose removal leading to both higher lignin content and better digestibility).

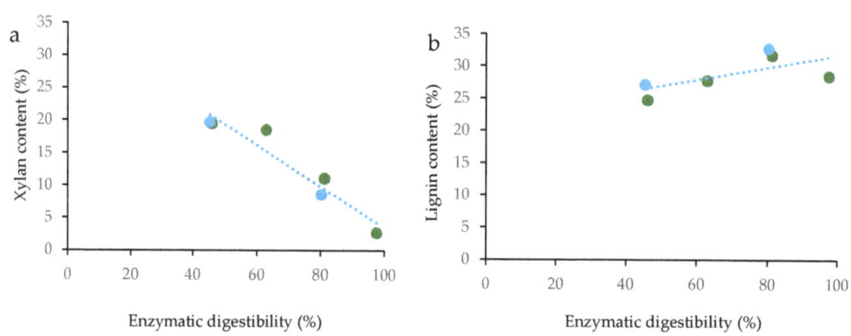

Figure 4. Correlation between enzymatic digestibility of glucan with the content of xylan (**a**) and lignin (**b**) in the pretreated solids. The green markers are for A-HTP and the blue markers are for SA-HTP. $R^2 = 0.92$ (xylan) and $R^2 = 0.45$ (lignin). The plotted lignin content was determined using Py-GC/MS.

In order to have a better picture of the effectiveness of different pretreatment conditions on the saccharification of wheat straw, the yield of sugars per ton of raw material was calculated. For A-HTP, the combined glucose yield (after pretreatment and enzymatic saccharification) increased almost proportionally with the temperature, giving the highest value (407 ± 13 kg/t) at 205 °C (Figure 5). The xylose yield reached a peak (127 ± 8 kg/t) at 190 °C, including the amount solubilized during pretreatment and that resulting from enzymatic hydrolysis. The xylose yield sharply decreased to only 21 ± 1 kg/ton raw wheat straw at 205 °C, mostly due to losses in the pretreatment step. For SA-HTP, glucose and xylose yields at 160 °C were higher than those observed for A-HTP, whereas at 190 °C, they were slightly lower. Putting together glucose and xylose, the highest total sugar yield, 480 ± 12 kg per ton of wheat straw, was observed for A-HTP at 190 °C. That yield was higher than the one achieved by SA-HTP at the same temperature.

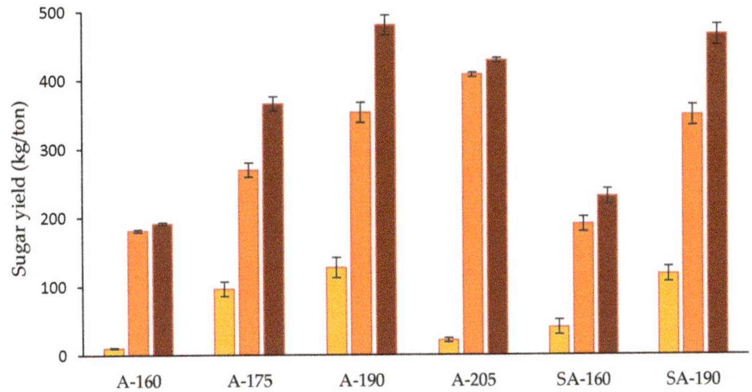

Figure 5. Yield of xylose (yellow bars), glucose (orange bars), and total sugar (brown bars) after pretreatment and enzymatic hydrolysis. Values are given in kg per ton raw wheat straw. Same codification of experimental conditions as in Table 1.

3.5. Fermentability of Pretreatment Liquids

For elucidating the effect of the inhibitors formed during pretreatment on ethanolic fermentation, the fermentability of the pretreatment liquids by *S. cerevisiae* was investigated. A strong inhibitory effect on yeast cell growth was observed for the A-HTP liquid from pretreatment at 205 °C (Figure 6a). Albeit at a lower degree, the cell growth in the medium with SA-190 was also inhibited. Some inhibition was also observed for the pretreatment liquid from A-HTP at 190 °C, but it was hardly noticeable after 12 h. The inhibitory effects exerted by the other pretreatment liquids were minor and restricted to the beginning of the fermentation.

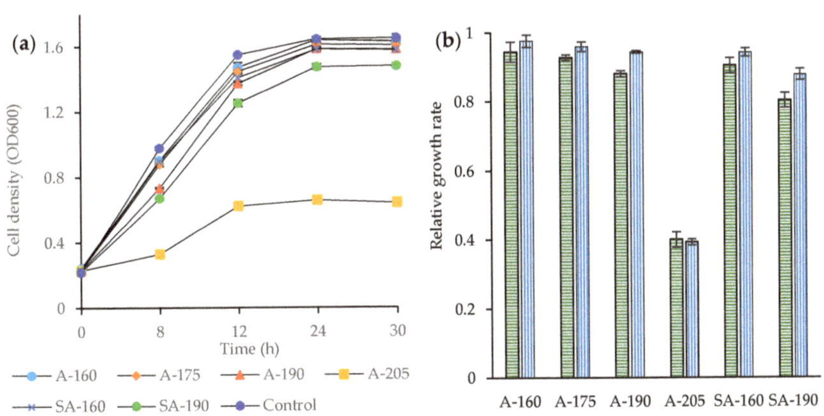

Figure 6. Cell growth of *S. cerevisiae* in the fermentation of the pretreatment liquids (**a**) and relative growth rate (**b**) after 12 h (horizontal green pattern bars) and 24 h (vertical blue pattern bars). Same codification of experimental conditions as in Table 1.

The inhibition pattern of the different pretreatment liquids is better revealed by comparing the relative growth rates (RGRs), which were calculated as the ratio between the growth rates in each pretreatment liquid and the reference. The remarkable inhibition of the 205 °C A-HTP liquid is highlighted by RGR values around 0.40 ± 0.06 after both 12 and 24 h of fermentation (Figure 6b). On the other hand, the fermentability of the 160 °C A-HTP liquid was comparable with that of the control, as indicated by a relative growth rate of 0.98 ± 0.06 after 24 h of fermentation. The other pretreatment liquids displayed RGR values between 0.88 ± 0.03 and 0.96 ± 0.04 after 24 h. Pretreatment liquids from SA-HTP were more inhibitory than those from A-HTP at the same temperature.

4. Discussion

4.1. Chemical Composition of Raw and Pretreated Solids

Although different issues related to hydrothermal pretreatment of wheat straw for bioconversion have already been well investigated, there are other aspects that remain to be addressed. The effects of pretreatment temperature, time and equipment configuration on the yield of solids, solubilization of the main components, and recovery of sugars have been widely discussed [32–35], but important details linking pretreatment conditions with byproduct formation, enzymatic digestibility, and inhibition of biochemical conversion are still lacking. The current study assesses the effects of different pretreatment temperatures and catalytic approaches on the formation of bioconversion inhibitors, the enzymatic digestibility of pretreated solids, and the inhibition of enzymatic saccharification and fermentation by pretreatment liquids.

Increasing the pretreatment temperature from 160 to 205 °C while holding the reaction time at 15 min resulted in increased solubilization of wheat straw constituents and led to a reduction of the gravimetric yield of pretreated solids (Figure 1). The high pretreatment yield (almost 88%) under low-severity pretreatment (160 °C, SF 2.9) indicates that under these conditions, wheat straw was only marginally affected, while the low yield (around 64%) under the most severe pretreatment (205 °C, SF 4.3) revealed major solubilization of the raw material. These values are within the typical range for wheat straw. Min et al. [33] showed a decrease in the yield of pretreated solids from 86% to 80% while increasing the SF from 3.2 to 3.9, and Chen et al. [32] reported yields of approximately 60% in pretreatment at 200 °C for 30 min (SF 4.4).

The analysis of the composition of pretreated solids and pretreatment liquids indicates that the main factor contributing to the reduction of the gravimetric yield is the solubilization of hemicelluloses. The increase of solubilization of hemicelluloses with the temperature range studied is in agreement with previous reports of hydrothermal pretreatment of wheat straw [34]. The fact that the yield of solids was more affected by the stepwise increase of temperature than by the use of sulfuric acid can be explained by the rather low acid-loading that was used.

The observed temperature-related increase of the glucan content in the pretreated solids was a consequence of hemicellulose solubilization (Table 1). Cellulose was not affected to any major extent by the different pretreatment conditions, as indicated by the high glucan recoveries achieved (Figure 2). The lack of substantial differences in glucan recovery between A-HTP and SA-HTP for similar temperatures might be explained by the rather low sulfuric acid-loading used in SA-HTP. In previous studies on hydrothermal pretreatment of sugarcane bagasse using a higher sulfuric acid loading (0.5 g/100 g of reaction mixture instead of 0.5 g/100 g dry biomass), glucan recovery was higher for A-HTP than for SA-HTP under the same temperature and time [35,36].

The decrease of xylan content with the increase of the temperature is a logical result, considering that a temperature-dependent increase of hemicellulose solubilization is normally expected for hydrothermal pretreatment [32]. In contrast to previous reports on studies in which higher sulfuric acid-loadings or longer reaction times were used [37], pretreatment at 160 °C was not effective for xylan solubilization, even when sulfuric acid was used (Table 1). At 190 °C, there was some sulfuric-acid-promoted xylan solubilization, as can be interpreted from the decrease of xylan recovery from around 41% with A-HTP to nearly 30% with SA-HTP (Figure 2).

The observed arabinose solubilization at 160 °C (Table 1) matches well with previous information on the hydrothermal deconstruction of different hemicellulosic components. Under mild conditions, arabinose side-chain moieties are cleaved faster than xylose units [38]. Xylan backbone is initially fragmented to a minor degree, and only after subsequent scissions, a major release of xylo-oligosaccharides and xylose occurs [39]. However, the solubilization of galactose and mannose at 160 °C is surprising considering that mannan and galactan typically exhibit comparable dissolution kinetics to xylan [40], and it would have been expected that they behave similarly under HTP at 160 °C.

The gradual increase of lignin content in the pretreated solids with the increase of the temperature (Table 1) is primarily a consequence of hemicellulose solubilization, as typically observed for hydrothermal processing and acid-based pretreatment methods [13]. As confirmed by combining TSSA and Py-GC/MS analysis, lignin recovery of over 100%, detected for A-HTP at 205 °C and for both SA-HTP experiments (Figure 1), was due to pseudolignin formation, a phenomenon that has not been well-studied previously and which is typically difficult to quantify. The lower lignin fraction found by using Py-GC/MS (Table 2) in comparison with the fraction determined by TSSA (Table 1) for A-205, SA-160, SA-190, and A-190 is due to the fact that TSSA does not distinguish differences between real lignin and Klason-positive partially degraded carbohydrates, i.e., pseudolignin. Py-GC/MS analysis will detect Klason-positive partially degraded carbohydrates as carbohydrates. The lignin recoveries exceeding 100% in Figure 1 can, with certainty, be attributed to pseudolignin formation. For estimating pseudolignin formation, we used the Δ_{Lignin} value, a factor we recently introduced based on TSSA analyses of lignin and biomass characterization using Py-GC/MS [35,36]. As shown by the negative value for raw wheat straw, TSSA and Py-GC/MS do not give identical values, and it is, therefore, a possibility that also some of the pretreated samples, with low negative Δ_{Lignin} values, contain small fractions of pseudolignin. As hydrothermal pretreatment mainly targets hemicelluloses, the lignin showed minor compositional changes, as shown by small changes observed for the G:S:H ratio. There was an evident decrease in the relative abundance of syringyl units (Table 2).

The high ash content of the raw material is within the range previously determined for wheat straw [33,34]. The ash content increased slightly with the temperature as other components were removed and minerals remained in the pretreated solids. These results are in agreement with previous studies showing an increase in ash content with the temperature in HTP of wheat straw [33,34].

The increase of the mass fraction of extractives observed at 190 and 205 °C (Table 1) might be due to the fact that the solvent used in the analytical procedure extracted not only native extractives present in untreated wheat straw but also lignin fragments deposited on the solid phase during pretreatment. This is further strengthened by the finding that at similar temperatures, the content of extractives was higher for SA-HTP, the pretreatment approach that is expected to cause more extensive lignin fragmentation, condensation, and redeposition [41] than for A-HTP, and that the difference was more obvious at 190 °C than at 160 °C.

4.2. Carbohydrates in the Pretreatment Liquids

The carbohydrate profiles of the pretreatment liquids, i.e., relatively high concentrations of xylose, xylo-oligosaccharides (Xylo-OS), glucose, and gluco-oligosaccharides (Gluco-OS) (Table 3), are as expected, considering the polysaccharide content of the raw material (Table 1) and considering that most of the glucan is cellulose, which is more resistant to pretreatment than hemicellulose. The increase of xylose and glucose concentrations with the increase of temperature is typical for hydrothermal pretreatment of wheat straw [31,32] and other herbaceous feedstocks such as sugarcane bagasse [42] and corn stover [43]. The discrepancy between the increase of xylose and the decrease of Xylo-OS for A-HTP at

205 °C compared to 190 °C point towards degradation reactions. This type of reaction has been discussed elsewhere [14].

The relatively high arabinose formation at 160 °C agrees well with the observations regarding the composition of pretreated solids (Section 4.1) and with the literature on the hydrolysis of hemicelluloses [38]. The fast dynamics of arabino- and galacto-oligosaccharides are in agreement with previous studies of HTP under mild conditions. Chen et al. [32] reported maximal arabinose and galactose formation from wheat straw at 160 °C, which is lower than the temperature of the maximum concentrations in the current work (175 °C), but they applied a longer pretreatment time (30 min). A similar trend has also been reported for hydrothermal pretreatment of sugarcane bagasse, which displayed maximum arabinose formation at 180 °C, followed by a sequential reduction of its concentration at 185, 190, and 195 °C [42].

It was unexpected that the amount of monosaccharides released by SA-HTP was almost as low as that of A-HTP and that the amount of oligosaccharides was rather high for both pretreatment approaches. Typically, SA-HTP liquids are rich in monosaccharides, and A-HTP liquids are rich in oligosaccharides. For instance, xylose concentrations around 19 g/L were detected in SA-HTP liquids of sugarcane bagasse at 175 °C for 3.9 min [35], while a yield of 61.7 g xylo-OS per kg raw material was reported for A-HTP of wheat straw at 180 °C for 30 min [32]. That, together with the observation that the yield of pretreated solids was more affected by temperature increase than by the use of sulfuric acid, can be attributed to the low acid-loading, which was not enough to cause any major hydrolysis of hemicelluloses. Although compositional analysis of solids pretreated at 190 °C revealed more extensive xylan solubilization for SA-HTP than for A-HTP (Table 1, Figure 2), the total concentration of xylose and Xylo-OS was comparable for both pretreatment approaches (Table 3). That indicates that a fraction of the solubilized xylan was degraded, and, therefore, it could not be quantified in the pretreatment liquids either as xylose or Xylo-OS.

4.3. *Effects of Pretreatment Conditions on the Formation of Bioconversion Inhibitors*

Degradation reactions leading to byproducts that are inhibitory to microorganisms and enzymes is a common problem for different pretreatment methods, including hydrothermal processing [14]. In the current work, the formation of inhibitory compounds was rather moderate for pretreatment temperatures up to 190 °C, but a clear increase was observed when the pretreatment temperature was raised to 205 °C or when sulfuric acid was used as a catalyst at 190 °C (Table 4).

Increased formation of furan aldehydes for SA-190 can partially explain the previously discussed (Section 4.2) sugars that were not accounted for by carbohydrate analyses. Dehydration of sugars to furan aldehydes is a typical phenomenon for pretreatments performed under acidic conditions and at high temperatures, and the furans can be further degraded if the pretreatment is very severe [14]. There was a sharp increase in sugar degradation when the A-HTP temperature was increased to 205 °C, as indicated by high concentrations of furfural and HMF after pretreatment at 205 °C. At 205 °C, using a method that has not been reported before for hydrothermal pretreatment of wheat straw, furfural was also detected in condensate from the gas phase. The increase of formic acid and levulinic acid at 205 °C indicates further degradation of furans, as is expected for harsh conditions [44]. Acetic acid is formed mainly by the hydrolysis of acetyl groups of hemicelluloses. For the A-HTP series, the concentration of acetic acid continued to increase even up to 205 °C (Table 4), which agrees with the observation that there was still plenty of xylan left in the solid fraction after treatment at 190 °C (Table 1).

The formation pattern of phenolic substances was found to be different depending on the length of the side chain, as phenols with one-carbon side chains increased over the whole temperature range, whereas phenols with two- or three-carbon side chains exhibited a maximum within the temperature range (Table 4). Thermal degradation of phenols with a two- or three-carbon side chain can contribute to increased formation of the corresponding phenolic benzaldehydes. For instance, degradation of coniferyl aldehyde and *p*-coumaraldehyde can result in the formation of vanillin and *p*-benzaldehyde, respectively. Vanillin has been shown to be formed from cleavage of the C_α-C_β bond of acetovanillone through radiolysis [45]. A similar trend was observed in studies of hydrothermal pretreatment of sugarcane bagasse, where the formation of vanillin, *p*-hydroxybenzaldehyde, and syringaldehyde increased with severity, while coniferyl aldehyde concentrations reached a maximum at intermediate severity (log R_0 = 3.8) and then decreased when the severity increased further [35].

4.4. Enzymatic Saccharification of Pretreated Wheat Straw

The reports on enzymatic hydrolysis of pretreated wheat straw are often limited to studies where pretreated solids are suspended in a buffer [33,46], which does not provide information on the effect of the pretreatment liquids on enzymatic conversion. The experimental setup used in this work, with pretreated solids suspended in either buffer or pretreatment liquids, allowed us to investigate the enzymatic digestibility of glucan contained in the pretreated solids and evaluate the inhibitory effect of the pretreatment liquids on the saccharification of glucan.

The results showed that (i) all pretreatment conditions greatly (2.5- to 5.5-fold) enhanced the enzymatic saccharification of wheat straw glucan; (ii) the enzymatic digestibility of pretreated solids increased with the pretreatment temperature and was found to be inversely correlated with the xylan content; (iii) the use of 0.5% (on a DW basis) sulfuric acid in the pretreatment did not have any major impact on saccharification in comparison to autocatalyzed pretreatment; (iv) the pretreatment liquids had a small (2–7%) but clear inhibitory effect on enzymatic saccharification; (v) the degree of inhibition increased with temperature, with no clear differences between SA-HTP pretreatment liquids and A-HTP pretreatment liquids (Figure 3). While some of these results agree with previous studies [41], the finding on the inhibition of enzymatic hydrolysis by the pretreatment liquids is an important contribution of this work to the knowledge on hydrothermal pretreatment. The relatively high enzyme inhibition observed for the pretreatment liquid from A-HTP at 205 °C (7%) can be attributed to its higher content of solubilized aromatics [16], expressed as total phenolic compounds and TAC (Table 4).

The correlations between the enzymatic convertibility of glucan and the content of xylan and lignin in the pretreated solids revealed that for achieving efficient enzymatic hydrolysis of wheat straw glucan, it is important that the pretreatment removes as much xylan as possible, while the effect of lignin is less important. That is in agreement with previous studies of pretreated lodgepole pine, implicating that removal of hemicelluloses is more important than removal of lignin in order to achieve efficient enzymatic saccharification [47].

Since different biorefinery applications, with glucose-based routes and xylose-based routes, can be considered for wheat straw, the total yields of sugars per ton of raw material are an important indicator for a pretreatment method of relevance. This work evaluates the effect of HTP on total sugar yield, including glucose, mainly formed during enzymatic saccharification, and xylose, mainly formed during the pretreatment stage. Although A-HTP at 205 °C was very effective for producing glucose in enzymatic hydrolysis, it led to extensive xylose degradation in the pretreatment stage. Instead, moderate pretreatment conditions (A-HTP, 190 °C) resulted in the highest xylose yield and, even though the glucose yield in the enzymatic saccharification step was lower than after pretreatment at 205 °C, the combined sugar yield was the highest. Thus, the investigation has revealed that since the highest yield was obtained with A-HTP at 190 °C, no sulfuric acid and no temperatures over 200 °C are required for maximizing sugar production from wheat straw.

4.5. Fermentability of the Pretreatment Liquids

The inhibitory effects on *S. cerevisiae* by the pretreatment liquids were low except for A-205. Additionally, with regard to microbial inhibition, A-190 turned out to be a good initial step for the biorefining of wheat straw. The chemical analyses indicated that the inhibitory impact on the fermentability of pretreatment liquid from A-205 corresponded with high values for TAC, total phenolics, TCAC, and furfural (Table 4). TAC has recently been found to be a useful indicator of inhibition in studies of hydrothermal pretreatment of sugarcane bagasse and Norway spruce [36]. TAC values are related to the values of total phenolics and furfural, both of which contribute to TAC. The most inhibitory (A-205) and second-most inhibitory (SA-190) pretreatment liquids also exhibited the highest and second-highest values for TAC, total phenolics, and furfural. Regarding TCAC, it is noteworthy that the values were always < 100 mM. Previous experiments with *S. cerevisiae* yeast indicate that inhibitory effects were achieved when the total concentration of aliphatic acids was above 100 mM [48]. Therefore, it is reasonable to assume that it was aromatic and heteroaromatic substances that caused the inhibitory effects rather than carboxylic acids. The absence or very low concentrations of newly discovered inhibitors such as formaldehyde and *p*-benzoquinone can tentatively be attributed to the relatively low lignin content of the raw material.

5. Conclusions

This investigation clarifies correlations between operational conditions during hydrothermal pretreatment and critical aspects of biochemical conversion of wheat straw, including hemicellulose removal, susceptibility to enzymatic saccharification, total sugar yields, byproduct formation, and inhibitory effects on enzymatic saccharification and ethanolic fermentation. Pseudolignin formation in autocatalyzed hydrothermal pretreatment at 205 °C and in sulfuric-acid-catalyzed hydrothermal pretreatment at lower temperatures was demonstrated. The importance of the removal of hemicelluloses for achieving high enzymatic digestibility of glucan was shown. Hydrothermal pretreatment without using mineral acids and without exceeding 200 °C was found to be the best approach to achieve high sugar yields while minimizing byproduct formation.

Author Contributions: D.I., conceptualization, methodology, investigation, data curation, writing—original draft preparation; S.S., methodology, validation, formal analysis; L.J.J., conceptualization, methodology, supervision, writing—review and editing, project administration, funding acquisition; C.M., conceptualization, methodology, supervision, writing—review and editing. All authors have read and agreed to the published version of the manuscript.

Funding: This research was funded by the Swedish Energy Agency (P41285-1 and P47516-1), Bio4Energy (www.bio4energy.se, accessed on 4 March 2021), and the Kempe Foundations.

Institutional Review Board Statement: Not applicable.

Informed Consent Statement: Not applicable.

Acknowledgments: The support provided by Guochao Wu in the preparation of the fermentability experiment, Junko Takahashi Schmidt in the Py-GC/MS analysis, and the statistical assistance given by Francesca Pilotto are gratefully acknowledged.

Conflicts of Interest: The authors declare no conflict of interest. The funding agencies had no role in the design of the study, in the interpretation of data, in the writing of the manuscript, or in the decision to publish the results.

References

1. Ragauskas, A.J.; Williams, C.K.; Davison, B.H.; Britovsek, G.; Cairney, J.; Eckert, C.A.; Frederick, W.J., Jr.; Hallett, J.P.; Leak, D.J.; Liotta, C.L.; et al. The path forward for biofuels and biomaterials. *Science* **2006**, *311*, 484–489. [CrossRef] [PubMed]
2. Zech, K.M.; Meisel, K.; Brosowski, A.; Toft, L.V.; Müller-Langer, F. Environmental and economic assessment of the Inbicon lignocellulosic ethanol technology. *Appl. Energy* **2016**, *171*, 347–356. [CrossRef]
3. Dias, M.O.S.; Lima, D.R.; Mariano, A.P. Techno-economic analysis of cogeneration of heat and electricity and second-generation ethanol production from sugarcane. In *Advances in Sugarcane Biorefinery*; Chandel, A.K., Silveira, M.H.L., Eds.; Elsevier: Amsterdam, The Netherlands, 2018; pp. 197–212.
4. Jiang, Y.; Xin, F.; Lu, J.; Dong, W.; Zhang, W.; Zhang, M.; Wu, H.; Ma, J.F.; Jiang, M. State of the art review of biofuels production from lignocellulose by thermophilic bacteria. *Bioresour. Technol.* **2017**, *245*, 1498–1506. [CrossRef] [PubMed]

5. Susmozas, A.; Martín-Sampedro, R.; Ibarra, D.; Eugenio, M.E.; Iglesias, R.; Manzanares, P.; Moreno, A.D. Process strategies for the transition of 1G to advanced bioethanol production. *Processes* **2020**, *8*, 1310. [CrossRef]
6. FAOSTAT—Food and Agriculture Organization Corporate Statistical Database. Available online: http://www.fao.org/faostat/en/#search/wheat (accessed on 24 September 2020).
7. Ericsson, K.; Nilsson, L.J. Assessment of the potential biomass supply in Europe using a resource focused approach. *Biomass Bioenerg.* **2006**, *30*, 1–15. [CrossRef]
8. Thomsen, M.H.; Thygesen, A.; Thomsen, A.B. Hydrothermal treatment of wheat straw at pilot plant scale using a three-step reactor system aiming at high hemicellulose recovery, high cellulose digestibility and low lignin hydrolysis. *Bioresour. Technol.* **2008**, *99*, 4221–4228. [CrossRef]
9. Thomsen, M.H.; Thygesen, A.; Thomsen, A.B. Identification and characterization of fermentation inhibitors formed during hydrothermal treatment and following SSF of wheat straw. *Appl. Microbiol. Biotechnol.* **2009**, *83*, 447–455. [CrossRef]
10. Talebnia, F.; Karakashev, D.; Angelidaki, I. Production of bioethanol from wheat straw: An overview on pretreatment, hydrolysis and fermentation. *Bioresour. Technol.* **2010**, *101*, 4744–4753. [CrossRef]
11. Galbe, M.; Wallberg, O. Pretreatment for biorefineries: A review of common methods for efficient utilisation of lignocellulosic materials. *Biotechnol. Biofuels* **2019**, *12*, 294. [CrossRef] [PubMed]
12. El-Zawawy, W.K.; Ibrahim, M.M.; Abdel-Fattah, Y.R.; Soliman, N.A.; Mahmoud, M.M. Acid and enzyme hydrolysis to convert pretreated lignocellulosic materials into glucose for ethanol production. *Carbohyd. Polym.* **2011**, *84*, 865–871. [CrossRef]
13. Hendriks, A.; Zeeman, G. Pretreatments to enhance the digestibility of lignocellulosic biomass. *Bioresour. Technol.* **2009**, *100*, 10–18. [CrossRef]
14. Jönsson, L.J.; Martín, C. Pretreatment of lignocellulose: Formation of inhibitory by-products and strategies for minimizing their effects. *Bioresour. Technol.* **2016**, *199*, 103–112. [CrossRef] [PubMed]
15. Gandla, M.L.; Martín, C.; Jönsson, L.J. Analytical enzymatic saccharification of lignocellulosic biomass for conversion to biofuels and bio-based chemicals. *Energies* **2018**, *11*, 2936. [CrossRef]
16. Silva, J.P.A.; Carneiro, L.M.; Roberto, I.C. Treatment of rice straw hemicellulosic hydrolysates with advanced oxidative processes: A new and promising detoxification method to improve the bioconversion process. *Biotechnol. Biofuels* **2013**, *6*, 23. [CrossRef]
17. Rasmussen, H.; Sørensen, H.R.; Meyer, A.S. Formation of degradation compounds from lignocellulosic biomass in the biorefinery: Sugar reaction mechanisms. *Carbohydr. Res.* **2014**, *385*, 45–57. [CrossRef]
18. Jönsson, L.J.; Alriksson, B.; Nilvebrant, N.O. Bioconversion of lignocellulose: Inhibitors and detoxification. *Biotechnol. Biofuels* **2013**, *6*, 16. [CrossRef]
19. Cavka, A.; Stagge, S.; Jönsson, L.J. Identification of small aliphatic aldehydes in pretreated lignocellulosic feedstocks and evaluation of their inhibitory effects on yeast. *J. Agric. Food Chem.* **2015**, *63*, 9747–9754. [CrossRef] [PubMed]
20. Martín, C.; Wu, G.; Wang, Z.; Stagge, S.; Jönsson, L.J. Formation of microbial inhibitors in steam-explosion pretreatment of softwood impregnated with sulfuric acid and sulfur dioxide. *Bioresour. Technol.* **2018**, *262*, 242–250. [CrossRef] [PubMed]
21. Stagge, S.; Cavka, A.; Jönsson, L.J. Identification of benzoquinones in pretreated lignocellulosic feedstocks and inhibitory effects on yeast. *AMB Express* **2015**, *5*, 62. [CrossRef] [PubMed]
22. Normark, M.; Pommer, L.; Gräsvik, J.; Hedenström, M.; Gorzsás, A.; Winestrand, S.; Jönsson, L.J. Biochemical conversion of torrefied Norway spruce after pretreatment with acid or ionic liquid. *Bioenergy Res.* **2016**, *9*, 355–368. [CrossRef]
23. Shinde, S.D.; Meng, X.; Kumar, R.; Ragauskas, A.J. Recent advances in understanding the pseudo-lignin formation in a lignocellulosic biorefinery. *Green Chem.* **2018**, *20*, 2192–2205. [CrossRef]
24. Toquero, C.; Bolado, S. Effect of four pretreatments on enzymatic hydrolysis and ethanol fermentation of wheat straw. Influence of inhibitors and washing. *Bioresour. Technol.* **2014**, *157*, 68–76. [CrossRef]
25. Overend, R.P.; Chornet, E. Fractionation of lignocellulosics by steam-aqueous pretreatments. *Phil. Trans. R. Soc. Lond.* **1987**, *A321*, 523–536.
26. Sluiter, A.; Hames, B.; Ruiz, R.; Scarlata, C.; Sluiter, J.; Templeton, D.; Crocker, D. *Determination of Structural Carbohydrates and Lignin in Biomass*; Technical Report NREL/TP-510-42618; National Renewable Energy Laboratory: Golden, CO, USA, 2012; p. 15.
27. Sluiter, A.; Ruiz, R.; Scarlata, C.; Sluiter, J.; Templeton, D. *Determination of Ash in Biomass*; Technical Report NREL/TP-510-42622; National Renewable Energy Laboratory: Golden, CO, USA, 2008; p. 12.
28. Gerber, L.; Eliasson, M.; Moritz, T.; Sundberg, B. Multivariate curve resolution provides a high-throughput data processing pipeline for pyrolysis-gas chromatography/mass spectrometry. *J. Anal. Appl. Pyrol.* **2012**, *95*, 95–100. [CrossRef]
29. Singleton, V.L.; Orthofer, R.; Lamuela-Raventós, R.M. Analysis of total phenols and other oxidation substrates and antioxidants by means of Folin-Ciocalteu reagent. *Methods Enzymol.* **1999**, *299*, 152–178.
30. R Core Team. *R: A Language and Environment for Statistical Computing*; R Foundation for Statistical Computing: Vienna, Austria, 2019.
31. Ambye-Jensen, M.; Thomsen, S.T.; Kádár, Z.; Meyer, A.S. Ensiling of wheat straw decreases the required temperature in hydrothermal pretreatment. *Biotechnol. Biofuels* **2013**, *6*, 116. [CrossRef]
32. Chen, X.; Li, H.; Sun, S.; Cao, X.; Sun, R. Co-production of oligosaccharides and fermentable sugar from wheat straw by hydrothermal pretreatment combined with alkaline ethanol extraction. *Ind. Crops Prod.* **2018**, *111*, 78–85. [CrossRef]

33. Min, D.; Wei, L.; Zhao, T.; Li, M.; Jia, Z.; Wan, G.; Zhang, Q.; Qin, C.; Wang, S. Combination of hydrothermal pretreatment and sodium hydroxide post-treatment applied on wheat straw for enhancing its enzymatic hydrolysis. *Cellulose* **2018**, *25*, 1197–1206. [CrossRef]
34. Merali, Z.; Ho, J.D.; Collins, S.R.A.; Le Gall, G.; Elliston, A.; Käsper, A.; Waldron, K.W. Characterization of cell wall components of wheat straw following hydrothermal pretreatment and fractionation. *Bioresour. Technol.* **2013**, *131*, 226–234. [CrossRef] [PubMed]
35. Ilanidis, D.; Stagge, S.; Jönsson, L.J.; Martín, C. Effects of operational conditions on auto-catalyzed and sulfuric-acid-catalyzed hydrothermal pretreatment of sugarcane bagasse at different severity factor. *Ind. Crops Prod.* **2021**, *159*, 113077. [CrossRef]
36. Ilanidis, D.; Wu, G.; Stagge, S.; Martín, C.; Jönsson, L.J. Effects of redox environment on hydrothermal pretreatment of lignocellulosic biomass under acidic conditions. *Bioresour. Technol.* **2021**, *319*, 24211. [CrossRef] [PubMed]
37. Ji, W.; Shen, Z.; Wen, Y. Hydrolysis of wheat straw by dilute sulfuric acid in a continuous mode. *Chem. Eng. J.* **2015**, *260*, 20–27. [CrossRef]
38. Mäki-Arvela, P.; Salmi, T.; Holmbom, B.; Willför, S.; Murzin, D.Y. Synthesis of sugars by hydrolysis of hemicelluloses—A review. *Chem. Rev.* **2011**, *111*, 5638–5666. [CrossRef] [PubMed]
39. Ibbett, R.; Gaddipati, S.; Davies, S.; Hill, S.; Tucker, G. The mechanisms of hydrothermal deconstruction of lignocellulose: New insights from thermal- analytical and complementary studies. *Bioresour. Technol.* **2011**, *102*, 9272–9278. [CrossRef] [PubMed]
40. Jara, R.; Lawoko, M.; van Heiningen, A. Intrinsic dissolution kinetics and topochemistry of xylan, mannan, and lignin during auto-hydrolysis of red maple wood meal. *Can. J. Chem. Eng.* **2018**, *97*, 649–661. [CrossRef]
41. Trajano, H.L.; Engle, N.L.; Foston, M.; Ragauskas, A.J.; Tschaplinski, T.J.; Wyman, C.E. The fate of lignin during hydrothermal pretreatment. *Biotechnol. Biofuels* **2013**, *6*, 110. [CrossRef]
42. Rocha, G.J.M.; Silva, V.F.N.; Martín, C.; Gonçalves, A.R.; Nascimento, V.M.; Souto-Maior, A.M. Effect of xylan and lignin removal by hydrothermal pretreatment on enzymatic conversion of sugarcane bagasse cellulose for second generation ethanol production. *Sugar Tech.* **2013**, *15*, 390–398. [CrossRef]
43. Lü, H.; Shi, X.; Li, Y.; Meng, F.; Liu, S.; Yan, L. Multi-objective regulation in autohydrolysis process of corn stover by liquid hot water pretreatment. *Chin. J. Chem. Eng.* **2017**, *25*, 499–506. [CrossRef]
44. Fengel, D.; Wegener, G. *Wood Chemistry, Ultrastructure, Reactions*; Walter de Gruyter: Berlin, Germany, 1989.
45. Madureira, J.; Leal, J.P.; Botelho, M.L.; Cooper, W.J.; Melo, R. Radiolytic degradation mechanism of acetovanillone. *Chem. Eng. J.* **2020**, *382*, 122917. [CrossRef]
46. Wang, R.; Yue, J.; Jiang, J.; Li, J.; Zhao, J.P.; Xia, H.H.; Wang, K.; Xu, J.M. Hydrothermal CO_2-assisted pretreatment of wheat straw for hemicellulose degradation followed with enzymatic hydrolysis for glucose production. *Waste Biomass Valor* **2021**, *12*, 1483–1492. [CrossRef]
47. Leu, S.Y.; Zhu, J.Y. Substrate-related factors affecting enzymatic saccharification of lignocelluloses: Our recent understanding. *Bioenerg. Res.* **2013**, *6*, 405–415. [CrossRef]
48. Larsson, S.; Palmqvist, E.; Hahn-Hägerdal, B.; Tengborg, C.; Stenberg, K.; Zacchi, G.; Nilvebrant, N.-O. The generation of fermentation inhibitors during dilute acid hydrolysis of softwood. *Enzyme Microb. Tech.* **1999**, *24*, 151–159. [CrossRef]

agronomy

MDPI

Article

Steam Refining with Subsequent Alkaline Lignin Extraction as an Alternative Pretreatment Method to Enhance the Enzymatic Digestibility of Corn Stover

Malte Jörn Krafft [1,†], Marie Bendler [2,†], Andreas Schreiber [1] and Bodo Saake [1,*]

1 Chemical Wood Technology, University of Hamburg, Haidkrugsweg 1, 22885 Barsbüttel, Germany; malte.krafft@uni-hamburg.de (M.J.K.); andreas.schreiber@uni-hamburg.de (A.S.)

2 Thünen Institute of Wood Research, Haidkrugsweg 1, 22885 Barsbüttel, Germany; marie.bendler@thuenen.de

* Correspondence: bodo.saake@uni-hamburg.de; Tel.: +49-40-822-459-206

† Authors of equal contribution.

Received: 7 April 2020; Accepted: 4 June 2020; Published: 8 June 2020

Abstract: Agricultural residues are promising and abundant feedstocks for the production of monomeric carbohydrates, which can be gained after pretreatment and enzymatic hydrolysis. These monomeric carbohydrates can be fermented to platform chemicals, like ethanol or succinic acid. Due to its high availability, corn stover is a feedstock of special interest in such considerations. The natural recalcitrance of lignocellulosic material against degradation necessitates a pretreatment before the enzymatic hydrolysis. In the present study, a novel combination of steam refining and alkaline lignin extraction was tested as a pretreatment process for corn stover. This combination combines the enhancement of the enzymatic hydrolysis and steam refining lignin can be gained for further utilization. Afterward, the obtained yields after enzymatic hydrolysis were compared with those after steam refining without alkaline extraction. After steam refining at temperatures between 160 °C and 210 °C and subsequent enzymatic hydrolysis with Cellic® CTec2, it was possible to enhance the digestibility of corn stover and to achieve 65.4% of the available carbohydrates at the lowest up to 89% at the highest conditions as monomers after enzymatic hydrolysis. Furthermore, the enzymatic degradation could be optimized with a subsequent alkaline lignin extraction, especially at low severities under three. After this combined pretreatment, it was possible to enhance the enzymatic digestibility and to achieve up to 106.4% of the available carbohydrates at the lowest conditions and up to 102.2% at the highest temperature as monomers after following enzymatic hydrolysis, compared to analytical acid hydrolysis. Regarding the utilization of the arising lignin after extraction, the lignin was characterized with regard to the molar mass and carbohydrate impurities. In this context, it was found that higher amounts and higher purities of lignin can be attained after pretreatment at severities higher than four.

Keywords: corn stover; pretreatment; steam refining; enzymatic hydrolysis; alkaline extraction; lignin

1. Introduction

Corn stover is a well-studied agricultural residue of the corn kernel production [1]. It is highly available and a significant amount of the produced straw is undervalued and until now not harvested [2]. Reasons, therefore, are the prevention of soil erosion or the maintaining of soil organic carbon [2,3]. Nonetheless, corn stover is one of the most promising feedstock candidates for large-scale lignocellulose biorefineries [4].

Because of the chemical composition of lignocelluloses, containing high amounts of cellulose and hemicelluloses, the enzymatic hydrolysis (EH) with cellulases is a suitable way to obtain monomeric carbohydrates for fermentation from the mentioned polysaccharides [3]. The hydrolysis of the glucoside linkage with cellulases (mostly from *T. reesei*) is highly specific to the β-(1→4) glycosidic bonds of the cellulose chain, resulting in high yields with nearly no undesired by-products formed in the EH. However, inhibitors are normally formed during pretreatment prior to EH.

The chemical composition of corn stover as a lignocellulosic feedstock involves also disadvantages, like the natural recalcitrance of the complex cellulose-hemicellulose-lignin-structure against microbial degradation and EH [5]. One important way to conquer this recalcitrance is the pretreatment of the raw material [6]. Over the years, different methods and applications have been invented and common classifications distinguish between biological, chemical, physical and (hydro)-thermal processes [6].

One additionally described category is the physicochemical pretreatment [7]. Well-described physicochemical pretreatments are steam pretreatments with and without (an acid) catalyst. They are reported as the most often investigated pretreatment methods and they combine the advantages of a simple process, low energy costs and no necessary recycling of process chemicals [7,8].

An alternative physicochemical pretreatment to the generally used steam explosion technique is steam refining [9–14]. Whereas the defibration of the fibers in steam explosion processes is achieved by pressure relief, some studies revealed that the explosion part of the steam explosion is nearly unnecessary for the enhancement of EH yields [15]. Therefore, the defibration can also be achieved by a refining step at the end of the steaming, then called the steam refining process by, e.g., Schütt et al. [12]. Nonetheless, in contrast to the defibration, the dimension of the used biomass particles has shown a much bigger impact on the results of the subsequent EH [12,15]. For steam-refined poplar wood, EH yields increase up to a severity of around four, which represents temperatures around 200 °C and holding times between 10 and 15 min. At severities higher than four the yields decline due to secondary carbohydrate degradation reactions [11]. Unfortunately, degradation of the pentoses to furfural and degradation of the hexoses to 5-hydroxymethylfurfural (5-HMF) occur in processes with increased temperature, like in steam processes. These components have an inhibitory effect on enzymes and it is essential to get knowledge on the best process conditions [16]. However, 5-HMF and furfural are platform chemicals by themselves and can be converted into several value-added products [17,18].

Today, industrial applications of cellulose systems are available at the commercial market for different purposes [19] and the use of nonfood raw material is of high interest. Further, the use of fermentation sugars and the production of platform chemicals from renewable resources are also highlighted in the literature, especially when the applied biorefinery approach is not subject to the food and fuel debate [20–22]. From this point of view, it is favorable to use undervalued agricultural residues, like corn stover, for the renewable production of highly valuable products.

Several authors describe the effect of the lignin content of the used raw material on the performance of the EH. It can be stated that lignin has a negative impact on the yields after EH [23]. Further, the formation of unproductive bindings between lignin and enzymes are described [24]. On the other hand, it was also reported that a nearly complete delignification below 5% lignin content of corn stover with acidified sodium chlorite results in a significant yield loss during the EH, whereas a softer dilute acid pretreatment improves the EH yields [25]. Suggested reasons are the aggregation of the cellulosic microfibrils after the near removal of lignin and xylan and a resulting decreased accessibility. Stücker et al. [13] report further about the utilization of alkaline-extracted poplar lignins in lignin–phenol–formaldehyde resins after a steam refining process. Due to the reported improvement of the enzymatic hydrolysis and the reported possibility of utilization, the influence of an alkaline lignin extraction should be tested for steam-refined corn stover, although alkaline treatment is reported as a preliminary treatment by itself [26].

The aim of the present study was to investigate the effect of different steaming severities on the EH of corn stover. The steamed fibers were subjected to EH with and without alkaline lignin extraction in order to differentiate the effect of lignin on the overall process balance. The degradation products of the carbohydrates were detected to get an overview of inhibitory compounds in the liquid fraction after the process. Furthermore, characterization of the extracted lignins was performed correlating steaming severities and lignin characteristics.

2. Materials and Methods

2.1. Raw Material

The used corn stover was harvested in 2018 in Fulda, Hesse (Germany). The material was separated into leaves, stalks, corn cobs and further impurities and was air-dried. After conditioning by air-drying to a stable dry matter content of 90.7%, the material was chopped with a garden chipper into segments with a length between 6 and 8 cm. For steam refining experiments, only leaves and stalks were used.

For raw material analysis, the ash content was measured according to TAPPI standard T 211 om-16. Extractives were determined by Accelerated Solvent Extraction (ASE) of milled (≤1 mm) material with an ASE 350 (Thermo Scientific™ Dionex™, Waltham, MA, USA). Three extraction steps for 10 min at 10 MPa with solvents of different polarities (petrol ether (70 °C); acetone/water 9:1 (70 °C); water (90 °C)) were conducted.

Two-step acid hydrolysis was performed subsequently for the determination of monomeric carbohydrates. Two hundred milligrams of dry material were prehydrolyzed for 60 min with 2 mL of 72% H_2SO_4. The reaction was stopped by the addition of 6 mL deionized water and the sample was transferred with 50 mL deionized water into a 100 mL volumetric flask. The second step of hydrolysis was conducted for 40 min at 120 °C and 0.12 MPa overpressure [27].

Afterward, the samples were cooled to room temperature and then filtered through a sintered glass frit (G4). The undiluted filtrate was used for further analysis, described in detail in Section 2.6. The acid-insoluble residue was washed, dried at 105 °C and gravimetrically weighed [12,27].

2.2. Steam Refining

The process of steam refining was conducted in a 10 L defibrator (Martin Busch & Sohn GmbH, Schermbeck, Germany). The input of raw material was 200 g dry material. The four blade-refiner-system (illustrated in [28]) in the reactor was rotated only in the final 30 s of steam treatment. The severity factor was calculated with Equation (1) according to Overend and Chornet [29]:

$$\log R_0 = \log\left(t \times e^{\frac{(T-100)}{14.75}}\right) \tag{1}$$

with time in minutes (t); temperature in °C (T).

For the further understanding of the used process, the different steps of the process are illustrated as a sequential process schematic in Figure 1.

After steaming and refining the raw material, the fiber fraction was washed and the yields were calculated after measuring the solid content. The liquid extract fraction including the wash water was separated and the amount of the combined aqueous extract fraction was gravimetrically measured for further calculations. For calculating the extract yield, the dilution and solid content of the combined extract were gravimetrically measured after freeze-drying. Further process steps will be explained in the following chapters. According to Equation (1), a trial design from 160 °C up to 210 °C was performed with different time steps as outlined in Table 1.

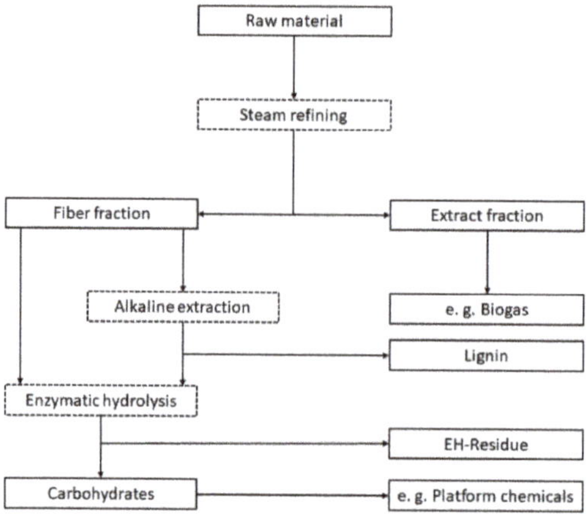

Figure 1. Sequential schematic of the used process. Dotted squares representing process steps, solid lines represent products.

Table 1. Reaction conditions and corresponding severity factors.

	Temperature (°C)	Time (min)	Severity Factor (log R_0)
1	160	10	2.77
2	170	10	3.06
3	180	10	3.36
4	180	15	3.53
5	190	10	3.65
6	190	15	3.83
7	190	20	3.95
8	200	10	3.94
9	200	15	4.12
10	200	20	4.25
11	210	10	4.24
12	210	15	4.41
13	210	20	4.54

2.3. *Acid Hydrolysis of the Extract and Fibers*

The acid hydrolysis of the freeze-dried extracts was conducted as a one-step hydrolysis. One hundred milligrams of dry extractive material were suspended with ultrasound in 10 mL deionized water. Then, 1.8 mL 2N H_2SO_4 were added and the suspension was hydrolyzed for 40 min at 120 °C and 0.12 MPa.

For carbohydrate analysis of fibers after steam refining, a two-step hydrolysis (see Section 2.1.) was performed on air-dried and fine milled samples.

2.4. *Alkaline Lignin Extraction*

The lignin extraction was performed according to Klupsch [9] and Schütt et al. [11]. Twenty grams of dry material was treated with 1.6 g (8% of dry raw material) sodium hydroxide and filled up with water to a consistency of 10% solid content. The treatment was carried out at 90 °C for 60 min. The filtered extract was acidified with 6 mL glacial acetic acid to a pH below 4. The precipitated lignin was separated by centrifugation at 18,000× *g* for 10 min.

The fiber residue was washed with hot deionized water and the solid content was determined for yield calculations. The lignin fraction was vacuum dried for 24 h and, afterward, the yields were determined gravimetrically.

2.5. Enzymatic Hydrolysis (EH)

The EH of the never-dried fibers was performed with 300 µL cellulases (Cellic® CTec2, Novozymes A/S, Bagsværd, Denmark) and 50 µL β-glucosidases (Novozyme 188, Novozymes A/S, Bagsværd, Denmark) per gram dry material for 72 h at 45 °C and a dry matter content of 4%. The dry matter content was adjusted with a pH 5 phosphate citrate (McIlvaine) buffer. Afterward, the hydrolysate and the fibers were transferred into a 250 mL volumetric flask and were filled up with deionized water. The suspension was filtrated through a sintered glass frit (G4) and the undiluted filtrate was used for carbohydrate analysis.

2.6. Analytical Methods

Monomeric carbohydrates in the hydrolysates were determined by borate–anion exchange chromatography (borate–AEC) with a Dionex™ UltiMate™ 3000 (Thermo Fisher Scientific™, Waltham, MA, USA) and MCI GEL® CA08F (Mitsubishi Chemical, Tokio, Japan) as anion exchange resin. Two potassium tetraborate/boric acid-buffers (pH 8.6 and pH 9.5) were used in different concentrations as mobile phase after post-column derivation at 65 °C. Carbohydrates were detected at 560 nm via UV/VIS-spectroscopy. More detailed information about the used borate–AEC was reported by Lorenz et al. [27].

For detection of furfural and 5-hydroxymethylfurfural, reversed phase-high-performance liquid chromatography (RP-HPLC) separation was performed with an AQUASIL™ C_{18} (250 × 4.6 mm; Thermo Fisher Scientific™, Waltham, MA, USA) column for 80 min with 10 µL extract at 25 °C. As a mobile phase, weak acidic water (A; 1 mM H_3PO_4) and acetonitrile (B; C_2H_3N) were used as eluents in different concentrations and a flow rate of 1 mL/min like shown in Table 2. The detection was conducted at 280 nm.

Table 2. Concentrations of the two eluents over time during RP-HPLC.

Time (min)	c (Eluent A) %	c (Eluent B) %
0	97.5	2.5
20	85	15
50	68	32
56	62	38
60	59	41
63	58	42
70	58	42
80	0	100

Size exclusion chromatography (SEC) was performed according to Podschun et al. [30] with a mixture of DMSO and 0.1% LiBr as eluent. One guard PolarGel-M column (50 × 7.5 mm; Agilent, Santa Clara, CA, USA) and two PolarGel-M columns (300 × 7.5 mm; Agilent, Santa Clara, CA, USA) were used with a flow rate of 0.5 mL/min^{-1} at 60 °C. Glucose and polyethylene glycol were used as standards with a refractive index detector (RI-501, Shodex™, Munich, Germany). The dissolved samples (c = 1 mg/mL^{-1}) were shaken for 24 h at room temperature into the eluent. The sample detection was made with a UV-2077 detector (JASCO, Pfungstadt, Germany) at 280 nm and phenol red as detector matching.

3. Results and Discussion

3.1. Raw Material Characterization

A comprehensive analysis of the raw material is needed to monitor the effect of steam refining. In the first step, the delivered corn stover was fractionated into its different components and existing impurities. The biggest fractions were leaves (44.4%) and stalks (38.6%). In nearly equal amounts, impurities (8.8%), mainly sand and corncobs (7.2%), occur. Further, minor amounts of corn silks (0.5%) and corn kernels (0.4%) were contained. In contrast to the determined corn stover composition, other studies report varying results. For example, Pordesimo et al. [31] show for corn stover from Tennessee, USA, a composition of 50.9% stalks, 21% leaves, 15.2% corn cobs and 12.9% husks after excluding the grain fraction. Further, they report good accordance with previous studies. However, corn stover is a natural product and its composition depends on the used variety of maize, the environmental conditions, the harvest time of the raw material and the harvesting technology applied [32].

As described before, the main components of corn stover (leaves, stalks and corncobs) represent around 90% of the delivered material. Not surprising, there is only a small amount of corn kernels left after harvesting the corn with a combine harvester. Nevertheless, a significant amount of sand is included as an impurity. For further investigations, the fractions were separated and the steam refining was performed with the leave and stalk fraction.

Hereafter, the chemical compositions of the used raw material and of pure leaves and stalks were analyzed. The results for the carbohydrate distribution, expressed as monomers, the lignin content, ash and the amounts of extracts, are listed in Table 3.

Table 3. Chemical composition of the used raw material in % based on raw material.

		Raw Material
Extractives	Petrol ether	0.8
	Acetone/Water (9:1)	8.1
	Water	7.3
	Σ	16.2
Carbohydrates	Glucose	35.6
	Xylose	19.5
	Arabinose	2.9
	Galactose	0.9
	Mannose	0.3
	Rhamnose	0.1
	Σ	59.3
Lignin	acid-insoluble [1]	17.1 [2]
	acid-soluble	2.2
	Σ	19.3 [2]
Ash		6.4

[1] Mainly analogs to Klason–Lignin [33,34]. [2] Proteins, e.g., from leaves can partly be detected as well in the acid-insoluble residue after hydrolysis.

As mentioned, the main characteristics of the original raw material mix were analyzed by ASE, two-step acidic hydrolysis with the following borate–AEC and determination of the ash content (Table 3). Nonetheless, due to the measuring method, proteins from the leaves may partly be detected in the acid-insoluble hydrolysis residue. Therefore, they are overestimating the detected acid-insoluble lignin content. The insoluble amounts of the raw material might be as well indicating slightly high lignin contents.

However, the found chemical composition is in good accordance with reference values for the different chemical fractions presented in the literature [35–37].

3.2. *Fractions after Steam Refining and Characterization*

3.2.1. Fiber and Extract Yields

Steam refining was conducted in a temperature range from 160 °C to 210 °C with severities from 2.77 to 4.54 (Table 1). To check the statistical scattering, a triplicate at the same severity was tested before. From this previous experiment is known that the standard deviation for the fiber and extract yields is below ± 1%. After steaming, the solid fiber fraction and the liquid extract fraction were separated, and the yields were calculated after measuring the moisture content of the fiber fraction and the solid content of the extract fraction (Figure 2).

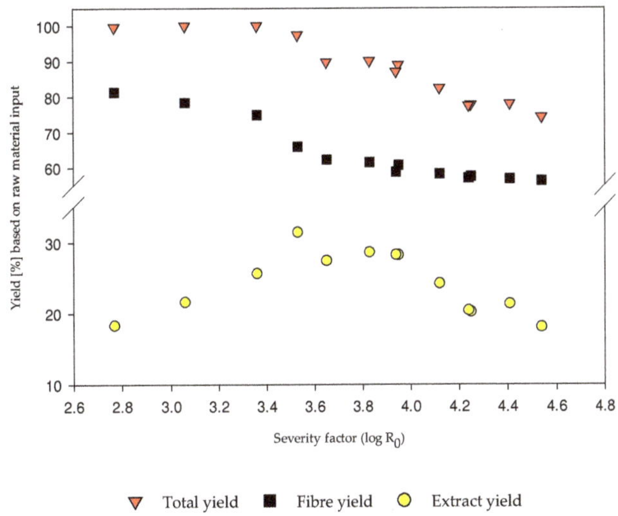

Figure 2. Fiber, extract and total yields after steam refining at different severities.

As illustrated in Figure 2 there is a clear tendency for the fiber, extract and total yield with increasing severity of the steam refining. While the fiber fraction yield decreases steadily with enhanced severity, the extract fraction yield shows a maximum at a severity around four. After that point, the extract yield decreases due to degradation reactions of the carbohydrates at severities higher than four. The total yield decreases continuously and at a severity of 4.54, it falls below 75% based on raw material input. The reduced recovery with increasing severity can be attributed to the formation of volatile components, which were not accounted for in the mass balance.

As stated for the fiber yield, extract yields and total yield, there is a decreasing tendency with increasing severity. For mild severity corn stover steam treatments for 10 min at 140 °C (log R_0 = 2.18) and 160 °C (log R_0 = 2.77), Takada et al. [38] report about slight decreasing yields, as shown in Figure 2. In the present study, only the extracts show a maximum yield at severities around 3.6. Schütt et al. [11] made similar findings for steam refining of poplar wood. In comparison, the fiber, extract and total yields are very similar; however, the described extract maximum occurs at higher severity.

3.2.2. Composition of Fibers and Extracts

The carbohydrate and lignin contents in the fiber after steam refining are important process characteristics and figured out in Figure 3. Therefore, the main carbohydrates glucose and xylose, the further hemicellulose monomers (arabinose, galactose, mannose, rhamnose) and as well the acid-insoluble hydrolysis residues were analyzed. The composition was calculated based on the fiber fraction (Figure 3a) and based on the original raw material (Figure 3b).

Regarding the fiber fraction (Figure 3a), it can be stated that with increasing severity the proportion of glucose in the fibers increases. In contrast to that, a strong decreasing tendency is visible for xylose, which is degraded or can be found in the extract fraction due to the preferential solubilization of hemicelluloses. This impact can also be seen for the hydrolysis residue, which is increasing in the fiber fraction due to the loss of xylose and the accumulation of lignin.

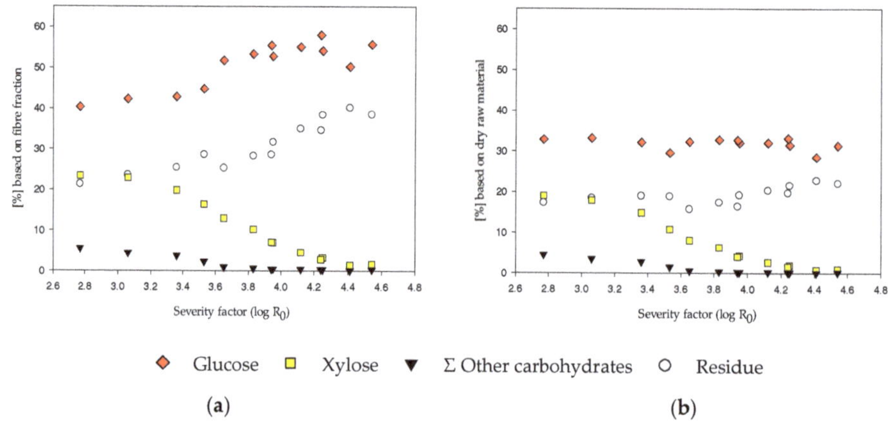

Figure 3. Comparison of carbohydrate and residue content of the steam-refined corn stover fiber fraction. The composition of fibers is referred to as fiber fraction (**a**) and raw material input (**b**).

If the obtained fiber yield is considered, the components can be calculated based on the dry material input. With this view, it can be seen that in contrast to the fiber-based view, the amounts of detected glucose and of the hydrolysis residue are nearly stable. Regarding the xylan content, the xylan is heavily degraded or dissolved (Figure 3b) and is thus decreasing.

The described tendencies for the carbohydrates in the fiber fraction, with nearly stable glucose contents and decreasing xylose amounts with increasing severity, are also described in the literature for steam refining of poplar at similar severity factors by Schütt [39]. For the mild steam explosion of corn stover, similar values of lignin, cellulose and hemicellulose are reported recently [38].

Referring to this, Bura et al. [40] described similar findings for steam treatment of corn stover with addition of SO_2 at low (log R_0 = 3), medium (log R_0 = 3.4) and high (log R_0 = 4.2) severities. They report increasing glucose yields, decreasing xylose yields and increasing lignin residue for the fiber fraction. Nevertheless, based on raw material the amounts of glucose are in comparison lower, whereas the amounts of xylose in the fiber fraction are much higher for corn stover experiments in contrast to poplar wood results [39].

The extract fraction was analyzed as well regarding the carbohydrate and residue content and is illustrated in Figure 4.

The composition of the extract is referred to as raw material input. When the yield data of the extract are referred to the raw material input (Figure 4), the maximum of xylose obtained between severities between 3.5 and 4 is, of course, less pronounced. This kind of presentation illustrated clearly that rather small quantities of the raw material components can be retrieved in the extract fraction, especially at severities higher than four. However, it is interesting to see that there is a gap between the raw material-based xylose yields from the fibers (Figure 3b) and the extract xylose yields in Figure 4, especially at severities higher than 3.5. This loss of hemicellulose can be explained with the formation of degradation products, like 5-HMF and furfural in the next chapter. Schütt [39] described the extract fraction of steam-refined poplar wood increasing xylose contents up to severities around 4. Subsequently, the xylose degradation occurred due to more intense pretreatment conditions. These findings are quite similar to the findings for the xylose content of the corn stover extracts after steam refining in the present study. Bura et al. [40] report slightly increasing glucose yields and increasing xylose yields in the extract fraction after steam pretreatment of corn stover at low (log R_0 = 3), medium (log R_0 = 3.4) and high (log R_0 = 4.2) pretreatment conditions. They did also report a slight increase of the xylose in the extract fraction from a severity of 3.4 to 4.2. It must be remarked that the authors used SO_2 as acid catalysts for the dataset. Additionally, the medium pretreatment condition was used without an acid catalyst. Higher yields of xylose in the extract fraction at medium conditions are clearly visible for the experiments with SO_2, but by calculating the total xylose mass balance there is a higher xylose loss visible compared with the results without a catalyst. However, the results without an acid catalyst are comparable with the results presented in this study.

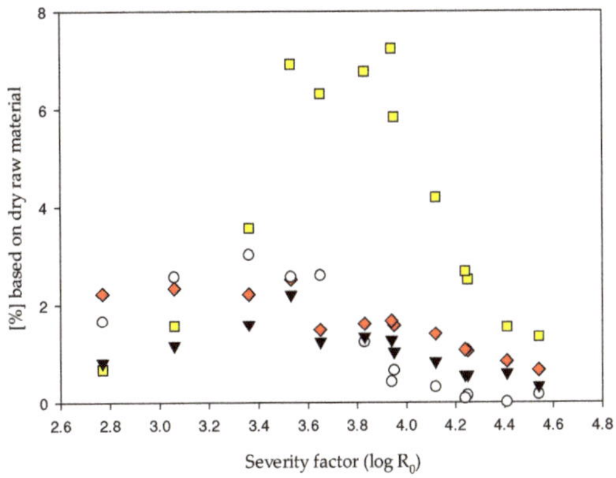

Figure 4. Comparison of carbohydrate and residue content of the steam-refined corn stover extract fraction.

3.2.3. Detection of 5-HMF, Furfural and pH Value

The negative influence of 5-HMF and furfural as main carbohydrate degradation products on enzymatic hydrolysis and subsequent fermentation is well described [41–44]. Therefore, 5-HMF and furfural were analyzed. As these compounds are unstable, the detection was performed directly in the extract obtained after the steam refining of the corn stover. The results are shown in Figure 5.

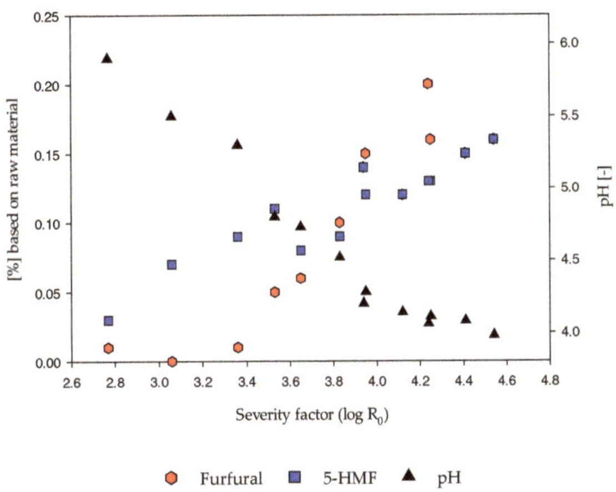

Figure 5. Effect of steaming at different severities on pH and the amounts of 5-HMF and furfural in the extract fraction.

Figure 5 shows that both, the content of furfural and 5-HMF in the extract fraction, increased steadily with increasing severity factors. This has particular importance for the approach to saccharify the fibers and the extract simultaneously. However, it is also important for subsequent processes such as EH of the fibers to fermentable carbohydrates or the production of biogas from the extract. The degradation of hexoses and pentoses to 5-HMF and furfural and the lowering of the pH value during steam explosion and steam refining is also well-described in the literature [11,45–49].

Ruiz et al. [49] report almost similar results for 5-HMF and furfural after steam explosion of sunflower stalks in comparison with the presented results for corn stover. Therefore, it can be assumed that the lower furan contents are typical for agricultural residues. Jacquet et al. [47] report results for really high severities up to 5.56 at extreme temperatures around 260 °C. For severities below severities of 4.5, they report similar data to the present data, although they used microcrystalline cellulose. For further results, a strong increase of 5-HMF, but not for furfural, is reported. In this context Um and van Walsum [50] report about the formation of furfural with increasing severity by a simultaneous decrease of the xylose contents in a corn stover dilute acid pretreatment. However, for severities above 4.43, they report a decrease and degradation of furfural contents and an increase of formate. In contrast to the present data, they received high severities by time, not by temperature.

However, the measured values for 5-HMF and furfural after steam refining of corn stover are much lower than the measured values for the furans after steam refining of poplar wood [11].

Figure 5 depicts as well the pH value in the extracts, which decreases with increasing severity. This is due to the formation of formic and acetic acid. Formic acid concentrations ranging from 0.61% at a severity factor of 3.65% up to 1.53% at a severity factor of 4.54%, all values based on raw material. The corresponding data for acetic are 2.02% and 2.89%, based on raw material.

Due to the formation of these organic acids, mainly caused by hemicellulose degradation, the pH value is reduced and autohydrolysis is intensified.

3.3. *Alkaline Lignin Extraction*

3.3.1. Fiber and Lignin Yields

The extraction of lignin with 8% NaOH (based on raw material input) was conducted for the fiber fraction of all severity grades. The loading of 8% NaOH is reported as the optimum for alkaline extractions after steam refining [9,11] and represents at a consistency of 10% a thin 0.8% *w*/*w* NaOH solution. The total recovery rate was calculated with the lignin and fiber recovery after extraction and is illustrated in Figure 6.

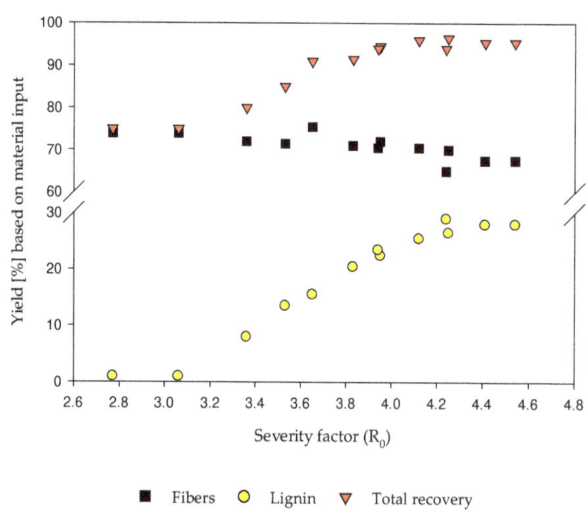

Figure 6. Illustration of the fiber and lignin yield and the total recovery rate, containing both mentioned fractions, after alkaline lignin extraction.

The fiber yield is slightly, but continuously decreasing with increasing severity (Figure 6). Simultaneously the lignin yield increases with the severity. Surprisingly, the reduction of fiber yield is significantly lower compared to the increase of lignin yield. This might be due to the condensation reactions of the lignin at high temperatures. The lignin with higher molar masses can be precipitated easier or more efficiently resulting in higher yields. However, when there is no covalent binding to the lignin, the precipitation is not sufficiently possible, resulting in lower yields.

As further illustrated, the combined recovery of lignin and fibers after extraction is strongly increasing with increasing severity of the pretreatment (Figure 6) from 75% up to 95% of the original raw material input. The increase in the total recovery rate can mainly be attributed to the higher lignin precipitation at severities above 4.

3.3.2. Analysis of the Lignins

The effect of steaming severity on the molar mass of the precipitated lignins is illustrated in Figure 7. It becomes evident that the molar mass of the lignins increased gradually with intensifying the steam refining conditions. Furthermore, the main peak of the molar mass distribution is visible between 6800 and 7700 g/mol (Figure 7).

At a severity of 3.65 (190 °C/10 min), components with low molar masses (100–250 g/mol) are visible in the distribution curves. They are significantly reduced at severities above four. The increasing shoulder at high molar masses is a clear indication for condensation reactions at higher severities. Thus, the SEC results confirm the assumption that lignin condensation occurs to a higher extent under harsh steaming conditions.

For considerations on the utilization potential of the extracted lignins, knowledge of the lignin characteristics and purity is essential. Therefore, their carbohydrate contents were determined after acid hydrolysis (Figure 8). With increasing severity, the carbohydrate impurities of lignin decrease. Nevertheless, after pretreatment at lower severities than 3.95, more carbohydrate impurities can occur in the alkaline extract. At severities above four, the carbohydrate content of lignins is negligible. For this finding, Schütt et al. [11] and Schütt [39] also report decreasing carbohydrate impurities with increasing severity for extracted lignins of steam-refined poplar wood.

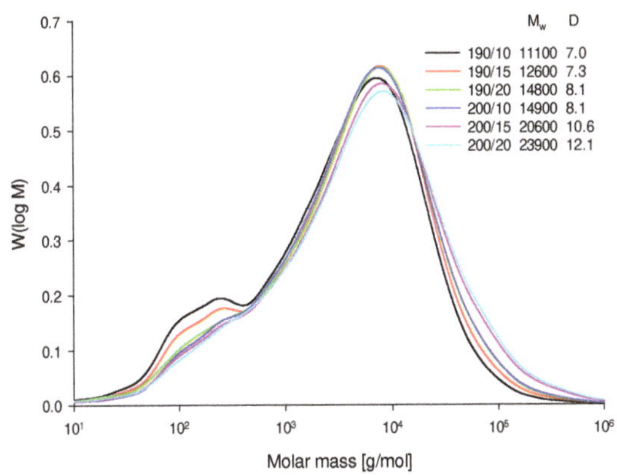

Figure 7. Molar mass distribution, molecular weight (M_W) and dispersity (D) of alkaline-extracted lignin.

Several authors discussed the alkaline extraction of steam-treated fibers and following the influence of lignin removal on EH. For the process of lignin extraction, Schütt et al. [11] and Schütt [39] reported higher molar masses and a higher dispersity of the extracted lignin with increasing severity. The influence of one or two steam explosion steps on the extraction behavior of the lignin and the influence of SO_2 as a catalyst were also investigated by Li et al. [51]. However, the reported findings are in good accordance with the illustrated results in the present study (Figures 7 and 8).

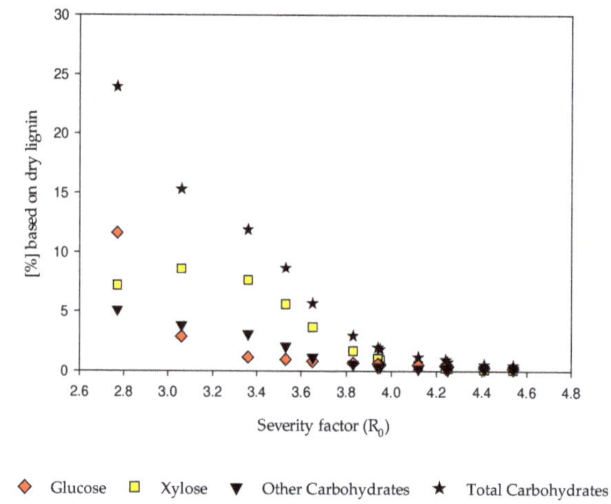

Figure 8. Effect of steaming severity on the carbohydrate content of alkaline-extracted lignins.

3.4. EH of the Fiber Fraction with and without Alkaline Extraction

The enzymatic hydrolysis of steam-refined fibers was compared with and without alkaline extraction. To enable a detailed comparison all carbohydrate yields were calculated on the theoretically available carbohydrates in the used fibers, detected after two-stage acidic hydrolysis (Figure 9). Therefore, the detected monomeric carbohydrates were concerned as a percentage based on the carbohydrate content in the fibers after an EH with Cellic® CTec2.

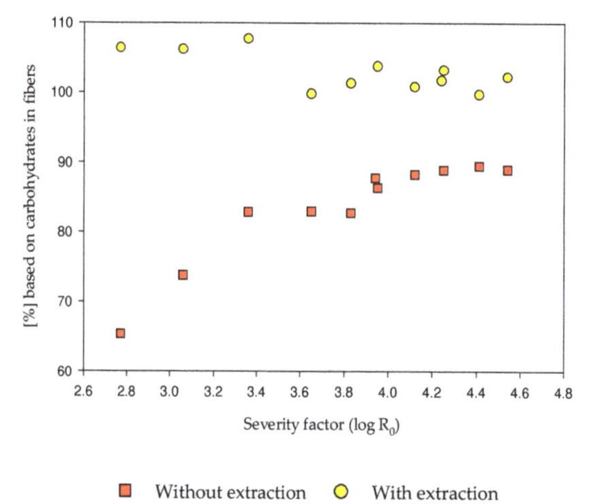

Figure 9. Effect of severity and alkaline extraction on the enzymatic hydrolysis (EH) of the fibers with Cellic® CTec2. The EH yields are calculated based on the theoretically available carbohydrates in the used fibers.

It is obvious that without extraction of the lignin, the yields after EH increases strongly with increasing steaming severity. After alkaline lignin extraction, there is no more positive effect of severity. There are slightly higher yields at severities lower than 3.5 and yields surpass constant 100%. This is because available carbohydrates in the fiber fraction were determined with a two-step acid hydrolysis. In this process, the carbohydrate content can be underestimated due to the secondary degradation of carbohydrates in the analytical method. This degradation will be more pronounced at severities below around 4 because more hemicelluloses and amorphous cellulose are present in the fiber. Hydrolysis with enzymes is more selective to carbohydrates and results in fewer degradation products. Therefore, higher yields are explainable by the high efficiency of EH. It can be concluded that after steam refining and alkaline extraction, all carbohydrates are available for the EH.

The described improvement of the EH yields by steam refining is an undisputed fact and pretreatments are often regarded as indispensable prerequisites for the economic implementation of such processes [6]. There are several parameters influencing the EH. Important characteristics for the impact on the EH are, e.g., the ratio of amorphous and crystalline regions, the degree of polymerization (DP) of the sample, the moisture content, available surface areas, the lignin content or the pore size of the samples [52]. Higher yields after EH are reported for wet substrate in contrast to dried biomass [53], as also found in this study (data not shown here). However, the effect of lignin removal is controversially discussed in the literature. Some authors complained about studies on wood having a reduced enzymatic digestibility after lignin extraction due to collapsing pores [54–56]. Schwalb et al. [57] reported no influence of the alkali extraction, whereas Excoffier et al. [58] and Schütt et al. [11] presented higher yields of the EH after lignin removal. Ishizawa et al. [25] report that a nearly complete delignification with acidified sodium chlorite below 5% lignin content decreases the yields after EH strongly. The authors report further that the influence of partial lignin and xylan extraction improves the EH. However, the xylan removal is suggested to be more significant for the improvement of the EH yields than lignin removal [25].

Regarding the discussions in the literature, a positive influence of the partial lignin removal on the EH results could also be stated in the present study for corn stover. Furthermore, the higher impact at severities below 4 and the fading influence at severities above 4 (Figure 10a,b) could be confirmed in accordance with Schütt et al. [11]. For more findings of the overall process efficiency, the glucose (Figure 10a) and xylose yield (Figure 10b) based on raw material were calculated for the process variants with and without lignin extraction prior to EH. Taking the yields after steam refining into account a strong influence of the extraction is apparent at severities below 4. This effect is disappearing at severities >4. After that point, no significant difference can be determined for the two process variants. Similar findings were reported for the steam refining of poplar by Schütt et al. [11]. The described influence at severities lower than 4 is also visible for the xylan yields. However, the influence is not that strong as for the glucose yields.

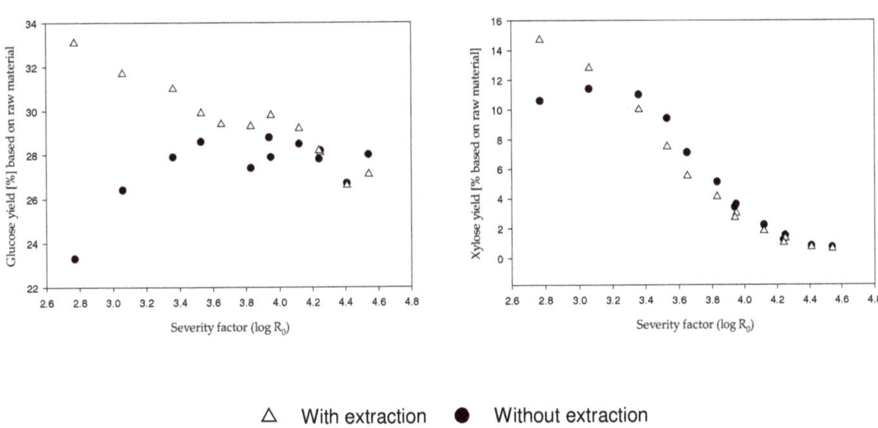

(a) (b)

Figure 10. Glucose yields (**a**) and xylose yields (**b**) after EH based on raw material for the process variants with and without alkaline extraction prior to EH.

3.5. Overall Process Balances

Finally, the glucose and xylose yields after EH of the fiber fraction and glucose and xylose yields after one-step acidic hydrolysis of the extract fraction were compared and calculated based on raw material. This consideration was made in order to evaluate the effect of alkaline extraction of the overall process balance and to monitor the carbohydrate losses during the whole process (Figure 11). Schütt [39] reports a consideration that a severity of around 4.5 is needed for gaining sufficient glucose rates after steam refining of poplar wood and enzymatic hydrolysis with Celluclast®.

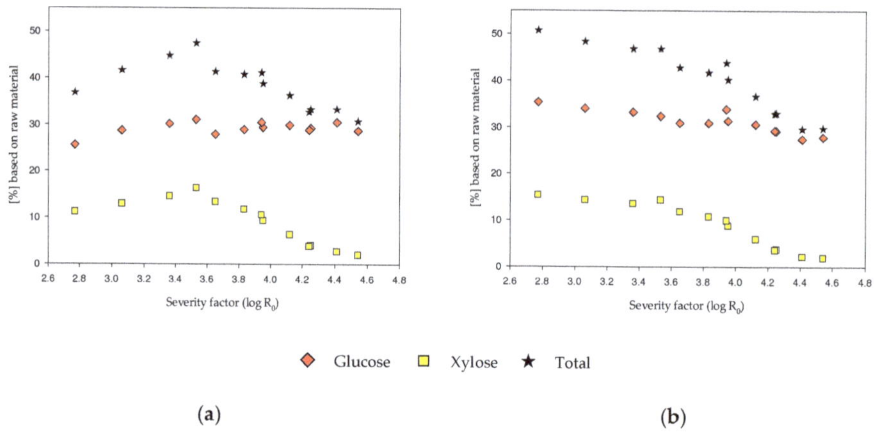

Figure 11. Summed glucose and xylose yields after EH of the fiber fraction and after one-step acidic hydrolysis of the extracts—without (**a**) and with (**b**) alkaline extraction; all based on raw material input.

A different behavior was found for the steam refining of corn stover and enzymatic hydrolysis with Cellic® CTec2 in this study. It must be remarked that the following comparisons and differences are mainly due to the used enzymes. From previous experiments (data not shown here) it is clearly known that Cellic® CTec2 shows a much higher activity to the used biomass than Celluclast®, used in previous days. However, not shown comparison experiments with Celluclast® and steam-treated corn stover also show the highest glucose yields at severities around 4.5.

As shown in Figure 11a the glucose is also available at severities below 3.5 and there is no need for steam refining in higher regions. There is a slight increase in the glucose yield with increasing severity with no significant optimum. In contrast to those findings, there is a clear optimum for the xylose yields at a severity of 3.2. Regarding Schütt [39], the findings for the xylose yields are nearly equal and the optimum is also in severity regions below 3.5. Figure 11b represents the overall process balance after alkaline extraction of the fibers and following EH. In contrast to the prior findings, the optimum of all yields is now located at a severity of 2.77. Due to these findings, the optimal steaming conditions for steam-refined and steam-extracted samples are clearly visible at severities below 3.5, mainly due to the higher reactivity of the used enzymes.

4. Conclusions

Due to the findings in the present study, several conclusions can be made. For steam refining experiments without subsequent alkaline extraction, the optimum of EH yields is located at a severity around 3.95. However, the optimum of total carbohydrate recovery from EH of the fibers and acidic hydrolysis of the extract fraction is at a severity around 3.4 with around 47.5% based on raw material.

For steam refining with subsequent alkaline extraction, different findings were made. For the EH yields around 100% of the theoretically available carbohydrates were found, even at severities below 3.5. However, also the total carbohydrate recovery shows the highest yields at these severities and it is known that low contents of carbohydrate degradation products are beneficial for further process steps. Nonetheless, for lignin utilization severities around 3.95 might be optimal due to fewer carbohydrate contaminations.

Due to these findings, it can be concluded that steam refining pretreatment especially at severities below 3.5 seems to be interesting for corn stover. It could further be stated that steam refining at different severities and alkaline lignin removal is enhancing the enzymatic digestibility for the supply of monomeric carbohydrates significantly. Higher fiber yields and good digestibility after alkaline extraction giving high yields of fermentable carbohydrates for further value-added products, like ethanol or dicarboxylic acids. Therefore, further undervalued agricultural residues should be tested in the future at severities below 3.5 and with alkaline extraction. Afterward, the results can be compared with the results for corn stover presented in this study.

Author Contributions: Conceptualization, B.S.; funding acquisition, B.S.; investigation, M.J.K., M.B. and A.S.; supervision, B.S.; visualization, M.J.K.; writing—original draft, M.J.K.; writing—review and editing, M.J.K., M.B., A.S. and B.S. All authors have read and agreed to the published version of the manuscript.

Funding: This research was funded by the Federal Ministry of Education and Research (BMBF) via Project Management Jülich (PtJ) in the PANDA project, grant number 031B0505B.

Acknowledgments: The Federal Ministry of Education and Research (BMBF) and Project Management Jülich (PtJ) are gratefully acknowledged for their financial support. The authors wish to thank Othar Kordsachia, who assisted in the proofreading of the manuscript. Martin Bellof, Autodisplay Biotech GmbH, is thanked for the raw material acquisition. Anna Knöpfle, Nicole Erasmy and Sascha Lebioda are also gratefully acknowledged for their technical support.

Conflicts of Interest: The authors declare no conflicts of interest.

References

1. Zhao, Y.; Damgaard, A.; Christensen, T.H. Bioethanol from corn stover—A review and technical assessment of alternative biotechnologies. *Prog. Energy Combust. Sci.* **2018**, *67*, 275–291. [CrossRef]
2. U.S. Department of Energy. *U.S. Billion-Ton Update: Biomass Supply for a Bioenergy and Bioproducts Industry*; U.S. Department of Energy: Washington, DC, USA, 2011.
3. Garlock, R.J.; Chundawat, S.P.S.; Balan, V.; Dale, B. Optimizing harvest of corn stover fractions based on overall sugar yields following ammonia fiber expansion pretreatment and enzymatic hydrolysis. *Biotechnol. Biofuels* **2009**, *2*, 29. [CrossRef] [PubMed]
4. Kadam, K.L.; McMillan, J.D. Availability of corn stover as a sustainable feedstock for bioethanol production. *Bioresour. Technol.* **2003**, *88*, 17–25. [CrossRef]
5. Himmel, M.E.; Ding, S.-Y.; Johnson, D.K.; Adney, W.S.; Nimlos, M.R.; Brady, J.W.; Foust, T.D. Biomass Recalcitrance: Engineering Plants and Enzymes for Biofuels Production. *Science* **2007**, *315*, 804–807. [CrossRef] [PubMed]
6. Yang, B.; Wyman, C.E. Pretreatment: The key to unlocking low-cost cellulosic ethanol. *Biofuels Bioprod. Biorefining* **2008**, *2*, 26–40. [CrossRef]
7. Sun, Y.; Cheng, J. Hydrolysis of lignocellulosic materials for ethanol production: A review. *Bioresour. Technol.* **2002**, *83*, 1–11. [CrossRef]
8. Galbe, M.; Zacchi, G. A review of the production of ethanol from softwood. *Appl. Microbiol. Biotechnol.* **2002**, *59*, 618–628. [CrossRef]
9. Klupsch, R. Untersuchungen zur Herstellung von Chemiezellstoff aus Aspen-und Buchenholz nach dem Dampfdruck-Extraktionsverfahren. Dissertation Thesis, University of Hamburg: Hamburg, Germany, 2000.
10. Klupsch, R.; Kordsachia, O.; Puls, J.; Karstens, T. Herstellung von Chemiezellstoffen nach dem Dampfdruck-Extraktionsverfahren. *Ipw Int. Pap. Das Pap.* **2001**, *55*, 73–79.
11. Schütt, F.; Puls, J.; Saake, B. Optimization of steam pretreatment conditions for enzymatic hydrolysis of poplar wood. *Holzforschung* **2011**, *65*, 453–459. [CrossRef]
12. Schütt, F.; Westereng, B.; Horn, S.J.; Puls, J.; Saake, B. Steam refining as an alternative to steam explosion. *Bioresour. Technol.* **2012**, *111*, 476–481. [CrossRef]

13. Stücker, A.; Schütt, F.; Saake, B.; Lehnen, R. Lignins from enzymatic hydrolysis and alkaline extraction of steam refined poplar wood: Utilization in lignin-phenol-formaldehyde resins. *Ind. Crop. Prod.* **2016**, *85*, 300–308. [CrossRef]
14. Poutanen, K.; Viikari, L. Biotechnical utilization of wood carbohydrates after steaming pretreatment. *Appl. Microbiol. Biotechnol.* **1985**, *22*, 416–423. [CrossRef]
15. Brownell, H.H.; Yu, E.K.C.; Saddler, J.N. Steam-explosion pretreatment of wood: Effect of chip size, acid, moisture content and pressure drop. *Biotechnol. Bioeng.* **1986**, *28*, 792–801. [CrossRef] [PubMed]
16. Kim, Y.; Ximenes, E.; Mosier, N.S.; Ladisch, M. Soluble inhibitors/deactivators of cellulase enzymes from lignocellulosic biomass. *Enzym. Microb. Technol.* **2011**, *48*, 408–415. [CrossRef]
17. Chinnappan, A.; Baskar, C.; Kim, H. Biomass into chemicals: Green chemical conversion of carbohydrates into 5-hydroxymethylfurfural in ionic liquids. *RSC Adv.* **2016**, *6*, 63991–64002. [CrossRef]
18. Mamman, A.S.; Lee, J.-M.; Kim, Y.-C.; Hwang, I.T.; Park, N.-J.; Hwang, Y.K.; Chang, J.-S.; Hwang, J.-S. Furfural: Hemicellulose/xylosederived biochemical. *Biofuels Bioprod. Biorefining* **2008**, *2*, 438–454. [CrossRef]
19. Himmel, M.E.; Adney, W.S.; Baker, J.O.; Nieves, R.A.; Thomas, S.R. Cellulases: Structure, Function, and Applications. In *Handbook on Bioethanol: Production and Utilization*; Wyman, C.E., Ed.; Routledge: Boca Raton, FL, USA, 1996; pp. 143–161, ISBN 1-56032-553-4.
20. Becker, J.; Lange, A.; Fabarius, J.; Wittmann, C. Top value platform chemicals: Bio-based production of organic acids. *Curr. Opin. Biotechnol.* **2015**, *36*, 168–175. [CrossRef]
21. Choi, S.; Song, C.W.; Shin, J.; Lee, S.Y. Biorefineries for the production of top building block chemicals and their derivatives. *Metab. Eng.* **2015**, *28*, 223–239. [CrossRef]
22. Rosenau, T. Advances in biorefinery research. *Holzforschung* **2018**, *73*, 1–2. [CrossRef]
23. Chang, V.S.; Holtzapple, M.T. Fundamental factors affecting biomass enzymatic reactivity. *Appl. Biochem. Biotechnol.* **2000**, *84*, 5–37. [CrossRef]
24. Pareek, N.; Gillgren, T.; Jönsson, L.J. Adsorption of proteins involved in hydrolysis of lignocellulose on lignins and hemicelluloses. *Bioresour. Technol.* **2013**, *148*, 70–77. [CrossRef] [PubMed]
25. Ishizawa, C.I.; Jeoh, T.; Adney, W.S.; Himmel, M.; Johnson, D.K.; Davis, M. Can delignification decrease cellulose digestibility in acid pretreated corn stover? *Cellulose* **2009**, *16*, 677–686. [CrossRef]
26. Chen, Y.; Stevens, M.A.; Zhu, Y.; Holmes, J.; Xu, H. Understanding of alkaline pretreatment parameters for corn stover enzymatic saccharification. *Biotechnol. Biofuels* **2013**, *6*, 8. [CrossRef] [PubMed]
27. Lorenz, D.; Erasmy, N.; Akil, Y.; Saake, B. A new method for the quantification of monosaccharides, uronic acids and oligosaccharides in partially hydrolyzed xylans by HPAEC-UV/VIS. *Carbohydr. Polym.* **2016**, *140*, 181–187. [CrossRef]
28. Bendler, M. Dampfdruck-Refiner-Aufschluss und Enzymatische Hydrolyse von Maisstroh. Bachelor's Thesis, Hamburg University of Applied Sciences, Hamburg, Germany, 2018.
29. Overend, R.P.; Chornet, E.; Gascoigne, J.A. Fractionation of lignocellulosics by steam-aqueous pretreatments. *Philos. Trans. R. Soc. Lond. Ser. A Math. Phys. Sci.* **1987**, *321*, 523–536. [CrossRef]
30. Podschun, J.; Saake, B.; Lehnen, R. Reactivity enhancement of organosolv lignin by phenolation for improved bio-based thermosets. *Eur. Polym. J.* **2015**, *67*, 1–11. [CrossRef]
31. Pordesimo, L.O.; Edens, W.C.; Sokhansanj, S. Distribution of Above Ground Biomass in Corn Stover. In Proceedings of the 2002 ASAE Annual International Meeting/CIGR XVth World Congress, Chicago, IL, USA, 28–31 July 2002; American Society of Agricultural and Biological Engineers: St. Joseph, MI, USA, 2002.
32. Templeton, D.; Sluiter, A.D.; Hayward, T.K.; Hames, B.R.; Thomas, S.R. Assessing corn stover composition and sources of variability via NIRS. *Cellulose* **2009**, *16*, 621–639. [CrossRef]
33. Deutschle, A. Charakterisierung und Anwendung von Kationischen Arabinoxylanen. Ph.D. Thesis, University of Hamburg, Hamburg, Germany, 2015.
34. Baar, J.; Paschová, Z.; Hofmann, T.; Kolář, T.; Koch, G.; Saake, B.; Rademacher, P. Natural durability of subfossil oak: Wood chemical composition changes through the ages. *Holzforschung* **2019**, *74*, 47–59. [CrossRef]
35. Kovačić, Đ.; Soybean, S. Corn Stover and Sunflower Stalk as Possible Substrates for Biogas Production in Croatia: A Review. *Chem. Biochem. Eng. Q.* **2017**, *31*, 187–198. [CrossRef]

36. Zhang, Y.-H.P.J.; Ding, S.-Y.; Mielenz, J.R.; Cui, J.-B.; Elander, R.T.; Laser, M.; Himmel, M.E.; McMillan, J.R.; Lynd, L.R. Fractionating recalcitrant lignocellulose at modest reaction conditions. *Biotechnol. Bioeng.* **2007**, *97*, 214–223. [CrossRef]
37. Saha, B.C.; Yoshida, T.; Cotta, M.; Sonomoto, K. Hydrothermal pretreatment and enzymatic saccharification of corn stover for efficient ethanol production. *Ind. Crop. Prod.* **2013**, *44*, 367–372. [CrossRef]
38. Takada, M.; Chandra, R.P.; Saddler, J.N. The influence of lignin migration and relocation during steam pretreatment on the enzymatic hydrolysis of softwood and corn stover biomass substrates. *Biotechnol. Bioeng.* **2019**, *116*, 2864–2873. [CrossRef] [PubMed]
39. Schütt, F. Dampfdruckaufschluss und Enzymatische Hydrolyse von Pappelholz. Ph.D. Thesis, University of Hamburg, Hamburg, Germany, 2012.
40. Bura, R.; Chandra, R.; Saddler, J.N. Influence of xylan on the enzymatic hydrolysis of steam-pretreated corn stover and hybrid poplar. *Biotechnol. Prog.* **2009**, *25*, 315–322. [CrossRef] [PubMed]
41. Palmqvist, E.; Hahn-Hägerdal, B.; Galbe, M.; Zacchi, G. The effect of water-soluble inhibitors from steam-pretreated willow on enzymatic hydrolysis and ethanol fermentation. *Enzym. Microb. Technol.* **1996**, *19*, 470–476. [CrossRef]
42. Palmqvist, E.; Hahn-Hägerdal, B. Fermentation of lignocellulosic hydrolysates. II: Inhibitors and mechanisms of inhibition. *Bioresour. Technol.* **2000**, *74*, 25–33. [CrossRef]
43. Martín, C.; Galbe, M.; Nilvebrant, N.-O.; Jönsson, L.J. Comparison of the fermentability of enzymatic hydrolyzates of sugarcane bagasse pretreated by steam explosion using different impregnating agents. *Appl. Biochem. Biotechnol.* **2002**, *98*, 699–716. [CrossRef]
44. Gurram, R.N.; Datta, S.; Lin, Y.J.; Snyder, S.; Menkhaus, T.J. Removal of enzymatic and fermentation inhibitory compounds from biomass slurries for enhanced biorefinery process efficiencies. *Bioresour. Technol.* **2011**, *102*, 7850–7859. [CrossRef]
45. Kaar, W.; Gutierrez, C.; Kinoshita, C. Steam explosion of sugarcane bagasse as a pretreatment for conversion to ethanol. *Biomass Bioenergy* **1998**, *14*, 277–287. [CrossRef]
46. Li, H.; Chen, H. Detoxification of steam-exploded corn straw produced by an industrial-scale reactor. *Process. Biochem.* **2008**, *43*, 1447–1451. [CrossRef]
47. Jacquet, N.; Quiévy, N.; Vanderghem, C.; Janas, S.; Blecker, C.; Wathelet, B.; Devaux, J.; Paquot, M. Influence of steam explosion on the thermal stability of cellulose fibres. *Polym. Degrad. Stab.* **2011**, *96*, 1582–1588. [CrossRef]
48. Han, G.; Deng, J.; Zhang, S.; Bicho, P.; Wu, Q. Effect of steam explosion treatment on characteristics of wheat straw. *Ind. Crop. Prod.* **2010**, *31*, 28–33. [CrossRef]
49. Ruiz, E.; Cara, C.; Manzanares, P.; Ballesteros, M.; Castro, E.; Castro, E. Evaluation of steam explosion pre-treatment for enzymatic hydrolysis of sunflower stalks. *Enzym. Microb. Technol.* **2008**, *42*, 160–166. [CrossRef] [PubMed]
50. Um, B.-H.; Van Walsum, G.P. Effect of Pretreatment Severity on Accumulation of Major Degradation Products from Dilute Acid Pretreated Corn Stover and Subsequent Inhibition of Enzymatic Hydrolysis of Cellulose. *Appl. Biochem. Biotechnol.* **2012**, *168*, 406–420. [CrossRef]
51. Li, J.; Gellerstedt, G.; Toven, K. Steam explosion lignins; their extraction, structure and potential as feedstock for biodiesel and chemicals. *Bioresour. Technol.* **2009**, *100*, 2556–2561. [CrossRef]
52. Hendriks, A.; Zeeman, G. Pretreatments to enhance the digestibility of lignocellulosic biomass. *Bioresour. Technol.* **2009**, *100*, 10–18. [CrossRef]
53. Liu, Z.-H.; Qin, L.; Jin, M.; Pang, F.; Li, B.-Z.; Kang, Y.; Dale, B.; Yuan, Y.-J. Evaluation of storage methods for the conversion of corn stover biomass to sugars based on steam explosion pretreatment. *Bioresour. Technol.* **2013**, *132*, 5–15. [CrossRef]
54. Donaldson, L.A.; Wong, K.K.Y.; Mackie, K.L. Ultrastructure of steam-exploded wood. *Wood Sci. Technol.* **1988**, *22*, 103–114. [CrossRef]
55. Ramos, L.P.; Breuil, C.; Saddler, J.N. Comparison of steam pretreatment of eucalyptus, aspen, and spruce wood chips and their enzymatic hydrolysis. *Appl. Biochem. Biotechnol.* **1992**, *34*, 37–48. [CrossRef]
56. Schell, D.; Nguyen, Q.; Tucker, M.; Boynton, B. Pretreatment of softwood by acid-catalyzed steam explosion followed by alkali extraction. *Appl. Biochem. Biotechnol.* **1998**, *70*, 17–24. [CrossRef]

57. Schwald, W.; Breuil, C.; Brownell, H.H.; Chan, M.; Saddler, J.M. Assessment of pretreatment conditions to obtain fast complete hydrolysis on high substrate concentrations. *Appl. Biochem. Biotechnol.* **1989**, *20*, 29–44. [CrossRef]
58. Toussaint, B.; Vignon, M.R. Saccharification of steam-exploded poplar wood. *Biotechnol. Bioeng.* **1991**, *38*, 1308–1317. [CrossRef]

© 2020 by the authors. Licensee MDPI, Basel, Switzerland. This article is an open access article distributed under the terms and conditions of the Creative Commons Attribution (CC BY) license (http://creativecommons.org/licenses/by/4.0/).

agronomy

MDPI

Article

Recovery of High Purity Lignin and Digestible Cellulose from Oil Palm Empty Fruit Bunch Using Low Acid-Catalyzed Organosolv Pretreatment

Kinanthi Mondylaksita [1,2], Jorge A. Ferreira [1], Ria Millati [2], Wiratni Budhijanto [3], Claes Niklasson [4] and Mohammad J. Taherzadeh [1,*]

1 Swedish Centre for Resource Recovery, University of Borås, 50190 Borås, Sweden; kinanthi.mondylaksita@hb.se (K.M.); jorge.ferreira@hb.se (J.A.F.)

2 Department of Food and Agricultural Product Technology, Universitas Gadjah Mada, Yogyakarta 55281, Indonesia; ria_millati@ugm.ac.id

3 Department of Chemical Engineering, Universitas Gadjah Mada, Yogyakarta 55281, Indonesia; wiratni@ugm.ac.id

4 Department of Chemistry and Chemical Engineering, Chalmers University of Technology, 41296 Gothenburg, Sweden; claesn@chalmers.se

* Correspondence: mohammad.taherzadeh@hb.se; Tel.: +46-(0)33-435-5908

Received: 11 April 2020; Accepted: 7 May 2020; Published: 11 May 2020

check for updates

Abstract: The lignocellulosic residue from the palm oil industry, oil palm empty fruit bunch (OPEFB), represents a challenge to both producing industries and environment due to its disposal difficulties. Alternatively, OPEFB can be used for the production of valuable products if pretreatment methods, which overcome OPEFB recalcitrance and allow tailored valorization of all its carbohydrates and lignin, are developed. Specifically, high-value applications for lignin, to increase its contribution to the feasibility of lignocellulosic biorefineries, demand high-purity fractions. In this study, acid-catalyzed organosolv using ethanol as a solvent was used for the recovery of high-purity lignin and digestible cellulose. Factors including catalyst type and its concentration, temperature, retention time, and solid-to-liquid (S/L) ratio were found to influence lignin purity and recovery. At the best conditions (0.07% H_2SO_4, 210 °C, 90 min, and S/L ratio of 1:10), a lignin purity and recovery of 70.6 ± 4.9% and 64.94 ± 1.09%, respectively, were obtained in addition to the glucan-rich fraction. The glucan-rich fraction showed 94.06 ± 4.71% digestibility within 18 h at an enzyme loading of 30 filter paper units (FPU) /g glucan. Therefore, ethanol organosolv can be used for fractionating OPEFB into three high-quality fractions (glucan, lignin, and hemicellulosic compounds) for further tailored biorefining using low acid concentrations. Especially, the use of ethanol opens the possibility for integration of 1st and 2nd generation ethanol benefiting from the separation of high-purity lignin.

Keywords: oil palm empty fruit bunch; lignin recovery; lignin purity; digestible cellulose; organosolv pretreatment

1. Introduction

As a result of being one of the largest crude palm oil producers in the world, Indonesia produces around 37 million tons of oil palm empty fruit bunch (OPEFB) yearly [1]. Indonesian oil palm plants are located and concentrated in some areas. For instance, more than 50% of the OPEFB is produced in 5 regions in Sumatra [2]. OPEFB accumulation can cause environmental problems including the proliferation of disease-causing pests and microorganisms, the requirement of extensive land for burying, and gas emissions originated from its combustion [3,4]. Therefore, handling and valorization of OPEFB are crucial, and at a starting point, the existence of areas with higher production of palm oil can be used as an advantage for valorization by decreasing transportation costs [2].

Being a lignocellulosic material, OPEFB is mainly composed of cellulose, hemicellulose, and lignin that account to 24–65 wt%, 21–34 wt%, and 14–31 wt% of the material, respectively [5]. The holocellulosic (cellulose and hemicellulose) fragments can be enzymatically converted into simple sugars [6] for further conversion into value-added products. However, the recalcitrance of OPEFB hinders easy access to its carbohydrate polymers. In general, enzymatic digestibility of lignocellulosic materials is limited by a number of factors such as the presence of lignin, cellulose crystallinity, degree of polymerization, acetyl groups bound to hemicellulose, surface area, and biomass particle size [7,8]. Hence, pretreatment is needed to open the lignocellulosic structure of OPEFB and have access to sugar polymers for further valorization.

Pretreatment of lignocelluloses can be performed by mechanical, chemical, enzymatic, and biological processes, or by a combination of these [9]. Chemical pretreatment is a strong and effective pretreatment to improve the digestibility of lignocellulosic biomass, applying acids, bases, or other catalysts such as hydrogen peroxide and ozone [10], with dissimilar end-results. For instance, in acid pretreatment, deconstruction of the lignocellulosic structure is carried out mainly through dissolution of hemicellulose, leaving lignin in the solid fraction together with cellulose [11]. The remaining lignin will interfere with the following enzymatic hydrolysis of cellulose. Alkali pretreatment can easily break the lignin bonds and enhance the solubilization of the polymer [12]. In this method, lignin is effectively removed; however, the lignin dissolved in the liquid is difficult to recover. Ozone pretreatment is safer than alkaline and acid pretreatments and leads to efficient lignin removal. Nonetheless, lignin recovery remains a hurdle and ozone generation is very expensive [13]. Hydrogen peroxide pretreatment is efficient towards the removal of lignin and xylan [14], but it is very toxic for the environment. Therefore, none of the pretreatment methods described can efficiently recover the lignin from the lignocellulose.

Lignin has normally been considered as a barrier to properly access carbohydrate polymers. As a low-value stream in lignocellulosic biorefineries, lignin is normally used for heat and power generation through combustion [15]. However, as a result of feasibility concerns of lignocellulosic biorefineries and increasing range of high-value applications for lignin, higher attention has been devoted to the full valorization of lignocellulose-derived polymers. In fact, high purity lignin can be utilized for the production of value-added products such as resin, flavor compounds, and nanofibers with antioxidant activity [16–18]. Hence, pretreatment methods that enable both efficient lignin removal from lignocelluloses and easy recovery into a high-purity lignin fraction, can positively contribute to the feasibility of lignocellulosic biorefineries.

The use of organic solvent (organosolv) for pretreatment is a promising strategy. The organic solvent is able to extract lignin and hydrolyses the hemicellulose [19]. A high-purity cellulose with only minor degradation and a high proportion of its amorphous phase, which is susceptible to enzymatic hydrolysis, can be recovered [20]. The extracted lignin can be further precipitated by dilution with water and recovered as a solid product, while hemicellulosic sugars remain in the liquid stream [19]. The organic solvent can easily be recovered using evaporation or distillation processes.

There are several parameters affecting the success of the fractionation process, especially delignification, during organosolv pretreatment such as solvent type and concentration, catalyst type and amount, temperature, retention time, and solid to liquid ratio (S/L ratio) [21]. Ethanol is a solvent that is frequently used for organosolv pretreatment of lignocellulosic biomass due to its low price, good solubility of lignin, lower toxicity compared to other alcohol-based solvents, its miscibility with water, and ease of recovery [22]. Moreover, the production of ethanol using sugars and starch-rich substrates is an industrially mature technology well distributed throughout the world; ca. 29,100 million gallons were produced in 2019 [23]. In the organosolv pretreatment, acids and bases can be added as catalysts to increase the delignification rate, whereas comparatively lower delignification rates (≤60%) are normally obtained during non-catalyzed organosolv pretreatment of lignocellulosic biomass [24,25]. Mineral acids have high reactivity and efficiency and sulfuric acid is the most studied catalyst for ethanol-organosolv pretreatment [26,27]. Higher S/L ratio is favorable because a smaller amount of solvent is employed and a better balance can be found among energy needed to carry out the pretreatment step and the amount of processed material. Overall, an organosolv pretreatment strategy that ensures high S/L ratio, minimum addition of acid and solvent, and optimum temperature and retention time, while attaining fractions that are rich in high-quality glucan, lignin, and hemicellulosic compounds, are of interest for the valorization of lignocellulosic materials.

The research landscape on organosolv pretreatment is characterized by a wide range of substrates and pretreatment conditions investigated, but by a lack of systematic studies to reveal unequivocally substrate-tailored pretreatment systems [28]. For instance, various organosolv strategies have been used for the pretreatment of OPEFB, as a result of extensive research on the development of efficient biorefinery systems for its valorization. These included the use of bisulfite, a mixture of acetic acid and ammonia, or ethylene glycol as solvents, where studies on ethanol organosolv pretreatment of OPEFB are scarce in literature [29–34]. Moreover, information about the influence of pretreatment parameters on lignin recovery and purity following organosolv pretreatment is still scarce in the literature, and it is common to all lignocellulosic substrates studied [28]. A recent study has shown that organosolv pretreatment conditions influence the recovery and purity of lignin from oat husks [35]. Therefore, the aim of the present study was to study the effect of acid-catalyzed ethanol organosolv pretreatment on the delignification of OPEFB. A range of parameters was studied in organosolv pretreatment of OPEFB, namely, acid type and concentration, temperature, retention time, and S/L ratio. Emphasis was given to lignin purity and lignin recovery as well as the digestibility of the glucan-rich fraction. The optimization steps carried out in this study led to high lignin purity and recovery. In addition, digestible glucan-rich fraction which has high glucan purity and recovery was obtained. The results obtained demonstrate the possibility to use very low acid concentration for deconstruction of OPEFB into high-quality fractions.

2. Materials and Methods

2.1. Materials

Oil palm empty palm fruit bunch (OPEFB) was obtained from a palm oil industry in Medan, Indonesia. It was sun-dried to achieve 7% moisture content. The dried OPEFB was then milled using a cutting mill (Retsch SM 100, Haan, Germany) using a screen with a pore size of 300 μm and resulted in the following particle size distribution: 44.24% of >500 μm; 17.96% of 250–500 μm; 23.81% of 100–250 μm; and 13.99% of 63–125 μm. The composition of OPEFB is presented in Table 1. The enzyme cocktail Cellic®® Ctec3 (Novozymes, Bagsværd, Denmark), with a cellulase activity of 222 filter paper units (FPU) /mL, was used for the hydrolysis of a mixture of glucan- and hemicellulosic compounds-rich fraction. The chemicals used were sulfuric acid, D-glucose, L-arabinose, D-maltose, D-xylose, and D-galactose from Sigma-Aldrich, and pure ethanol (100%) and glacial acetic acid from Scharlau.

Table 1. Composition of oil palm empty palm fruit bunch (OPEFB).

Component.	OPEFB (% Dry Weight)
Lignin	21.77 ± 0.27
Glucan	40.09 ± 0.01
Hemicellulose	23.94 ± 0.02
Protein *	4.18 ± 0.51
Ash	3.72 ± 0.07
Others	6.30

* A nitrogen-to-protein conversion of 6.25 was used.

2.2. Organosolv Pretreatment

Oil palm empty fruit bunch (OPEFB), solvent ethanol 50% (v/v), and acid catalysts were added to 150 mL stainless steel tubular reactors (Swagelok, El Paso, TX, USA). The raw material and solvent were added according to the required solid loading. The reactors were sealed and placed into an oil bath (Julabo, Seelbach Germany) at specific temperature and retention time. After the desired retention time, the reactors were removed from the oil bath and quenched directly in an ice bath. The mixture was filtered using a 250 μm sieve for separation of the solid (glucan-rich) and the liquor (solvent, lignin-rich, and hemicellulosic compounds-rich) fractions after cooling down. The solid was washed using 28.3 mL of solvent/g of dry sample. The solvent was being collected and added to the liquor. To induce precipitation of lignin, 56.6 mL of water/g of dry sample was added to the liquor which was then centrifuged at 3360× *g* for 5 min to separate the lignin from the solvent and hemicellulosic compounds-rich fraction. Then, lignin-rich fraction, glucan-rich fraction, and hemicellulosic compounds-rich fractions were kept in the refrigerator at 4 °C until use. Series of pretreatments were carried out by varying four parameters, namely, acid type and concentration, temperature, retention time, and S/L ratio. To study the effect of catalyst, two strategies were used: (a) sulfuric acid and acetic acid were added to the solvent to reach the pH of 3; and (b) acetic acid was added in the equal amount as that of sulfuric acid added to reach the pH of 3. The pretreatment temperatures studied were 180, 210, and 220 °C; the retention times were 30, 60, 90, and 120 min; and the S/L ratios were 1:20, 1:10, and 1:5. The experiments were carried out using a one factor-at-time strategy.

2.3. Enzymatic Hydrolysis

The glucan-rich and hemicellulosic compounds-rich fractions obtained at the best conditions found in the study were mixed and evaporated using a rotary evaporator (LABO ROTA 20, Heidolph, Schwabach, Germany) at 110 °C, 40 rpm, and at a vacuum pressure of 100 mPa. The concentrated slurry had a glucan loading of 1.37% (w/v) and 1.85% (w/v) of total solid. Cellic®® Ctec3 enzyme solution was prepared by 10× dilution with ultra-pure water followed by sterile filtration using disposable disc filters with a pore size of 0.2 μm (GVS, Findlay, OH, USA). Diluted enzyme solution was added to the slurry based on the enzyme activity and glucan content (10, 15, 20, and 30 FPU/g glucan). Enzymatic hydrolysis was performed in 20 mL Erlenmeyer flasks, containing 10 mL of slurry adjusted to pH 5.2, that were incubated in a water-bath at 50 °C and shaking at 125 rpm for 24 h. Samples of 1.5 mL were withdrawn at 18 h and 24 h for chromatographic analysis. The percentage of glucan hydrolysis was then calculated based on the ratio between the amount of glucose released after hydrolysis and the theoretical maximum. Untreated OPEFB and evaporated hemicellulosic compounds-rich fraction were also used for enzymatic hydrolysis for comparison purposes.

2.4. Analytical Methods

The moisture content of untreated OPEFB and of the glucan-rich and lignin-rich fractions, to determine recovery yields, was quantified through sample drying in an oven at 70 °C until constant weight. The total solids of the evaporated mixture of glucan-rich and hemicellulosic compounds-rich fractions were determined according to Sluiter et al., (2008) [36]. The lignin-rich fraction, glucan-rich fraction, and untreated OPEFB were also analyzed for carbohydrates, lignin, and ash according to the methods described by Sluiter et al., (2008) [37], in order to determine the purity of the fractions. The percentage of lignin recovery was calculated using the following Equation (1)

$$(\%)\text{ Lignin recovery} = \frac{\text{Lop} \times (\%)\text{ lignin purity}}{\text{Lf}} \times 100\% \tag{1}$$

Lop = weight of lignin-rich fraction obtained after organosolv pretreatment; Lf = weight of lignin on OPEFB feed

Derived sugars from the acid treatment of samples for compositional analysis and glucose released during enzymatic hydrolysis were measured using High Performance Liquid Chromatography (HPLC) (Waters, Milford, MA, USA). The system was equipped with a hydrogen-based column (Aminex HPX-87P, Bio-Rad, Milford, MA, USA) operating at 60 °C and using 0.6 mL/min of 5 mM H_2SO_4 as the eluent. A refractive index (RI) detector (Waters 2414) was used to quantify the compounds.

The crystallinity of the untreated and pretreated OPEFB was analyzed using a Fourier Transform Infrared (FTIR) spectrometer using Nicolet OMNIC 4.1 software (Impact 410 iS10, Nicolet Instrument Corp., Madison, WI, USA). The spectral data were obtained with an average of 64 scans and resolution of 4 cm^{-1}, in the range of 400–4000 cm^{-1}. The total crystallinity index was calculated by the absorbance ratio of wavenumbers 1428 cm^{-1} and 897 cm^{-1} [38].

The nitrogen content of OPEFB was analyzed using the Kjeldahl method according to Mahboubi et al., (2017) [39] and the crude protein was obtained by using a nitrogen-to-protein conversion factor of 6.25.

2.5. Statistical Analysis

All experiments and analyses were carried out in duplicate. All intervals and error bars reported represent two times the standard deviation. The data acquired were statistically analyzed using MINITAB®® 17 (Minitab Ltd., Coventry, UK). A general linear model with a confidence interval of 95% was applied for the analysis of variance (ANOVA); statistical differences were identified at p-value <0.05. To have a better understanding of the extent of differences between results obtained, pairwise comparisons according to the Bonferroni test were performed.

3. Results

Oil palm empty fruit bunch was pretreated using acid-catalyzed ethanol organosolv to achieve deconstruction into three high-quality streams, namely, glucan-rich, lignin-rich, and hemicellulosic compounds-rich fractions. The effect of pretreatment factors, namely, acid catalyst type and concentration, temperature, retention time, and S/L ratio on lignin purity and recovery were studied. A mixture of evaporated glucan and hemicellulosic compounds-rich fractions obtained at the best condition of pretreatment was then enzymatically hydrolyzed in order to investigate the effect of organosolv pretreatment on glucan digestibility. The untreated OPEFB was also enzymatically hydrolyzed as a control.

3.1. Effect of Acid Catalyst

Catalyst Type and Concentration

In order to evaluate acetic acid and sulfuric acid as catalysts, ethanol organosolv pretreatment of OPEFB was carried out at 210 °C for 120 min with S/L ratio of 1:20. Both acids were added to the solvent until an initial pH of 3 was achieved. The amount of acetic acid added was 0.32 g acid/g substrate, while the use of sulfuric acid was much lower, i.e., 0.0013 g acid/g sample. The effect of catalyst type on lignin purity and recovery is presented in Figure 1a. The addition of acetic acid and sulfuric acid did not have a significant effect on the lignin purity, which were 65.70 ± 4.10 and 68.25 ± 3.50%, respectively, based on the dry weight. Lignin recovery from the two pretreatment conditions also showed no significant difference (53.02 ± 0.46% for acetic acid and 48 ± 2.91% for sulfuric acid). Since the amount of acetic acid added was higher than that of sulfuric acid, sulfuric acid is thus considered more efficient as a catalyst.

A further experiment was carried out where the amount of acetic acid added to the solvent was equal to the amount of sulfuric acid added to reach pH 3. The amount of both acids added was thus 0.0013 g/g sample. The final pH of the solvent with acetic acid was 4.66. Results from this experiment are presented in Figure 1b. Organosolv pretreatment with sulfuric acid showed better performance with significantly higher lignin purity (68.25 ± 3.50%) and lignin recovery (48.00 ± 2.91%) than those obtained when acetic acid was the used catalyst. The lignin purity of OPEFB treated with acetic acid as catalyst was lower by 9.17% while the recovery was lower by 17.85%. As sulfuric acid was more efficient regarding lignin purity and recovery, it was therefore employed for all subsequent pretreatment experiments.

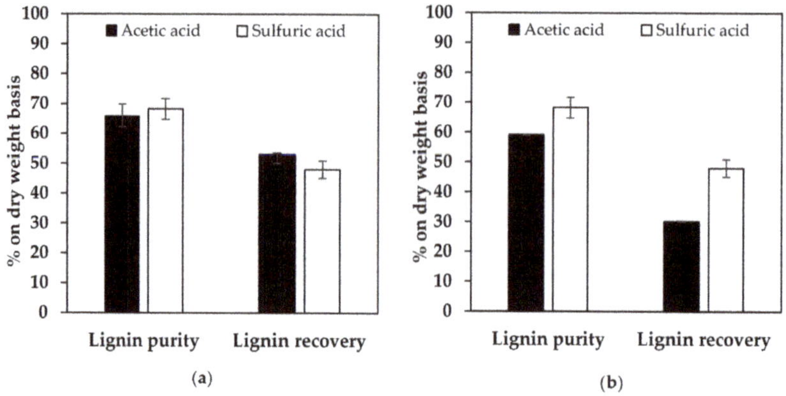

Figure 1. Effect of (**a**) catalysts type (added in different concentration to reach pH 3) and (**b**) catalyst concentration (added in the same concentration, 0.13% (w/w)) on lignin purity (p-value = 0.444 and p-value = 0.035, respectively) and recovery (p-value = 0.076 and p-value = 0.007, respectively). Error bars represent ± 2× the standard deviation.

3.2. *Effect of Temperature and Retention Time*

Three different temperatures, namely, 180 °C, 210 °C, and 220 °C were employed for organosolv pretreatment of OPEFB for 120 min with S/L ratio of 1:20 and using sulfuric acid as catalyst. The lignin purity and recovery were significantly higher when the temperature of the pretreatment was increased from 180 °C to 210 °C (Figure 2a). However, when the temperature was 220 °C, no further improvement on lignin purity and recovery were observed. Hence, 210 °C was chosen for further pretreatment experiments.

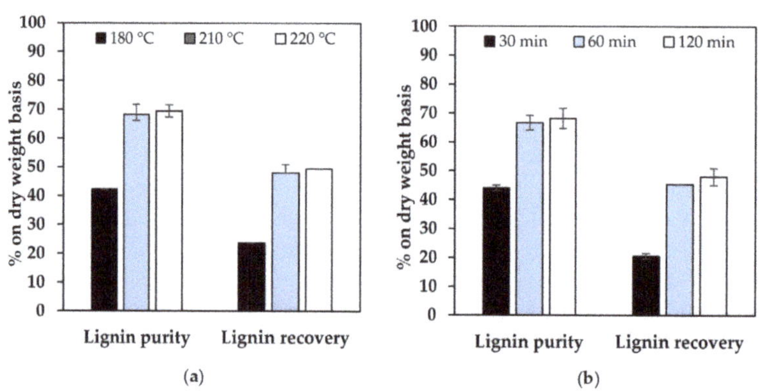

Figure 2. Effect of (**a**) temperature (for 120 min and using solid-to-liquid (S/L) ratio of 1:20) and (**b**) retention time (at temperature of 210 °C and S/L ratio of 1:20) on lignin purity (p-value = 0.001 and p-value = 0.001, respectively) and recovery (p-value = 0 and p-value = 0, respectively). Error bars represent ± 2× the standard deviation.

To investigate the effect of retention time, the pretreatment was conducted for 30, 60, and 120 min. Increasing the retention time from 30 to 60 min resulted in higher lignin purity (from 44.09 ± 0.41 to 66.73 ± 2.55) and recovery (from 20.52 ± 1.27 to 45.35 ± 0.00%) (Figure 2b). However, no significant improvements were observed by extending the pretreatment to 120 min.

3.3. Effect of S/L Ratio

The S/L ratio is the ratio of solid phase (biomass) and liquid phase (solvent plus water) in the organosolv pretreatment. In the first approach, different S/L ratios, namely, 1:20, 1:10, and 1:5, were applied for 60 min at 210 °C. The lignin purity obtained from pretreatment experiments at different S/L ratios of 1:20, 1:10, and 1:5 was 66.73 ± 2.55, 77.60 ± 2.80, and 80.26 ± 2.09%, respectively. Changing the S/L ratio from 1:20 to 1:10 increased the purity of lignin, however, no further improvement on lignin purity was achieved at S/L ratio of 1:5 (Figure 3a).

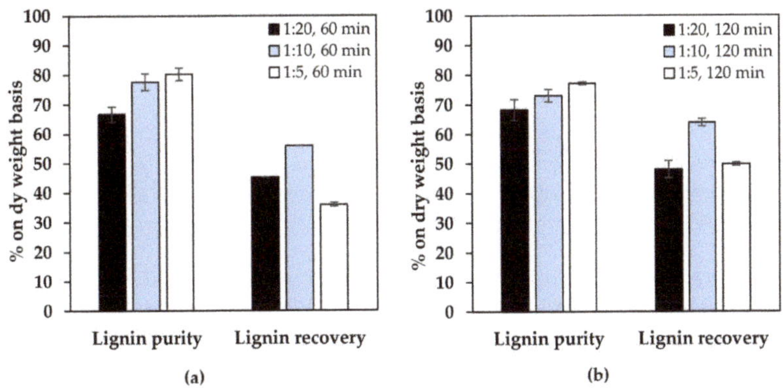

Figure 3. Effect of solid to liquid ratio and retention time of (**a**) 60 min and (**b**) 120 min (pretreatment at 210 °C) on lignin purity (p-value = 0.009 and p-value = 0, respectively) and recovery (p-value = 0 and p-value = 0.002, respectively). Error bars represent ± 2× the standard deviation.

However, such a trend was not observed for lignin recovery (Figure 3a), where a significant decrease was observed when the S/L ratio was changed from 1.10 to 1:5 (55.86 ± 0.06% vs. 35.99 ± 0.63%, respectively). Accordingly, the retention time was extended to 120 min (Figure 3b); similar trends were found for lignin purity and recovery although an overall increase in the later was observed. These results together with those obtained throughout the optimization approach, culminating in higher purity and recovery, indicate the interaction between S/L ratio and retention time which should be the focus of future research.

In a further optimization experiment, a retention time of 90 min was also investigated for pretreatment at an S/L ratio of 1:10. This retention time resulted in a significantly higher lignin recovery of 64.94 ± 1.09% than that obtained from the pretreatment at 60 min which was 55.85 ± 0.06%. A similar lignin recovery value of 63.94 ± 1.25% was obtained at 120 min (Figure 4); therefore, shorter retention times can be used while achieving similar recovery and purity of lignin. In summary, the best conditions found among those tested in this work, for the recovery and purity of lignin, include sulfuric acid as catalyst, temperature of 210 °C, retention time of 90 min, and S/L ratio of 1:10. In these conditions, lignin purity of 70.56 ± 4.48% and lignin recovery of 64.94 ± 1.09% were obtained.

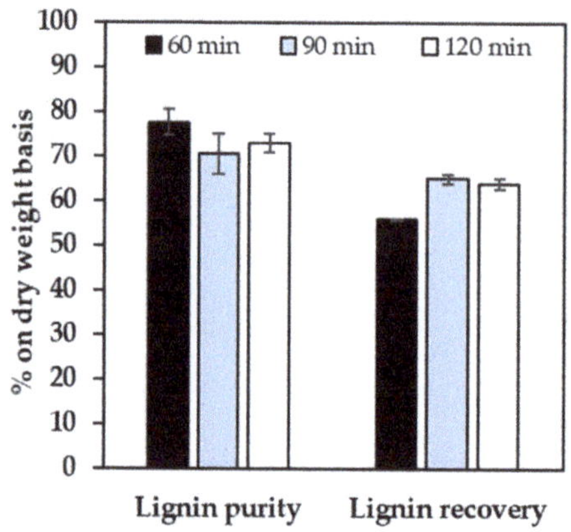

Figure 4. Effect of retention time at S/L ratio of 1:10 on lignin purity (p-value = 0.118) and recovery (p-value = 0.002). Error bars represent ± 2× the standard deviation.

Analysis of the composition of the lignin-rich fraction and the glucan-rich fraction, obtained under the best conditions mentioned, was carried out and the result is presented in Table 2. Based on the recovery of lignin in the glucan-rich fraction (around 10%), delignification of about 90% was obtained which shows some material loss during recovery through centrifugation.

Table 2. Composition of glucan-rich fraction and lignin-rich fraction obtained after acid-catalyzed organosolv pretreatment at 210 °C, for 90 min, and using an S/L ratio of 1:10.

Compound	Glucan-Rich Fraction	Lignin-Rich Fraction
Glucan	74.16 ± 0.52%	10.11 ± 0.08%
Glucan recovery	80.00 ± 0.02%	5.06 ± 0.08%
Lignin	5.09 ± 0.48%	70.56 ± 4.48%
Lignin recovery	10.44 ± 0.78%	64.94 ± 1.09%
Hemicellulose	8.24 ± 0.06%	7.79 ± 0.06%
Hemicellulose recovery	15.40 ± 0.11%	6.54 ± 0.16
Ash	3.23 ± 0.34%	1.87 ± 0.00%
Others	9.28%	9.67%

3.4. *Crystallinity of Cellulose*

The effect of organosolv pretreatment on the crystallinity of OPEFB was determined by calculating the total crystallinity index (TCI) value (absorbance ratio at wavenumbers 1248 and 897 cm^{-1}). Crystallinity index can be used to observe changes in cellulose digestibility after pretreatment. Glucan-rich fraction spectra showed a higher absorption band at 897 cm^{-1} and a lower absorption band at 1428 cm^{-1} (Figure 5). The result shows an increase in amorphous cellulose and a decrease in crystalline cellulose after pretreatment which indicates an increase in cellulose digestibility. The TCI value of the untreated OPEFB was 1.25 ± 0, whereas after pretreatment in the best conditions, i.e., 0.07% H_2SO_4, 210 °C, 90 min, and a solid-to-liquid ratio of 1:10, the TCI was 0.95 ± 0.03, representing a reduction of 24.48%.

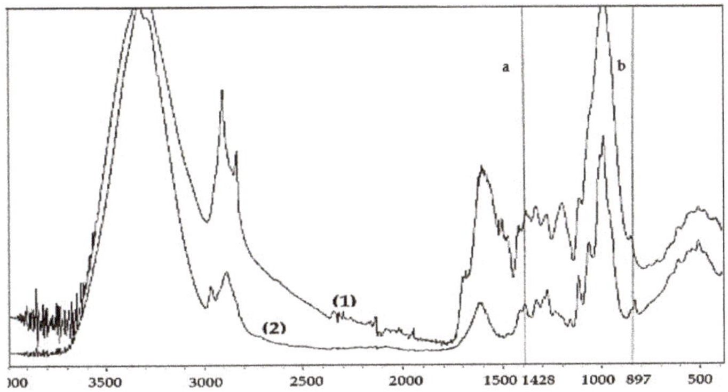

Figure 5. FTIR spectra of **(1)** untreated oil palm empty fruit bunches (OPEFB) and **(2)** glucan-rich fraction from the best organosolv pretreatment condition (0.07% H_2SO_4, 210 °C, 90 min, and solid-to-liquid ratio of 1:10). Bar **(a)** shows the absorbance at 1428 cm^{-1} whereas bar **(b)** shows the absorbance at 897 cm^{-1}.

3.5. Enzymatic Hydrolysis

Untreated OPEFB and a mixture of evaporated glucan-rich fraction and hemicellulosic compounds-rich fraction were digested using 10 FPU of Cellic®® Ctec3 enzyme per gram of glucan, for 24 h. The use of organosolv pretreatment led to an increase of 4.56-fold in glucan digestibility (Figure 6a). Enzymatic hydrolysis of evaporated hemicellulosic compounds-rich fraction resulted in 0.25 g/L glucose after 18 h. Since an enzyme loading of 10 FPU/g glucan could not achieve 100% glucan digestibility, higher enzyme loadings were used (Figure 6b). There was no difference on glucan digestibility between 18 h and 24 h of enzymatic hydrolysis. Therefore, enzymatic hydrolysis for 18 h resulted in glucan digestibility of 36.29 ± 2.84%, 55.45 ± 3.72%, 82.66 ± 3.09%, and 94.06 ± 4.71% at enzyme loading of 10, 15, 20, and 30 FPU/g glucan, respectively.

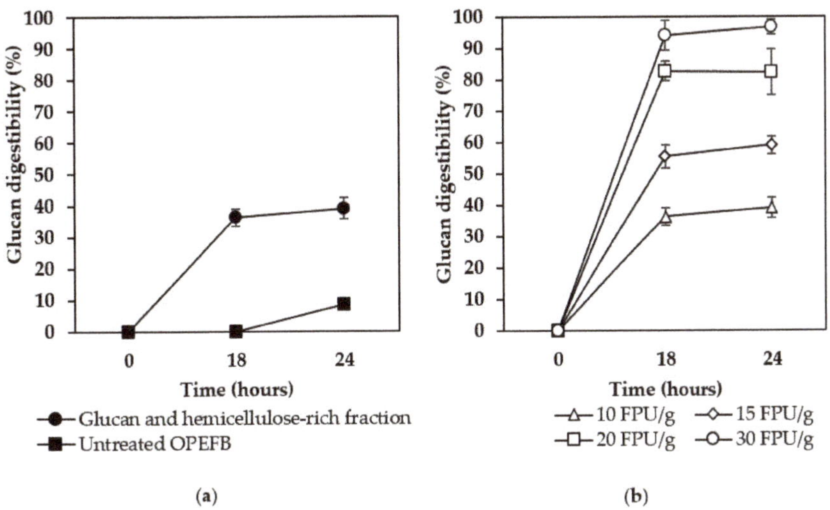

Figure 6. Glucan digestibility of (**a**) untreated OPEFB and of a mixture of evaporated glucan-rich and hemicellulosic compounds-rich fractions at enzyme loading of 10 FPU/g glucan and (**b**) a mixture of evaporated glucan-rich and hemicellulosic compounds-rich fractions using different enzyme loadings. Error bars represent ± 2× the standard deviation.

4. Discussion

Organosolv pretreatment uses organic solvents to remove lignin and hemicellulose, leaving a high purity glucan-rich fraction after solid-liquid separation. The fine tuning of organosolv pretreatment can consider a manifold of parameters including the solvent type and its concentration, catalyst addition, catalyst type and its concentration, temperature, retention time, S/L ratio, biomass particle size, and pretreatment vessel design. Substantial research on organosolv pretreatment has been carried out using a wide range of organosolv systems, applied to a wide range of lignocellulosic materials. The emphasis has been on the optimization of the pretreatment system in order to recover a high-purity and highly digestible glucan fraction, while the effect of pretreatment parameters on the recovery and purity of lignin has been comparatively less investigated [28]. Moreover, research studies using acid-catalyzed ethanol organosolv for the pretreatment of OPEFB are scarce in the scientific literature.

In this work, we demonstrated that parameters including catalyst type and its concentration, temperature, retention time, and S/L ratio had an effect on the recovery and purity of lignin following ethanol organosolv pretreatment of OPEFB. Through a series of optimization steps, where factor interaction is hypothesized to take place, OPFEB was decomposed into glucan-, lignin-, and hemicellulosic compounds-rich fractions. A delignification of ca. 90% was obtained at the following conditions: 0.07% H_2SO_4 (g/g substrate), 50% ethanol, 210 °C, 90 min, and S/L ratio of 1:10. The purity and recovery of glucan was 74% and 80%, respectively, and the corresponding values for lignin were 71% and 65%, respectively.

Lignin is the most abundant aromatic polymer in nature. Lignin is composed by coniferyl alcohol and sinapyl alcohol, with a small amount of *p*-coumaryl alcohol. Currently, most of the lignin (in low-purity) is produced by Kraft pulping process which reaches 50 million tons/year as low-value residuals [40]. Low-quality lignin limits its range of applications. Lignin produced through organosolv pretreatment of lignocellulose materials can be of high-purity, widening the range of potential applications. High-purity lignin can be converted into various polymers such as surfactants, plasticizers, superabsorbent gels and hydrogels, coating, and stabilizing agents [41–43]. Lignin also has the potential to be utilized as raw material for nanofiber with antioxidant capacity, resins, and vanillin synthesis [16–18]. The compositional analysis of the glucan-rich fractions revealed that loss of lignin takes place during the precipitation and centrifugation steps. This can be related to the intrinsic limitations of the recovery methods used or to the chemical characteristics of the resulting lignin. It has been reported that the dissolved lignin is polydispersed and, therefore, the used solvent for precipitation might be specific for certain lignin fractions. Reduced efficiency of water as anti-solvent has been reported for lower-molecular weight lignins with high contents of methoxyl and phenolic hydroxyl groups [44]. Therefore, establishing relationships among organosolv pretreatment parameters, lignin molecular weight and surface characteristics, and lignin recovery yields, need to be considered in future studies. Furthermore, lignin surface and compositional characteristics will influence the end-application and, consequently, influence the economic contribution of lignin to lignocellulosic biorefineries.

Among the organic solvents employed for organosolv pretreatment of lignocellulosic materials, ethanol is the most studied solvent due to its low price, good solubility of lignin, lower toxicity compared to other alcohol-based solvents, its miscibility with water, and ease of recovery [45]. Furthermore, the use of ethanol as solvent opens the possibility for integration of 1st and 2nd generation ethanol plants. Such strategy has been proposed as a result of techno-economic bottlenecks that still need to be overcome for commercialization of lignocellulosic biorefineries [46]. Accordingly, the glucan-rich fraction could be directed to yeast fermentation into ethanol, while the hemicellulosic compounds-rich fraction could be added to the yeast fermentation left-overs, following distillation, for e.g., biogas production through anaerobic digestion. Another strategy would be to take advantage of the capacity of filamentous fungi to consume pentose sugars so additional ethanol and fungal biomass for feed applications could be produced [47]. The ethanol as an end-product or as a pretreatment solvent could be recirculated through the system through already installed distillation columns and evaporators. Such integration strategy contributes to a needed holistic approach for cost-effective organosolv pretreatment processes, where optimization of pretreatment parameters leading to efficient material fractionation, need to be coupled with efficient fraction separation and solvent recycling methods [48]. By leading to lower energy consumption, the use of distillation is preferable to the use of evaporation for solvent recycling in systems containing water/low-boiling point organic solvent such as ethanol [48]. Reasonably, the extent of dilution of the black liquor will influence the energy consumption during distillation; therefore, studies on optimization of the amount of water added to induce lignin precipitation are also of relevance.

In this study, the crystallinity of glucan-rich fraction was 24.48% lower than that of the raw OPEFB. A decrease in crystallinity was also observed for pretreated Loblolly pine using 65% (v/v) ethanol with 1.1% (w/w) H_2SO_4 as the catalyst at 170 °C for 1 h [49]. Following OPEFB delignification and decreased crystallinity, 94.06 ± 4.71% glucan digestibility was achieved with an enzyme loading of 30 FPU/g within 18 h. Previous studies showed that high digestibility yield (>75%) of glucan-rich fraction from organosolv pretreatment of lignocellulosic material was obtained with enzyme loading higher than 10 FPU/g for 42–78 h of hydrolysis [50–53]. In addition to its use for ethanol production, glucan-rich fraction has also been proposed for production of acetone-butanol-ethanol (ABE) solvent, bio succinic acid, and fat-rich biomass [28].

A catalyst is usually added during organosolv pretreatment. The main effect in the addition of a catalyst is an increase in the rate and extent at which hydrolysis of hemicellulose and the cleavage of lignin-lignin bonds (α- and β-aryl ether linkages) occur [54]. Two types of acid catalysts i.e., acetic acid and sulfuric acid were employed in this study. The lignin recovery and purity of pretreatment using sulfuric acid as catalyst were significantly higher than those achieved when acetic acid was added at the same amount. To achieve similar lignin purity and recovery, about 270 times more acetic acid was needed. Even though acetic acid is more environmentally friendly, the high amount of acetic acid added can become problematic in the further utilization of lignocellulose fractions, that is, through microbial conversion where acetic acid can act as an inhibitor [55]. The higher lignin purity and its recovery after pretreatment with sulfuric acid as the catalyst is more likely due to the higher reactivity and efficiency of sulfuric acid on breaking the lignin-carbohydrate and lignin-lignin bonds than acetic acid [21]. The result from this study is in agreement with previous research by de la Torre et al., (2013) [56] who studied lignin recovery yields with different catalysts including acetic acid and sulfuric acid. Wheat straw was pretreated with 50% ethanol as a solvent and 0.001 N of each catalyst was tested for 30 min pulping. The organosolv pretreatment using acetic acid as catalyst led to 51% lignin recovery, whereas pretreatment using sulfuric acid led to 61% lignin recovery. Therefore, the results of this study support the need for alternative and effective catalysts. The catalyst should have a comparatively lower environmental footprint than that of sulfuric acid, to be used in organosolv pretreatment systems, a research gap previously identified [28].

The observations made on lignin purity and its recovery while varying the temperature and retention time agree with those reported by other studies. In this study, when the temperature was increased from 180 °C to 210 °C, lignin purity and recovery were increased by ca. 62% and 103%, respectively. This result is in agreement with an existing strong relationship between lignin solubility and temperature [57]. High delignification (>85%) was achieved during organosolv pretreatment of Silver birch wood chips and Norway spruce using ethanol as solvent and 1% sulfuric acid as catalyst at 200 °C [51,58]. The purity of the lignin from organosolv pretreatment of Silver birch wood chips was higher than that obtained in this work though, which was 96% [58]. A study by Goh et al., (2011) [59] reported 52% of lignin recovery from organosolv pretreatment of OPEFB using ethanol 65% (v/v) as solvent and 1.63% of sulfuric acid as the catalyst at a temperature of 190 °C. Generally, to obtain a better delignification rate, longer retention times are required. Results from this study showed that the use of retention time of 30 min was only able to recover 20% of lignin with a purity of less than 50%. When the pretreatment time was prolonged to 60 min, 45% of lignin was able to be recovered with 66% purity. A retention time of 60 min was reported to be necessary in order to obtain high delignification

(≥80%) on organosolv pretreatment of mixed sawmill and beechwood [60,61]. A study by Alio et al., (2019) [60] showed that 57% of lignin was recovered from organosolv pretreatment sawdust mixture of softwood species using sulfuric acid as a catalyst for 60 min. Furthermore, a retention time >60 min was needed in order to recover >30% of lignin during sulfuric acid-catalyzed organosolv pretreatment of OPEFB [59].

The S/L ratio is another key parameter for the commercialization of organosolv pretreatment. Higher S/L ratio is more favorable due to higher concentrations of glucan, hemicellulosic compounds, and lignin per unit volume. The higher concentration of substrate per volume leads concomitantly to the reduction of the amount of catalyst and solvent used per gram of substrate. However, the S/L ratio plays a role in organosolv pretreatment performance by affecting the contact between biomass and solvent. An S/L ratio higher than 1:10 was found to be detrimental to lignin purity and its recovery. Among research works applying organosolv systems for the pretreatment of lignocellulosic materials, such S/L ratio predominates [38]. The use in this study of an S/L ratio of 1:10 represented a concentration of 0.07% H_2SO_4 (gram acid per gram of substrate), which is almost 10× lower than that normally used for organosolv pretreatment [38], while attaining similar delignification yields together with high lignin recovery and purity. For instance, a study carried out by Goh et al., (2011) [59] showed that in order to recover 52% of the lignin from organosolv pretreatment of OPEFB using ethanol 65% as solvent at 190 °C for 75 min, 1.63% sulfuric acid was needed as a catalyst. Such a lower amount of acid used can have impacts on corrosion potential, economic feasibility, environmental footprint, and on the application of edible fungal biomass as feed ingredients. Altogether, the results of this work show the potential of acid-catalyzed ethanol organosolv pretreatment for fractionation of OPEFB and achieved a set of conditions leading to high lignin recovery and purity and digestible glucan. Moreover, the research work demonstrates the possibility to use much lower concentrations of acid with potential economic and environmental impacts. However, the system still can benefit from more environmentally-friendly catalysts and further system optimization that allow higher solid loading to be used, lower water usage during lignin precipitation, and clear relationships between lignin characteristics and end-application. Combining these optimization and characterization strategies with integration of OPEFB in 1st generation ethanol plants can create a beneficial biorefinery system for a such readily available lignocellulosic material in Indonesia, and in other countries such as Malaysia, the second largest worldwide producer of palm oil.

5. Conclusions

The purity and the recovery of both glucan and lignin from OPEFB could be obtained through a series of sequential steps of ethanol organosolv pretreatment. Lignin purity and recovery of ca. 71% and 65%, respectively, were obtained, opening the potential for higher-value applications. Furthermore, high-purity glucan fraction was also obtained that showed ca. 94% digestibility within 18 h of enzymatic treatment. The study shows that high delignification of OPEFB (of ca. 90%) and consequent recovery of high-purity lignin- and glucan-rich fractions can be obtained at remarkably lower acid concentration (0.07%) in comparison to previous studies within the area as result of an increase in the solid loading used.

Author Contributions: Conceptualization, R.M., W.B., C.N., and M.J.T.; methodology, K.M., M.J.T. and J.A.F.; formal analysis, K.M. and J.A.F.; investigation, K.M. and J.A.F.; writing—original draft preparation, K.M. and J.A.F.; writing—review and editing, M.J.T., R.M., W.B., C.N.; supervision, R.M., W.B., C.N., J.A.F., and M.J.T.; project administration, R.M., W.B., C.N., and M.J.T.; funding acquisition, R.M., C.N., and M.J.T. All authors have read and agreed to the published version of the manuscript.

Funding: This research was funded by the Program Menuju Doktor Sarjana Unggul (PMDSU) Batch III of the Ministry of Research, Technology and Higher Education of the Republic of Indonesia (Kemenristekdikti RI) (grant number 5844/DUN1.DITLIT/DIT-LIT7LT/2018) and Swedish Reseach Council (VR).

Acknowledgments: The authors would like to acknowledge PMDSU Scholarship from the Ministry of Research, Technology and Higher Education, Indonesia and Swedish Reseach Council (VR) for the administrative and funding support.

Conflicts of Interest: The authors declare no conflict of interest.

References

1. USDA Foreign Agricultural Service. *Table 11: Palm Oil: World Supply and Distribution*; USDA Foreign Agricultural Service: Washington, DC, USA, 2017.
2. Ditjenbun. *Tree Crop Estate Statistics of Indonesia Palm Oil (2017–2019)*; Ministry of Agriculture of Indonesia: Jakarta, Indonesia, 2018.
3. Ekpo, F.E.; Okey, E.N. Effect of oil palm empty fruit bunches (OPEFB) amendments in crude oil polluted soil on germination and growth performance of white mangrove species (Laguncularíaa racemosa). *Eur. Environ. Sci. Ecol. J.* **2014**, *1*, 19–28.
4. Geng, A. Conversion of oil palm empty fruit bunch to biofuels. In *Liquid, Gaseous and Solid Biofuels—Conversion Techniques*; Zhen, F., Ed.; IntechOpen: London, UK, 2013; pp. 479–487.
5. Chang, S.H. An overview of empty fruit bunch from oil palm as feedstock for bio-oil production. *Biomass Bioenergy* **2014**, *62*, 174–181. [CrossRef]
6. Zulkiple, N.; Maskat, M.Y.; Hassan, O. Pretreatment of oil palm empty fruit fiber (OPEFB) with aquaeous ammonia for high production of sugar. In *Procedia Chemistry*; Hase, T., Kurisu, G., Lusida, M.I., Dijkstra, B.W., Dixon, N., Eds.; Elsevier: Surabaya, Indonesia, 2016; pp. 155–161.
7. Kumar, R.; Wyman, C.E. Does change in accessibility with conversion depend on both the substrate and pretreatment technology? *Bioresour. Technol.* **2009**, *100*, 4193–4202. [CrossRef] [PubMed]
8. Kumar, R.; Wyman, C.E. Effects of cellulase and xylanase enzymes on the deconstruction of solids from pretreatment of Poplar by leading technologies. *Biotechnol. Prog.* **2009**, *25*, 302–314. [CrossRef] [PubMed]
9. Kumar, A.K.; Sharma, S. Recent updates on different methods of pretreatment of lignocellulosic feedstocks: A review. *Bioresour. Bioprocess.* **2017**, *4*, 1–19. [CrossRef] [PubMed]
10. Chen, H.; Liu, J.; Chang, X.; Chen, D.; Xue, Y.; Liu, P.; Lin, H.; Han, S. A review on the pretreatment of lignocellulose for high-value chemicals. *Fuel Process. Technol.* **2017**, *160*, 196–206. [CrossRef]
11. Li, X.; Mupondwa, E.; Panigrahi, S.; Tabil, L.; Sokhansanj, S.; Stumborg, M. A review of agricultural crop residue supply in Canada for cellulosic ethanol production. *Renew. Sustain. Energy Rev* **2012**, *16*, 2954–2965. [CrossRef]
12. Kim, J.S.; Lee, Y.Y.; Kim, T.H. A review on alkaline pretreatment technology for bioconversion of lignocellulosic biomass. *Bioresour. Technol.* **2016**, *199*, 42–48. [CrossRef]
13. Garcia-Cubero, M.A.; Gonzalez-Benito, G.; Indacoechea, I.; Coca, M.; Bolado, S. Effect of ozonolysis pretreatment on enzymatic digestibility of wheat and rye straw. *Bioresour. Technol.* **2009**, *100*, 1608–1613. [CrossRef]
14. Kucharska, K.; Rybarczyk, P.; Holowacz, I.; Lukajtis, R.; Glinka, M.; Kaminski, M. Pretreatment of lignocellulosic materials as substrates for fermentation processes. *Molecules* **2018**, *23*, 2937. [CrossRef]
15. Miliotti, E.; Dell'Orco, S.; Lotti, G.; Rizzo, A.; Rosi, L.; Chiaramonti, D. Lignocellulosic ethanol biorefinery: Valorization of lignin-rich stream through hydrothermal liquefaction. *Energies* **2019**, *12*, 723. [CrossRef]
16. Bassett, A.W.; Breyta, C.M.; Honnig, A.E.; Reilly, J.H.; Sweet, K.R.; La Scala, J.J.; Stanzione, J.F. Synthesis and characterization of molecularly hybrid bisphenols derived from lignin and CNSL: Application in thermosetting resins. *Eur. Polym. J.* **2019**, *111*, 95–103. [CrossRef]
17. Fache, M.; Boutevin, B.; Caillol, S. Vanillin production from lignin and its use as a renewable chemical. *ACS Sustain. Chem. Eng.* **2015**, *4*, 35–46. [CrossRef]
18. Kai, D.; Ren, W.; Tian, L.; Chee, P.L.; Liu, Y.; Ramakrishna, S.; Loh, X.J. Engineering poly(lactide)–lignin nanofibers with antioxidant activity for biomedical application. *ACS Sustain. Chem. Eng.* **2016**, *4*, 5268–5276. [CrossRef]
19. Zhao, X.; Cheng, K.; Liu, D. Organosolv pretreatment of lignocellulosic biomass for enzymatic hydrolysis. *Appl. Microbiol. Biotechnol.* **2009**, *82*, 815–827. [CrossRef]
20. Hallac, B.B.; Ray, M.; Murphy, R.J.; Ragauskas, A.J. Correlation between anatomical characteristics of ethanol organosolv pretreated *Buddleja davidii* and its enzymatic conversion to glucose. *Biotechnol. Bioeng.* **2010**, *107*, 795–801. [CrossRef]
21. Borand, M.N.; Karaosmanoğlu, F. Effects of organosolv pretreatment conditions for lignocellulosic biomass in biorefinery applications: A review. *J. Renew. Sustain. Energy* **2018**, *10*. [CrossRef]

22. Zhao, X.; Li, S.; Wu, R.; Liu, D. Organosolv fractionating pre-treatment of lignocellulosic biomass for efficient enzymatic saccharification: Chemistry, kinetics, and substrate structures. *Biofuels Bioprod. Biorefin.* **2017**, *11*, 567–590. [CrossRef]
23. Annual World Fuel Ethanol Production. Available online: https://ethanolrfa.org/statistics/annual-ethanol-production/ (accessed on 28 January 2020).
24. Kim, S.; Um, B.; Im, D.; Lee, J.; Oh, K. Combined ball milling and ethanol organosolv pretreatment to improve the enzymatic digestibility of three types of herbaceous biomass. *Energies* **2018**, *11*, 2457. [CrossRef]
25. Kim, T.; Im, D.; Oh, K.; Kim, T. Effects of organosolv pretreatment using temperature-controlled bench-scale ball milling on enzymatic saccharification of *Miscanthus* × *giganteus*. *Energies* **2018**, *11*, 2657. [CrossRef]
26. Park, N.; Kim, H.Y.; Koo, B.W.; Yeo, H.; Choi, I.G. Organosolv pretreatment with various catalysts for enhancing enzymatic hydrolysis of pitch pine (*Pinus rigida*). *Bioresour. Technol.* **2010**, *101*, 7057–7064. [CrossRef] [PubMed]
27. Wildschut, J.; Smit, A.T.; Reith, J.H.; Huijgen, W.J. Ethanol-based organosolv fractionation of wheat straw for the production of lignin and enzymatically digestible cellulose. *Bioresour. Technol.* **2013**, *135*, 58–66. [CrossRef] [PubMed]
28. Ferreira, J.A.; Taherzadeh, M.J. Improving the economy of lignocellulose-based biorefineries with organosolv pretreatment. *Bioresour. Technol.* **2020**, *299*, 122695. [CrossRef] [PubMed]
29. Chin, D.W.K.; Lim, S.; Pang, Y.L.; Lim, C.H.; Lee, K.M. Two-staged acid hydrolysis on ethylene glycol pretreated degraded oil palm empty fruit bunch for sugar based substrate recovery. *Bioresour. Technol.* **2019**, *292*, 121967. [CrossRef] [PubMed]
30. Kim, D.Y.; Kim, Y.S.; Kim, T.H.; Oh, K.K. Two-stage, acetic acid-aqueous ammonia, fractionation of empty fruit bunches for increased lignocellulosic biomass utilization. *Bioresour. Technol.* **2016**, *199*, 121–127. [CrossRef]
31. Mardawati, E.; Badruzaman, I.; Nurjanah, S.; Bindar, Y. Effect of organosolv pretreatment on delignification for bioethanol feedstock from oil palm empty fruit bunch (OPEFB). In *IOP Conference Series: Earth and Environmental Science*; IOP Publishing: Queensland, Australia, 2018.
32. Ong, H.C.; Jan, B.M.; Tong, C.W.; Fauzi, H.; Chen, W.-H. Effects of organosolv pretreatment and acid hydrolysis on palm empty fruit bunch (PEFB) as bioethanol feedstock. *Biomass Bioenergy* **2016**, *95*, 78–83. [CrossRef]
33. Tan, L.; Yu, Y.; Li, X.; Zhao, J.; Qu, Y.; Choo, Y.M.; Loh, S.K. Pretreatment of empty fruit bunch from oil palm for fuel ethanol production and proposed biorefinery process. *Bioresour. Technol.* **2013**, *135*, 275–282. [CrossRef]
34. Tan, Y.T.; Ngoh, G.C.; Chua, A.S.M. Evaluation of fractionation and delignification efficiencies of deep eutectic solvents on oil palm empty fruit bunch. *Ind. Crop. Prod.* **2018**, *123*, 271–277. [CrossRef]
35. Chopda, R.; Ferreira, J.A.; Taherzadeh, M.J. Biorefining oat husks into high-quality lignin and enzymatically digestible cellulose with acid-catalyzed ethanol organosolv pretreatment. *Processes* **2020**, *8*, 435. [CrossRef]
36. Sluiter, A.; Hames, B.; Hyman, D.; Payne, C.; Ruiz, R.; Scarlata, C.; Sluiter, J.; Templeton, D.; Wolfe, J. *Determination of Total Solids in Biomass and Total Dissolved Solids in Liquid Process Samples*; National Renewable Energy Laboratory: Golden, CO, USA, 2008.
37. Sluiter, A.; Hames, B.; Ruiz, R.; Scarlata, C.; Sluiter, J.; Templeton, D.; Crocker, D. *Determination of Structural Carbohydrates and Lignin in Biomass: Laboratory Analytical Procedure*; National Renewable Energy Laboratory: Golden, CO, USA, 2008.
38. Nelson, M.L.; O'Connor, R.T. Relation of certain infrared bands to cellulose crystallinity and crystal latticed type. Part, I. Spectra of lattice types I, II, III and of amorphous cellulose. *J. Appl. Polym. Sci.* **1964**, *8*, 1311–1324. [CrossRef]
39. Mahboubi, A.; Ferreira, J.A.; Taherzadeh, M.J.; Lennartsson, P.R. Value-added products from dairy waste using edible fungi. *Waste Manag.* **2017**, *59*, 518–525. [CrossRef] [PubMed]
40. Ferreira, J.A.; Agnihotri, S.; Taherzadeh, M.J. Waste Biorefinery. In *Sustainable Resource Recovery and Zero Waste Approaches*; Elsevier: Amsterdam, The Netherlands, 2019; pp. 35–52.
41. Cerrutti, B.M.; de Souza, C.S.; Castellan, A.; Ruggiero, R.; Frollini, E. Carboxymethyl lignin as stabilizing agent in aqueous ceramic suspensions. *Ind. Crop. Prod.* **2012**, *36*, 108–115. [CrossRef]
42. Faria, F.A.C.; Evtuguin, D.V.; Rudnitskaya, A.; Gomes, M.T.S.R.; Oliveira, J.A.B.P.; Graça, M.P.F.; Costa, L.C. Lignin-based polyurethane doped with carbon nanotubes for sensor applications. *Polym. Int.* **2012**, *61*, 788–794. [CrossRef]

43. Wang, H.; Zou, J.; Shen, Y.; Fei, G.; Mou, J. Preparation and colloidal properties of an aqueous acetic acid lignin containing polyurethane surfactant. *J. Appl. Polym. Sci.* **2013**, *130*. [CrossRef]
44. Zhu, W.; Westman, G.; Theliander, H. Investigation and characterization of lignin precipitation in the lignoboost process. *J. Wood Chem. Technol.* **2014**, *34*, 77–97. [CrossRef]
45. Zhou, Z.; Lei, F.; Li, P.; Jiang, J. Lignocellulosic biomass to biofuels and biochemicals: A comprehensive review with a focus on ethanol organosolv pretreatment technology. *Biotechnol. Bioeng.* **2018**, *115*, 2683–2702. [CrossRef]
46. Lennartsson, P.R.; Erlandsson, P.; Taherzadeh, M.J. Integration of the first and second generation bioethanol processes and the importance of by-products. *Bioresour. Technol.* **2014**, *165*, 3–8. [CrossRef]
47. Ferreira, J.A.; Lennartsson, P.R.; Taherzadeh, M.J. Production of ethanol and biomass from thin stillage byNeurospora intermedia: A pilot study for process diversification. *Eng. Life Sci.* **2015**, *15*, 751–759. [CrossRef]
48. Thoresen, P.P.; Matsakas, L.; Rova, U.; Christakopoulos, P. Recent advances in organosolv fractionation: Towards biomass fractionation technology of the future. *Bioresour. Technol.* **2020**, *306*, 123189. [CrossRef]
49. Sannigrahi, P.; Miller, S.J.; Ragauskas, A.J. Effects of organosolv pretreatment and enzymatic hydrolysis on cellulose structure and crystallinity in Loblolly pine. *Carbohydr. Res.* **2010**, *345*, 965–970. [CrossRef]
50. Choi, J.-H.; Jang, S.-K.; Kim, J.-H.; Park, S.-Y.; Kim, J.-C.; Jeong, H.; Kim, H.-Y.; Choi, I.-G. Simultaneous production of glucose, furfural, and ethanol organosolv lignin for total utilization of high recalcitrant biomass by organosolv pretreatment. *Renew. Energy* **2019**, *130*, 952–960. [CrossRef]
51. Patel, A.; Matsakas, L.; Rova, U.; Christakopoulos, P. Heterotrophic cultivation of *Auxenochlorella protothecoides* using forest biomass as a feedstock for sustainable biodiesel production. *Biotechnol. Biofuels* **2018**, *11*, 169. [CrossRef] [PubMed]
52. Yao, L.; Yang, H.; Yoo, C.G.; Pu, Y.; Meng, X.; Muchero, W.; Tuskan, G.A.; Tschaplinski, T.; Ragauskas, A.J. Understanding the influences of different pretreatments on recalcitrance of *Populus* natural variants. *Bioresour. Technol.* **2018**, *265*, 75–81. [CrossRef] [PubMed]
53. Yuan, Z.; Wen, Y.; Kapu, N.S.; Beatson, R. Evaluation of an organosolv-based biorefinery process to fractionate wheat straw into ethanol and co-products. *Ind. Crop. Prod.* **2018**, *121*, 294–302. [CrossRef]
54. Zhang, Z.; Harrison, M.D.; Rackemann, D.W.; Doherty, W.O.S.; O'Hara, I.M. Organosolv pretreatment of plant biomass for enhanced enzymatic saccharification. *Green Chem.* **2016**, *18*, 360–381. [CrossRef]
55. Van der Pol, E.C.; Vaessen, E.; Weusthuis, R.A.; Eggink, G. Identifying inhibitory effects of lignocellulosic by-products on growth of lactic acid producing micro-organisms using a rapid small-scale screening method. *Bioresour. Technol.* **2016**, *209*, 297–304. [CrossRef]
56. De la Torre, M.J.; Moral, A.; Hernández, M.D.; Cabeza, E.; Tijero, A. Organosolv lignin for biofuel. *Ind. Crop. Prod.* **2013**, *45*, 58–63. [CrossRef]
57. Zhang, X.; Yuan, Z.; Wang, T.; Zhang, Q.; Ma, L. Effect of the temperature on the dissolution of corn straw in ethanol solution. *RSC Adv.* **2016**, *6*, 102306–102314. [CrossRef]
58. Matsakas, L.; Karnaouri, A.; Cwirzen, A.; Rova, U.; Christakopoulos, P. Formation of lignin nanoparticles by combining organosolv pretreatment of birch biomass and homogenization processes. *Molecules* **2018**, *23*, 1822. [CrossRef]
59. Goh, C.S.; Tan, H.T.; Lee, K.T.; Brosse, N. Evaluation and optimization of organosolv pretreatment using combined severity factors and response surface methodology. *Biomass Bioenergy* **2011**, *35*, 4025–4033. [CrossRef]
60. Alio, M.A.; Tugui, O.C.; Vial, C.; Pons, A. Microwave-assisted Organosolv pretreatment of a sawmill mixed feedstock for bioethanol production in a wood biorefinery. *Bioresour. Technol.* **2019**, *276*, 170–176. [CrossRef] [PubMed]
61. Kalogiannis, K.G.; Matsakas, L.; Lappas, A.A.; Rova, U.; Christakopoulos, P. Aromatics from beechwood organosolv lignin through thermal and catalytic pyrolysis. *Energies* **2019**, *12*, 1606. [CrossRef]

© 2020 by the authors. Licensee MDPI, Basel, Switzerland. This article is an open access article distributed under the terms and conditions of the Creative Commons Attribution (CC BY) license (http://creativecommons.org/licenses/by/4.0/).

agronomy

MDPI

Article

Formosolv Pretreatment to Fractionate Paulownia Wood Following a Biorefinery Approach: Isolation and Characterization of the Lignin Fraction

Elena Domínguez [1], Pablo G. del Río [2,3], Aloia Romaní [4], Gil Garrote [2,3], Patricia Gullón [2,*] and Alberto de Vega [5]

1 Technological Centre of Multisectorial Research (CETIM), Business Park of Alvedro, 15181 Culleredo, Spain; edominguez@cetim.es

2 Department of Chemical Engineering, Faculty of Science, Universidade de Vigo (Campus Ourense), As Lagoas, 32004 Ourense, Spain; pdelrio@uvigo.es (P.G.d.R.); gil@uvigo.es (G.G.)

3 Environmental Technology and Assessment Laboratory, Campus da Auga-Campus Ourense, Universidade de Vigo, 32004 Ourense, Spain

4 Center of Biological Engineering, University of Minho, Campus of Gualtar, 4710 057 Braga, Portugal; aloia@ceb.uminho.pt

5 Group of Environmental Chemical Engineering (EnQA), Centro de Investigacións Científicas Avanzadas (CICA), Dept. of Chemistry, Faculty of Science, Univerdade da Coruña, As Carballeiras, s/n, 15071 A Coruña, Spain; alberto.de.vega@udc.es

* Correspondence: pgullon@uvigo.es

Received: 28 July 2020; Accepted: 13 August 2020; Published: 17 August 2020

check for updates

Abstract: Paulownia is a rapid-growth tree with a high biomass production rate per year and low demand of water, which make it very suitable for intercropping systems, as it protects the crops from adverse climatic conditions, benefiting the harvest yields. Moreover, these characteristics make Paulownia a suitable raw material able to be fractionated in an integrated biorefinery scheme to obtain multiple products using a cascade conversion approach. Different delignification pretreatments of biomass have been purposed as a first stage of a lignocellulosic biorefinery. In this study, the formosolv delignification of Paulownia wood was investigated using a second order face-centered factorial design to assess the effects of the independent variables (concentrations of formic and hydrochloric acids and reaction time) on the fractionation of Paulownia wood. The maximum delignification achieved in this study (78.5%) was obtained under following conditions: 60 min, and 95% and 0.05% formic and hydrochloric acid, respectively. In addition, the remained solid phases were analyzed to determine their cellulose content and cooking liquors were also chemically analyzed and characterized. Finally, the recovered lignin by precipitation from formosolv liquor and the pristine lignin (milled wood lignin) in Paulownia wood were characterized and compared by the following techniques FTIR, NMR, high-performance size-exclusion chromatography (HPSEC) and TGA. This complete characterization allowed verifying the capacity of the formosolv process to act on the lignin, causing changes in its structure, which included both phenomena of depolymerization and condensation.

Keywords: organosolv; fractionation; lignin; characterization; Paulownia

---

1. Introduction

The current economic system, based on a linear model, has led to humankind's overreliance on non-renewable fossil resources causing its depletion, besides entailing harmful consequences for the environment, society, economy and health [1]. This global reality is provoking an unsustainable situation with consequences still incalculable. In an attempt to mitigate this situation, in recent decades, an unceasing search for alternative strategies has been encouraged to find new and suitable production systems founded in the use of renewable resources as raw material within a biorefinery context. This would suppose a transition from the traditional linear economic model to a circular economy, more efficient and greener, moving the current trend towards a global and sustainable bioeconomy [2].

In this scenario, lignocellulosic materials are promising candidates as feedstock to obtain biofuels, building blocks, bio-chemicals, food additives, adhesives, or cosmetics, among others [3,4]. Lignocellulosic materials (LCM) present a three-dimensional and recalcitrant structure mainly composed of cellulose (homopolymer made up of glucose units), hemicelluloses (heteropolymer made up of different sugars) and lignin (aromatic polymer). The integral use of LCM involves a selective separation of its components through fractionation treatments following the biorefinery concept [5]. Therefore, the selection of the adequate fractionation process is key to achieve an efficient utilization of all fractions.

Organic solvent pretreatment (also known as organosolv) is an emerging alternative to conventional pulping processes, since it allows fractionating the LCM into cellulose, lignin, and hemicellulose, with a high purity of all fractions opening the possibility of its integral exploitation [6]. The organosolv fractionation process provides an efficient and clean way of transforming lignocellulose into valuable products, facilitating a subsequent recovery of all the fractions obtained [7]. This kind of treatment uses a mixture of an aqueous organic solvent and water, with or without the addition of a mineral acid, which dissolves most part of the lignin and hemicellulose from the raw material [8]. Moreover, these procedures present interesting advantages such as: easy solvent recovery and recyclability, free sulfur, low-cost investment and environmentally friendly [9].

Short-chain organic acids have been considered good solvents for lignin in the delignification of LCM [10]. Particularly, formic acid has attracted considerable attention as a delignification agent, due to its ability to achieve a selective fractionation of the biomass showing high efficiency, both non-wood, hardwood and softwood biomass [9]. The mixture of LCM and concentrated aqueous solutions of formic acid at boiling temperature, with the addition of small quantities of hydrochloric acid employed as a catalyst, is known as formosolv. In this work, the formosolv treatments were performed at atmospheric pressure in all cases, that is, depending on the composition of the cooking liquor, in the range 105–109 °C. Nevertheless, the temperature and pressure can be raised in order to reduce the time of reaction [9]. This process yields a cellulose rich pulp, an aqueous fraction rich in hemicellulosic sugars and a lignin fraction [11]. Among the LCM, Paulownia species are rapid-growth trees with a high biomass production rate per year (50 t/ha/yr) and a low demand of water, which make it very suitable for intercropping systems as it protects the crops from adverse climatic conditions, benefiting the harvest yields. In addition to its uses as wood to build from plywood for musical instruments, other applications suggested for Paulownia wood include its exploitation as a source for pulp due to its fast development and uniform growth [12].

Several works have reported the use of Paulownia species in a biorefinery framework. For example, Domínguez et al., (2017) [13] subjected *P. tomentosa* to hydrothermal pretreatment to solubilize the hemicellulosic fraction yielding a solid fraction that was evaluated to obtain bio-ethanol. Gong and Bujanovic (2014) [14] purposed a sequence based on hot water extraction to solubilize most of the hemicelluloses followed by the delignification in acetone/water in the presence of oxygen for the production of cellulose and lignin from *P. tomentosa* and *P. elongata*. However, to the best of our knowledge, scarce research has been performed to fractionate Paulownia by formosolv pulping to recover cellulose and lignin.

The resulting cellulose of the fractionation process can be used for the production of pulp, derivatives, nanofibrillated cellulose or fermentable glucose (after the cellulose hydrolysis) depending on its physicochemical properties [14].

Lignin is a polyphenolic amorphous material originated from the random oxidative coupling of three main *p*-hydroxycinnamyl alcohol monomers (*p*-coumaryl, coniferyl, and sinapyl alcohols), which are representative of the *p*-hydroxyphenyl (H-units), guaiacyl (G-units) and syringyl (S-units) phenylpropanoid units, respectively [15,16]. Due to its polyphenolic chemical structure, it can be employed in the manufacture of adhesives, epoxy, phenolic resins, and polyolefins, as well as in a variety of novel applications.

The objective of this work was the systematic study of the operational variables of formosolv cooking (concentrations of formic and hydrochloric acids and reaction time) to obtain the highest yield of delignification of Paulownia wood. In addition, a secondary target was to identify the main changes caused during the delignification process on the Paulownia lignin, by means of different analytical methods such as FTIR, NMR, high-performance size-exclusion chromatography (HPSEC) and TGA, comparing with pristine lignin used as reference.

2. Materials and Methods

2.1. Raw Material

The raw material used in this study was wood coming from a dihybrid species of Paulownia (*Paulownia elongata x fortunei*), and was provided by Maderas Álvarez Oroza, S.L. located in Foz (Lugo, NW, Spain). The wood was milled in a Wiley mill until the size of a particle smaller than 8 mm, homogenized, and stored in containers with aeration in a cool, dry and dark place until its use.

2.2. Analysis of Raw Material

Samples taken from the homogenized lot were milled to a size smaller than 0.5 mm and analyzed (composition shown in Table 1) using the following methods: extractives (National Renewable Energy Laboratory/Technical Procedure NREL/TP-510-42619, 2008), moisture (NREL/TP-510-42621, 2008), ashes (NREL/TP-510-42622, 2008), and quantitative acid hydrolysis (NREL/ TP-510-42618, 2008).

Table 1. Chemical composition of *Paulownia elongata x fortunei* wood.

Component	g/100 g Oven-Dried Paulownia ± Standard Deviation
Glucan	39.7 ± 0.97
Xylan + Galactan + Mannan	14.7 ± 0.56
Arabinan	1.09 ± 0.05
Acetyl groups	3.30 ± 0.01
Klason lignin	21.9 ± 0.50
Extractives	5.60 ± 0.01
Ashes	0.50 ± 0.05
Uronic acids (expressed as glucuronic acid)	1.30 ± 0.30

The liquid phase from quantitative acid hydrolysis was analyzed by high-performance liquid chromatography (HPLC) to quantify the monosaccharides, acetic acid and formic acid (detector, refractive index at 30 °C; column, Aminex HPX-87H; mobile phase, 0.01 M H_2SO_4; flow rate, 0.6 mL/min; column temperature 50 °C). In this column, xylose, galactose and mannose were co-eluted and therefore, these monosaccharides were quantified together using the notation (Xyl + Gal + Man). The concentrations of glucose, Xyl + Gal + Man, arabinose and acetic acid, before and after quantitative acid hydrolysis (121 °C, 60 min, 4% H_2SO_4), were used to calculate the equivalent content of glucan, Xylan + Galactan + Mannan, arabinan and acetyl groups, respectively. The insoluble phase from the quantitative acid hydrolysis was gravimetrically quantified and reported as Klason lignin. Uronic acids were determined using a colorimetric method [17]. Analyses were carried out in quadruplicate.

2.3. Formosolv Fractionation Process

Mixtures of Paulownia wood, water and formic acid (80–95%) at 8 g liquid/g dry Paulownia wood, were heated to boiling point in 250 mL Pyrex flasks. When the boiling started, hydrochloric acid was added, taking this as time zero. The selected concentrations of hydrochloric acid were in the range of 0.05 to 0.10%, according to the best results found in previous work [18,19]. The mixture was stirred at atmospheric pressure along a desired time (30–60 min). Reactions were stopped by removing the flasks from the heating plates and by fast filtration through medium-pore Gooch crucibles to separate the pulp from black liquor. In order to prevent the reprecipitation of lignin on the solids, the pulps were subsequently washed with two formic acid solutions, the first one at the same concentration employed in the pretreatment, and the second one at half of its concentration. Finally, the solids were washed with distilled water until neutral pH (named Paulownia formosolv pulp, PFP). Pulp yield (PY; g oven-dried pulp/100 g oven-dried Paulownia wood) was determined gravimetrically after being oven-dried. The solids were analyzed as described in Section 2.2. and in addition, they were subjected to kappa number (KN) (Tappi T236) and, intrinsic viscosity (VIS) (Tappi T230) determinations.

The aliquots of the black liquors were subjected to posthydrolysis (121 °C, 40 min, 4% H_2SO_4) in order to quantify the solubilized hemicelluloses. In addition, another aliquot was employed in theh precipitation assays of the lignin by adding a specific amount of water to the black liquor. Precipitated solids were separated by centrifugation (4200 rpm, 10 min), repeatedly washed with water, and centrifuged until the supernatant was neutral. Afterwards, the precipitated lignin (Paulownia formosolv lignin, PFL) was lyophilized and used for structural characterization experiments. The quantitative acid hydrolysis of lignin was carried out following the same procedure as in the previous section (NREL/TP-510-42618, 2008).

2.4. Experimental Design and Statistical Analysis

In this study, 17 runs (experiments 1–17 collected in Table 2) were performed in accordance with a face-centered factorial design with three replicates at the central point [20] so as to investigate the effects of the variables: X_1, cooking time (30–60 min), X_2, formic acid (FA) concentration (80–95%) and X_3 hydrochloric acid (HA) concentration (0.05–0.10%). Following the normalization of the independent variables to the range (−1, 1), quadratic response surfaces were fitted by least-squares multiple regression (Equation (1)) using an excel statistical module (DOE PRO XL, SigmaZone):

$$DV = b_O + \sum\nolimits_{J=1}^{3} b_j X_j + \sum\nolimits_{j=1, j \leq k}^{3} b_{jk} X_j X_k a \quad (1)$$

where DV denotes each of the dependent variables (system responses) studied, and X_j and X_k are the previously defined normalized variables. The values of b_0, b_i, and b_{jk} represent the fitting parameters calculated by multiple regression between the system responses (DV) and the normalized variables.

Table 2. Design of the experiments used to evaluate the delignification process by the formosolv of Paulownia wood and the main results (all data expressed in an oven dry basis).

Experimental Variables				Pulp Composition						Liquor Composition
Exp. no.	**Time (min)**	**FA (%) [1]**	**HA (%) [1]**	**PY (%) [1]**	**KN**	**VIS (mL/g)**	**GnP (%) [2]**	**FAP (%) [2]**	**AGP (%) [2]**	**Xyl+Gal+Man Dis (%) [3]**
1	60	95	0.1	45.2	32.6	605	70.4	3.2	0.9	17.4
2	30	80	0.05	54.1	42.2	361	63.5	1.9	1.9	17.9
3	60	80	0.05	49.6	38.7	607	73	1.8	1.1	18.1
4	30	80	0.1	55.1	47.9	467	67.8	2.1	2	17.7
5	60	80	0.1	52.1	41.8	529	60.6	1.8	1.2	18
6	30	95	0.1	49.6	37.1	653	63.5	3.6	1.1	17
7	45	87.5	0.075	48.9	39.7	613	65	2.3	1.1	17.8
8	45	87.5	0.075	48	40	611	65.1	2.3	1	17.8
9	45	87.5	0.075	49.4	40.4	613	65.2	1.5	1.2	18.3
10	30	95	0.05	56	52.8	600	66.7	6.9	2.3	14.3
11	60	95	0.05	47	27.7	626	71.7	5.8	2.4	15.9
12	30	87.5	0.075	52.3	39.9	686	64.7	2.7	1.6	17.4
13	60	87.5	0.075	50.6	45.7	640	69.4	2.4	0	17.7
14	45	80	0.075	58.3	48.9	532	61.8	6.1	0	14.8
15	45	95	0.075	51.8	40.4	617	69.6	2.2	2.9	17.7
16	45	87.5	0.05	55.7	52.7	595	67.3	7.3	0	14.2
17	45	87.5	0.1	54.4	46.8	556	69.8	6.8	0	14.6

PY: Pulp yield KN: kappa number; VIS: viscosity; FA: formic acid concentration; **HA**: hydrochloric acid concentration; **GnP**: glucan remaining in pulp; **FAP**: formic acid remaining in pulp; **AGP**: acetyl groups remaining in pulp; **Xyl+Gal+Man dis**: xylose, galactose, and mannose dissolved in liquor.; [1] kg pulp/100 kg raw material; [2] kg component/100 kg pulp; [3] percentage over content of component in Paulownia wood.

2.5. Milled Wood Lignin (MWL) Isolation

To compare the lignin recovered by the formosolv process with the pristine lignin in Paulownia wood, the milled wood lignin was prepared according to methods as previously described [21,22]. Milled Paulownia wood from an IKA WERKE MF 10 grinder (particle size < 1 mm) was subjected to several steps, as follows: (i) extraction with acetone for 8 h in Soxhlet and with hot water (100 °C) during 3 h, (ii) milling in a Retsch PM 400 centrifugal ball mill with toluene using an agate jar and balls, (iii) three consecutive extractions (12 h) with dioxane:water (9:1, *v*/*v*) (25 mL of solvent/g of milled wood), (iv) the separation of the supernatant by centrifugation and evaporation at 40 °C at reduced pressure until dryness, (v) the residue obtained (raw MWL) was redissolved in a mixture of acetic acid:water (9:1, *v*/*v*) (20 mL of solvent/g of raw MWL), (vi) precipitated in water and separated by centrifugation, (vii) milled in an agate mortar, (viii) dissolved in a solution of 1,2-dicloroethane:ethanol (1:2, *v*/*v*), (ix) the mixture was then centrifuged to eliminate the insoluble material, precipitating the supernatant in diethyl ether and separating the residue obtained by centrifugation, (x) this residue was then resuspended in petroleum ether and centrifuged again to obtain the final purified MWL fraction, which was dried under a N_2 current. The final yield was ca. 20% of the original Klason lignin content.

2.6. Infrared Spectroscopy (FTIR-ATR)

FTIR spectra were obtained on a Perkin-Elmer Spectrum 2000 instrument by the attenuated total reflectance (ATR) technique. The spectra were recorded in the 4000–600 cm^{-1} range with 16 scans at a resolution of 4.0 cm^{-1} and an interval of 1.0 cm^{-1} [23]. All the spectra were baseline corrected and normalized in an area, in order to make adequate comparisons.

2.7. Nuclear Magnetic Resonance (NMR) Spectroscopy

^{1}H NMR, ^{13}C NMR and HSQC (heteronuclear single quantum correlation) solution state analysis were performed on lignin samples (~40 mg) dissolved on 1 mL of DMSO-d_6. The spectra were recorded at 25 °C on a Bruker Advance instrument at 500 MHz. Standard Bruker programs such as zg30 for ^{1}H NMR, zgpg60 for ^{13}C NMR, and hsqcetgp for HSQC, were used.

For ^{31}P NMR, the samples were derivatized with a phosphitylation reagent according to the procedures previously described [24–27]. Lignin (20 mg) was first dissolved in 500 μL of a mixture of pyridine and deuterated chloroform (1.6:1 *v*/*v*). Afterwards, 100 μL of cholesterol (10 mg/mL) and 100 μL of chromium (III) acetylacetonate solution (5 mg/mL) were added, as an internal standard and relaxation agent, respectively. Finally, the solution was mixed with 100 μL of 2-chloro-4,4,5,5-tetramethyl-1,3,2-dioxaphospholane (agent of phosphitylation) for about 10 min and transferred into a 5 mm NMR tube for subsequent NMR analysis at 25 °C on a Bruker Avance 400 MHz instrument using the pulse program zgig (inverse gated decoupling).

In the aliphatic oxygenated region of the HSQC spectra, the relative abundance of interunit linkages were estimated using the volume integrals from the C_α–H_α correlations. In the aromatic region, the $C_{2,6}$–$H_{2,6}$ correlations from S units and the C_2–H_2 from G units were used to estimate the S/G ratio according with previous works [28].

2.8. High-Performance Size-Exclusion Chromatography (HPSEC)

The average molecular weight (M_w), the number average (M_n) and the polydispersity index (M_w/M_n) of MWL and the PFL were analyzed by HPSEC. The chromatography equipment employed was a Jasco LC Net II/ADC (Analog data Can) equipped with a refractive index detector provided with two Varian Polymer Laboratories PolarGel-M columns in series (300 mm × 7.5 mm). Isolated lignins were dissolved in the mobile phase which consisted in 0.1% of lithium bromide in dimethylformamide. The chromatographic conditions were as follows: injection volume 20 μL, a flow of 0.7 mL/min and a temperature of 40 °C. The calibration of the SEC was carried out using polystyrene standards with different molecular weights (between 62,500 and 266 g/mol) provided by Sigma Aldrich.

2.9. Thermogravimetric Analysis (TGA)

TGA was carried out on a NETZSH STA 449 F3 Jupiter instruments to investigate the general characteristics of lignin thermal decomposition. A sample of approximately 2.5 mg lignin was loaded into alumina pans, where it was gradually heated from 25 to 900 °C at a heating rate of 10 °C/min under a nitrogen atmosphere with a constant flow of 60 mL/min. When the Klason lignin was subjected a calcination process, the content of the ashes was negligible.

2.10. Elemental Analysis

The C, H, and N contents of organosolv lignins were determined in triplicate by elemental analysis in a Flash EA1112 (ThermoFinnigan). The oxygen content was calculated from the difference between the sample weight and the C, H, N and S contents.

3. Results and Discussion

3.1. Formosolv Process: Delignification of Paulownia Wood

The formic acid delignification was carried out with the objective to obtain: (i) a high glucan content in the solid phase, potentially useful in later enzymatic hydrolysis stages, (ii) a high delignification degree and (iii) a high solubilization of hemicellulosic sugars, leading to a solid phase mainly composed of glucan.

Paulownia wood was treated with different concentrations of formic and hydrochloric acid. In order to obtain a quantitative description of the effects of these variables on the fractionation process, the experiments were carried out according to a second order face-centered factorial design. The structure of the factorial design, analysis of the pulps and liquors are shown in Table 2, and regression parameters for the variables' solid yield, viscosity and glucan in the pretreated pulp are displayed in Table 3.

Table 3. Regression parameters for the dependent variables: solid yield, viscosity and glucan.

	Yield	Viscosity	Gn
b_0	51.680	619.59	65.992
b_1, Time	−2.260 [c]	23.94 [c]	1.893 [b]
b_2, FA	−1.960	60.48 [a]	1.525 [c]
b_3, HA	−0.600	2.12	−1.026
b_{12}	−0.737	−41.31 [b]	1.195
b_{13}	0.763	−32.20 [c]	−1.872 [c]
b_{23}	−1.463	0.52	0.452
b_{123}	0.388	13.96	2.321 [b]
b_{11}	−2.415	38.04	0.404
b_{22}	1.185	−50.75 [c]	−0.973
b_{33}	1.185	−49.45 [c]	1.893
R^2	0.6738	0.918	0.8351

[a] Coefficients significant at 99% confidence level based on the Student's t-test; [b] coefficients significant at 95% confidence level based on the Student's t-test; [c] coefficients significant at 90% confidence level based on the Student's t-test.

Formic acid treatment produced an important solubilization of Paulownia wood, ranging the pulp yield between 45.2 (in experiment 1, carried out at the highest values of the three independent variables, time = 60 min, FA: 95%, HA: 0.1%) and 58.3% (in experiment 14, time = 45 min, FA: 80%, HA: 0.075%). According to the data of Table 2, the yield decreased when the independent variables increased, with more influence of time and FA concentration, according to the data of Table 3.

The highest delignification (78.5% of lignin removal) was achieved under the conditions of experiment 11 (see Table 2), corresponding to a kappa number value of 27.7 (10.6% Klason lignin).

The most significant variables on the results of the KN were the duration of the treatment and the concentration of formic acid, but none of them were significant at the 95% confidence level.

Regarding the hemicellulosic sugars recovery in liquors, significant differences were found between the operational conditions, achieving a concentration of hemicellulosic sugars varying in the range of 14.3–18.3 g of sugar/100 g of Paulownia. The operational range employed allowed to reach very similar percentages of dissolution among the experiments (about 17% of hemicelluloses were solubilized from Paulownia wood during the pretreatment) (Table 2). In the optimum conditions to remove the maximum lignin content (experiment 11), the distribution of sugars in black liquor was largely dominated by xylooligosaccharides (66.9% of total dissolved carbohydrates, 2.2 g/L), followed by glucooligosaccharides (14.3% of total dissolved carbohydrates, 0.47 g/L) and arabinooligosaccharides (7.5% of total dissolved carbohydrates, 0.24 g/L). The presence of monosaccharides was reduced up to 11.0% by mass, while their products of decomposition, furfural and hydroxymethylfurfural, only accounted for 0.3% of the mixture of dissolved carbohydrates.

The high glucan content is directly related to the cellulose purity. In our case, the values were very close to 70% in the best experimental combinations (experiments 1, 3, 11). The regression equation of this variable indicates that long treatment times and high FA concentrations promote a high percentage of glucan in pulp. However, the HA concentration must be low as to partially prevent the hydrolytic effects on the cellulose. Figure 1 shows the predicted response surface for the content of glucan in the pulp showing the influences of the more significant variables.

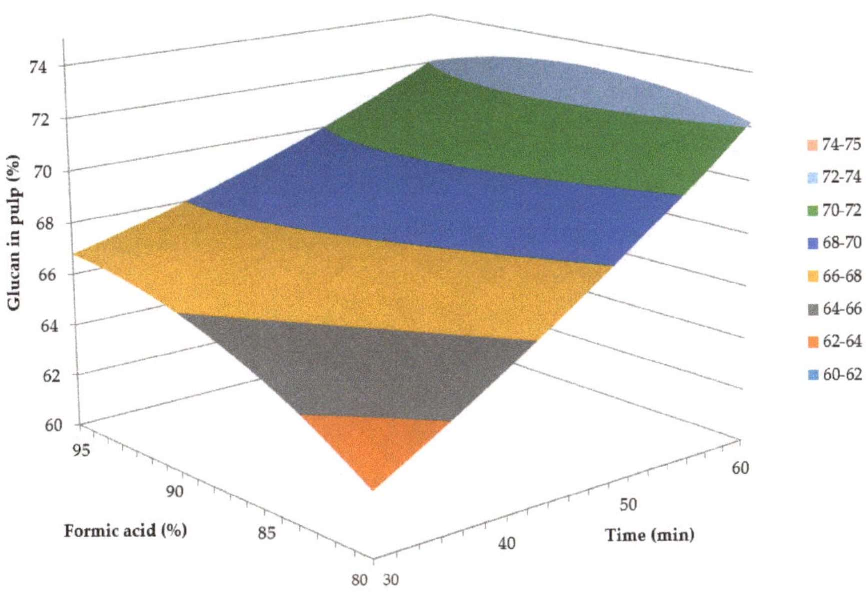

Figure 1. Response surface of glucan in pulp at a constant value of 0.05% HA.

In order to achieve a more precise information about the severity of the treatment over cellulose, the glucose content of the black liquors was analyzed, as well as the viscosity of the pulps, which is a measure of the chain length of the cellulose. The maximum value of monomeric glucose solubilized from glucan in the liquor was 7.2% (the average was 4.0%), indicating a probable hydrolysis of cellulose to some extent. On the other hand, the viscosities of the pulps, under conditions in which the highest delignification was achieved, were around 600 mL/g, a relatively low value compared with other pulps with the same KN. It was confirmed that the cellulose suffered a certain attack resulting in a reduction of the chain length.

It should be noted that the analysis of the formic acid content of the pulps (after quantitative acid hydrolysis) reached a significant mass percentage (~5%), probably due to the incorporation of formates in the lignin molecule. This fact was determined by several researchers in different treatments with formic acid [11,29] and is confirmed, in this work, with the spectroscopic data shown below.

A mathematical optimization allowed the definition of the optimal conditions of fractionation introducing, with the same weight, the following constraints for the composition of the pulp: the highest possible content in glucose, the lowest KN and the lowest content in hemicelluloses. These optimal conditions are a time of 60 min, FA concentration of 95%, and a HA concentration of 0.05% (experiment 11).

3.2. Precipitation of the Lignin from Formosolv Black Liquor

In order to collect an adequate amount of lignin for the study of its physicochemical properties, a larger reactor was used at the optimum conditions. The black liquor was treated with different proportions of water to evaluate the precipitation rate as a function of the dilution factor. The curve soon reached an asymptotic value at a dilution factor of 3 mL water/mL liquor. The recovery, gravimetrically measured, was 104% of the lignin dissolved. This value exceeds 100% because the solids consisted of not only pure lignin, but also associated carbohydrates (11%) and FA (1.6%).

3.3. FTIR of Paulownia Wood and Residual Solid

FTIR spectroscopy is a fast and simple technique that allows extracting valuable information about the existence of functional groups and their changes during chemical treatments. Figure 2 shows the fingerprint region of the FTIR spectra of Paulowia wood (P), and the solid residue after the formosolv treatment (PFP). The assignment of the bands was made based on data published elsewhere [9,28].

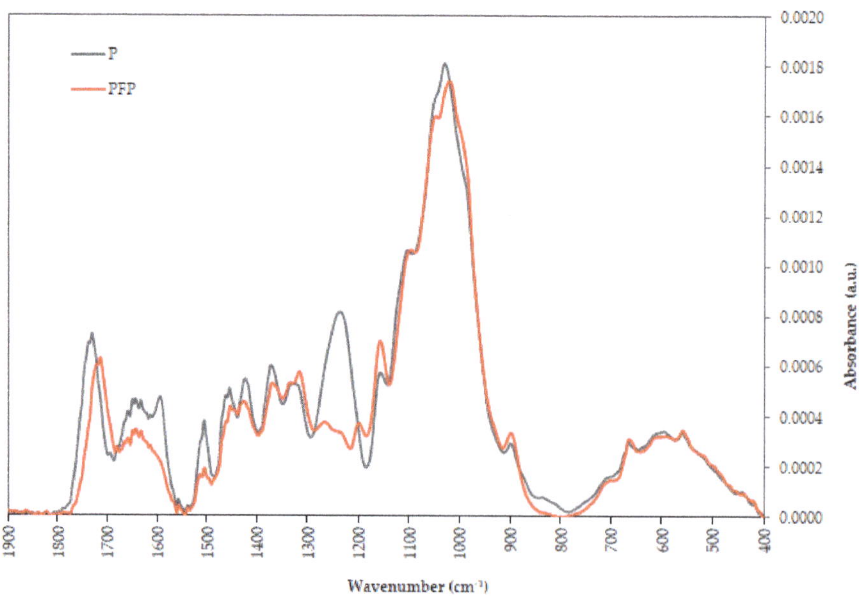

Figure 2. FTIR spectra of Paulowia wood (P) and Paulownia formosolv pulp (PFP).

Several typical bands of lignin can be clearly seen in the spectra, which involve aromatic skeleton vibrations, which disappeared or were greatly diminished in the residual solid of the formosolv treatment due to extensive delignification. For example: 1594, 1504, 1420 cm^{-1}, and the broad band around 1230 cm^{-1} that contains the contributions of vibration modes from G and S units. In addition, the band located around 1650 cm^{-1} reflects some loss of C=O conjugate aryl ketones in lignin. The absorption band in the Paulownia wood spectrum at 1730 cm^{-1}, attributable to the presence of acetyl groups of the hemicellulose, disappeared after organosolv treatment, indicating the solubilization of the majority of the hemicellulose. In addition, a new absorption band appeared in the spectrum of formosolv pulp, at lower wavelength (1710 cm^{-1}) that is assigned to the incorporation of formate groups to the propyl chain of lignin [11]. Finally, the bands located at 1325 and 1265 cm^{-1}, originated by the stretching of C–O bonds in S and G-units, respectively, modified their relative intensity in an evident way. Although it is not possible to confirm the exclusive contribution of G-units in the band at 1265 cm^{-1}, its great decrease in intensity seems to indicate an important loss of guaiacyl structures [30].

3.4. *Structural Characterization of Lignins*

^{1}H NMR Spectroscopy of MWL and Formosolv Lignin

Figure S1 shows the ^{1}H NMR spectra of MWL and formosolv Paulownia lignin. The ^{1}H NMR spectra of MWL and PFL showed that the proton signals in non-oxygenated carbons (methyl and methylene groups, δ = 0.8–1.2 ppm) were fewer, and less intense in the organosolv residue, which indicates structural modifications in the side chain of the phenylpropane units. The methyl acetate protons resonated in the range of 1.8–2.1 ppm and disappeared almost completely after the treatment, due to the hydrolytic effect of the acidic medium in the black liquor. Peaks at 2.50 and 3.33 ppm were due to residual proton signals in DMSO-d_6 and water, respectively [31].

Between 4 and 5.5 ppm, a large number of signals originated from several protons of the propyl side chain of lignin. In this region of the spectrum, resonances due to the different protons of the hemicelluloses components also appear, but their contribution should be very low. The areas of these signals are smaller in the organosolv residue, which indicates the changes in the functionalities of the side chain. Specifically, the signal at 4.44 ppm (part per million), which originated from the γ protons in β-O-4′ aryl ether units, decreased drastically in the spectrum of PFL due to delignification.

The range between 6.3 and 7.5 ppm showed the different signals of the aromatic protons, and between 8 and 9 ppm (part per million), those of the phenolic hydroxyl protons. The sharp peak at 8.1 ppm originated from carboxylic functionalities (formic acid) incorporated into a lignin molecule.

3.5. *^{13}C NMR Spectroscopy*

^{13}C NMR spectra of MWL and PFL (Figure 3) provide interesting information about several functional features of MWL and the changes suffered after the organosolv treatment. The assignment of signals was made based on previous published works [32–35].

The spectrum of MWL showed the characteristic peaks assigned to methoxy groups (55.9 ppm), carbonyl carbon in acetates (169.6 ppm), carbons 3 and 5 ($C_{3,5}$) in etherified syringyl (S) units (152.2 ppm), C_3 in etherified guaiacyl (G) units (149.1 ppm), C_4 in etherified G units and $C_{3,5}$ in non-etherified S units, and C_4 in S units (138.2 ppm). The aromatic carbons 2, 5 and 6 of the G structures resonated at 111.1, 114.7 and 119.1 ppm, respectively, while the $C_{2,6}$ in the S units produced a signal around 104 ppm.

Between 70 and 90 ppm, the signals of C_β in β-O-4′, and C_α in β-5 and β-β′ were overlapped with those belonging to the carbons of the residual carbohydrates, complicating an accurate assignment, but probably the sharpest signal in this area (71.9 ppm) could be due to C_α of β-O-4′ linkages. Other identifiable peaks were: C_γ in units with oxidized C_α (65.6 ppm), C_γ in substructures β-O-4′ (62.9 ppm), methylene carbons in the α,β positions of the side chain (29.1 ppm), the methyl carbon in acetates (21.1 ppm), and the primary carbon in methyl groups of the propyl chain (15.3 ppm). The absence of typical *p*-hydroxyphenyl group signals (167, 161, 158 ppm) agrees with those found in the literature [36,37] and reflects that the Paulownia hardwood lignin is GS-type [35].

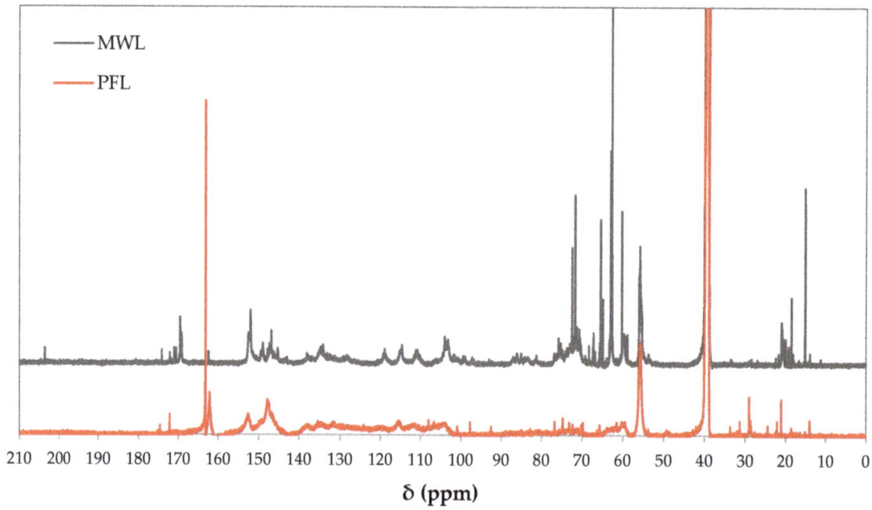

Figure 3. ^{13}C NMR spectra of milled wood lignin (MWL) and Paulownia formosolv lignin (PFL).

After the formosolv treatment, the lignin recovered from the liquor (PFL) showed important changes compared to MWL. The acetate groups were almost completely eliminated from the structure (absence of signals at 169.6 and 21.1 ppm). At the same time, two intense signals (163.4 and 162.4 ppm), not appearing in the MWL spectrum, reflected the introduction of formate groups into the molecule [38]. The effects of delignification are clearly reflected in ^{13}C NMR spectrum of PFL. The depolymerization of lignin through the disruption of β-O-4′ bonds and to some extent, β-5′ and β-β′, was reflected by the decrease in the intensity of many of the typical signals of these substructures in the 90–52 ppm range: specifically, 71.9, 65.6, and 62.9, whose assignment was mentioned above. Additionally, the solubilization of a large proportion of the remaining carbohydrates in the MWL causes this area of the spectrum to have less signals than MWL. The distribution of aliphatic carbon signals is modified, appearing more and as different types of signals: more methylenic carbons and less methyl groups.

In the region of aromatic carbons (166–102 ppm), the changes produced by delignification were clearly visible. The signals of carbons 2, 5 and 6 in the G-units (111.1, 114.7 and 119.1 ppm) were broader and with a larger area, which is compatible with a partial recondensation of lignin. In addition, the signals $C_{3,5}$ in the S-units (152.2 ppm) and C_3 in the G-units (149.1 ppm) increased their intensity. However, this increase was more important in the signal due to guaiacyl groups, indicating a greater tendency of these groups to undergo recondensation processes. The integration of the clusters due to the signals of aryl bonds [33] that is C–O (161–141 ppm), C–C (141–124 ppm) and C–H (124–100 ppm) indicated different percentage between the values of MWL and PFL of −2%, 31% and −21%, respectively. These values can be explained by the formation of a higher proportion of C–C links and a loss of C–H, which provide an additional symptom of the existence of lignin condensation reactions.

Table S1 (Supplementary Material) collects the assignments of carbon chemical shifts (δ, ppm) in the ^{13}C NMR spectrum of MWL and formosolv lignin.

3.6. ^{1}H-^{13}C HSQC NMR Spectroscopy of Lignins

The HSQC spectrum of the MWL of Paulownia wood (Figure 4a,c) shows some prominent signals that allowed to extract information about the most important features on how phenylpropanoid units are linked together. The most important correlations correspond to:

- The C–H correlations in aromatic methoxyl between δ_C–δ_H: 56.1/3.2–56.2/3.7;

- Aryglycerol β-aryl ether with β-O-4′ links (label A), which can be seen at δ_C–δ_H: 72.5/4.9 and 71.8/4.7 (C_α–H_α in S and G-units respectively), 84.4/4.3 and 86.5/4.1 (C_β–H_β in S and G-units, respectively) and a broad signal, overlapping with others, in 60.2/3.7–60.4/3.2 (C_γ–H_γ);
- Resinol type units (α-O-γ′, β-β′, and γ-O-α′ linkages, label B): their different C–H correlations are evident at δ_C–δ_H: 85.5/4.7 (C_α–H_α), 54.3/3.1 (C_β–H_β), 71.6/3.8 and 71.7/4.2 (C_γ–H_γ);
- Phenylcoumaran type structures (α-β-5′, α-O-4′ linkages, label C): only two types of correlations are visible in the spectrum: C_α–H_α at δ_C–δ_H: 87.5/5.5, and C_β–H_β at δ_C–δ_H: 53.9/3.5, they are partially overlapped with the correlations of the aromatic methoxyl groups;
- Spirodienone type structures (β-1′, α-O-α′ linkages, label D): they present less intense signals at δ_C–δ_H: 81.80/5.04 (C_α–H_α correlations), and 60.4/2.7 (C_β-H_β);
- A weak signal, due to C_β–H_β correlation on dibenzodioxocin-like structures (label E), also appears in the spectrum at δ_C–δ_H 86.6/3.8—however, the one corresponding to the C_α–H_α correlation (83.4/4.8) can only be observed at very low contour levels;
- Low-intensity correlations corresponding to cynnamyl alcohol end-groups (129.15/6.45, 128.89/6.23, and 62.06/4.10, correlations between carbonhydrogen α, β and γ, respectively, and label I) and cynnamyl aldehyde end-groups (C_β–H_β: 126.72/6.77. Label J) are also present;
- In the area of the spectrum corresponding to aromatic C–H correlations the signals belonging to S (carbons 2 and 6 δ_C–δ_H 104.4/6.7. label $S_{2,6}$) and G-units (111.6/7.0, carbon 2, label G2; 115.5/6.7-115.8/6.9, carbon 5, label G5, and 119.7/6.8, carbon 6, label G6) are predominant—other less intense signals belonging to S-units oxidized in the α-carbon ($S_{2,6'}$) are also observable at δ_C–δ_H 107.1/7.3 and 107.0/7.2;
- Minor signals can be seen due to the cinnamaldehyde end-groups in S-units ($J_{2,6\text{-}S}$) at 106.7/7.0 and H-units ($H_{2,6}$) at 128.9/7.2;
- Finally, it is worth mentioning that the presence of associated carbohydrate signals, which are indicated in the figure corresponding to the anomeric positions and some other with acetyl substituents [39].

Table S2 (Supplementary Material) collects the assignments of the ^{13}C–^{1}H correlation signals in the HSQC NMR spectrum of the obtained lignin fractions.

HSQC spectrum of PFL (Figure 4b,d) reflects important structural changes after the treatment. The more apparent difference is the presence of a broad signal (around δ_C–δ_H 63.5/4.3) which corresponds to the β-O-4′ structures with acylated γ-OH, probably as formates. The formylation of the γ-OH was referred by other researchers [9,28] treating LCM with FA solutions of different concentrations and at different temperatures. The ^{13}C NMR spectrum reflected this fact (peaks at 163.4 and 162.4 ppm already mentioned), but also the HSQC spectrum of the PFL showed a signal at δ_C-δ_H 162.0/8.2 that confirms the formylation of the lignin.

Table 4 presents the quantification of the different types of linkages according to Sun and coworkers (Sun et al., 2014). In the MWL, β-O-4′ units were predominant and accounted for 57.6% of all identified linkages, followed by resinols (21.4%), phenylcoumarans (9.4%), and spirodienones (3.7%). In addition, minor proportions of *p*-hydroxycinnamyl with alcohol (2.2%) and aldehyde (2.3%) end-groups were quantified in the spectrum.

All the correlations corresponding to the different linkages between phenylpropane units were drastically diminished in the spectrum of PFL. The effect of delignification, through the destruction of the linkages identified in MWL, is very important. The quantification of this breakdown of bonds was 87% of the β-aryl ethers, 78% of the resinol type subunits, 65% of the phenylcoumarans and 70% of the spirodienones. The cinnamyl alcohol or aldehyde-type units completely disappeared in PFL.

Moreover, the signals derived from the aromatic nucleus (G_2, G_5, G_6 and $S_{2,6}$) were less intense and extended towards smaller chemical shifts in both dimensions, which is indicative of the existence of condensation in the lignin [9,40].

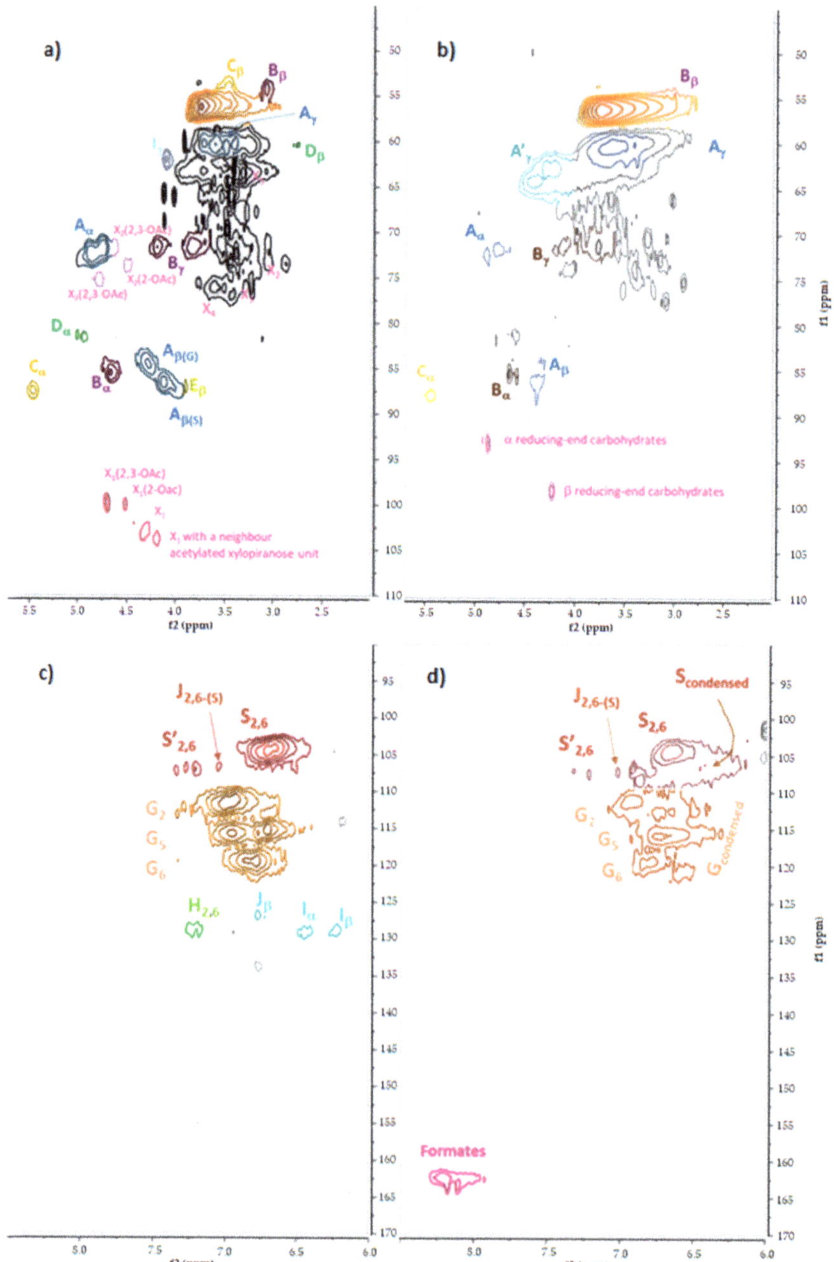

Figure 4. HSQC spectra of MWL (**a**,**c**) and PFL (**b**,**d**).

Table 4. Properties of MWL and PFL.

		MWL	PFL
Structural units/C9 by HSQC	Arylglycerol β-O-4′ aryl ethers	0.41	0.13
	Resinols	0.15	0.11
	Phenylcoumarans	0.07	0.07
	Spirodienones	0.04	0
	Dibenzidioxocins	0.02	0
	p-OH-cinnamyl alcohol end-groups	0.03	0
	p-OH-cinnamyl aldehyde end-groups	0.02	0
	S/G	0.63	0.83
OH groups by ^{31}P NMR (mmol/g)	Aliphatic OH	5.66	2.21
	OH in 5-substituted units	0.17	0.83
	Guaiacyl OH	0.34	0.57
	Total OH	6.16	3.6
	Carboxyl (COOH)	0.01	0.23
Elemental analysis (%)	C	55.04	59.99
	H	5.96	5.33
	O	35.67	34.37
	N	0.33	0.31
	C_9 formulae	$C_9H_{10.5}O_{3.5}N_{0.05}(OH_{Ph.})_{0.10}$ $(OH_{Al.})_{1.11}(COOH)_{0.002}$	$C_9H_{8.9}O_{3.2}N_{0.04}(OH_{Ph.})_{0.27}$ $(OH_{Al.})_{0.38}(COOH)_{0.05}$
SEC (Da)	M_n	9145	3589
	M_w	22792	44458
	D (M_n/M_w)	2.5	12.4
TGA/DTG *	$T_{5\%}$ (°C)	171	251
	Residue at 900°C (% mass)	35.2	53
	DTG_{max} (%/°C)	0.31	0.29
	T of $DTG_{max.}$ (°C)	291	370

* DTG: first derivative of the *TGA* curve.

Table 4 also shows the S/G ratio calculated based on the volume integrals of signals $S_{2,6}$ and G_2, according to previously published works [28]. The value of S/G for MWL (0.63) was in accordance to that measured by Rencoret et al. [21] for *Paulownia fortunei* (0.66). The corresponding value for PFL (0.83) indicated that G-units have suffered more degradation than S-units.

Figure S2 (Supplementary Material) shows the main substructures identified in MWL and PFL of Paulownia.

3.7. ^{31}P NMR Spectroscopy of Lignins

Figure 5 shows the ^{31}P NMR spectra of the phosphitylated MWL and PFL, and Table 4 shows the calculated distribution of the different OH groups. According to the bibliography [25–27], the signals can be classified according to the different resonances of each type of the phosphitylated hydroxyl groups present in the molecule (Figure 5). Thus, the signals corresponding to aliphatic OH (150.6–145.2 ppm) of the PFL decreased very significantly (61%) after the organosolv treatment. Several mechanisms of esterification (formylation) and acid-catalyzed dehydration can explain this decrease [6,7,41]. The signals produced by G and S-type units were much broader in the PFL due to recondensation, as was also reflected in the HSQC spectra. In fact, the area of signals increased almost four times after the formosolv treatment due to the presences of 5-substituted structures which include syringyl, and β-5′, 4-O-5′ and 5-5′ linked units. The signals due to G-units also showed condensation processes since the spectrum reflected similar qualitatively similar characteristics to those of the S-units. At 135.8 ppm a signal, absent in MWL, appeared due to the formic acid carboxyl group [42]. The spectrum area corresponding to the carboxyl groups (136.3–134.21 ppm) revealed more intense signals in PFL. Table 4 reflects the content of the different hydroxyl, in mmol OH/mg lignin, depending on the content of the internal standard (cholesterol) and according to the methods already published [27].

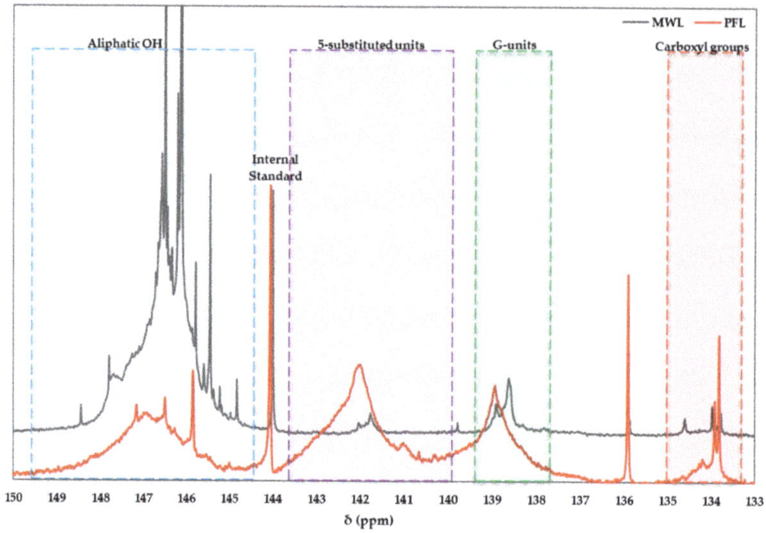

Figure 5. ^{31}P NMR spectra of MWL and PFL.

3.8. Molecular Weights of Lignins

The SEC analysis on the lignin samples (Figure 6) was used to evaluate its molecular size distribution. Table 4 shows the most important parameters calculated from the chromatograms. Compared with MWL, the molar mass distribution of PFL showed a lower M_n but a higher M_w. In order to explain this, in PFL a larger number of small and big molecules should be present in relation to MWL; that is, a much higher polydispersity than that of MWL. In fact, it is almost five times higher (2.5 vs. 12.4), according to the SEC data. This points out that there was a concurrent depolymerization and recondensation of lignin in the reaction medium. Taking as reference the simple C9 formulas (Table 4) to calculate the molecular mass of each C9 unit and the value of M_w as an average representative of the lignin molecular size, values of 116.5 and 246.2 units C9/lignin molecule for MWL and PFL, respectively, were calculated.

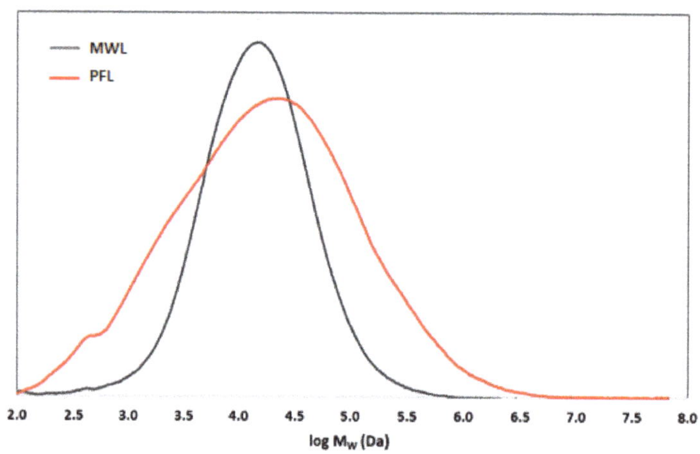

Figure 6. SEC chromatograms of MWL and PFL (the molecular weight is estimated relative to polystyrene standards).

3.9. Thermogravimetric Analysis of Lignins (TGA)

The TGA was used to determine the basic behavior and thermal stability of the MWL and PFL. The processes of pyrolytic decomposition can be divided into three stages: the release of water and volatiles (from approximately 100 to 200 °C), the rapid decomposition of the bonds that form the structure of lignin (200 to 500 °C), and a last slow carbonization to form charry residue (500 to 900 °C). The TGA and DTG curves are shown in Figure S3 (Supplementary Material).

The first step in the thermal decomposition was qualitatively similar for both samples, although the process in MWL was faster and produced greater weight loss. The weight loss in this stage came from the release of volatile compounds and water from the samples, and their values were 7.2% and 2.8% for MWL and PFL, respectively. The most important mass change corresponded to the temperatures between 200 and 500 °C, approximately, which accounted for the additional weight loss values of 32.2% for MWL and 37.5% for PFL. In this stage, the bonds forming the structure of the lignin molecule were progressively modified, starting with the fragmentation of the interunit links [43] that were broken to release volatile oxygen compounds (CO, CO_2, formaldehyde, formic acid, and some simple ethers and alkanes, among others) leaving a solid with a higher degree of unsaturation [44,45]. In this period, the maximum weight loss rate was located at 291 °C (0.31%/°C) for MWL and 375 °C (0.29%/°C) for PFL. The significant difference between these values clearly reflected the structural changes already mentioned, suffered by the lignin in the process of delignification. The residual phase of pyrolysis (500–900 °C) was very slow and generated a very low rate of volatile compounds: in this phase, mass losses of 6.8 and 6.7% were measured for MWL and PFL, respectively. The percentage of non-volatilized material at 900 °C was higher in the case of PFL (53.0%) than in MWL (35.2%). A greater amount of non-volatile residue was related to a more branched structure (more condensed) which agreed with the spectroscopic data of the lignins studied in this work [45,46].

3.10. High Added-Value Applications for Lignin

As an overview, the lignin obtained from the formosolv fractionation of Paulownia wood results in a high decrease in labile ether bonds, with an increase in its polydispersity, and also reflecting recondensation phenomena. However, some works reflected the chance of employing this kind of lignin for several purposes. Some applications, via a further upgrading of the lignin, may be its use as an adsorbent for heavy metal such as Pd or Cr [47] in the form of lignin nanoparticles, as well as for the manufacture of polymeric resins, polyurethanes and polyesters [48,49]. In addition, this lignin may also be used as a flocculant, cement additive, etc. On the other hand, if the lignin is employed for consequent catalytic depolymerization, it can be subjected to pyrolysis or other kinds of treatment for the production of aromatic monomers, or for the obtainment of biofuels [47]. All of these may be attractive applications for the exploitation of lignin.

4. Conclusions

Formosolv treatments of Paulownia, under optimum conditions, produced a cellulose-enriched pulp (80%) and a solubilization of the 78.5% of the initial lignin in the solid. The physicochemical and spectroscopic characteristics of the formosolv lignin and its comparison with the MWL showed that the lignin underwent depolymerization phenomena but also recondensation, leading to a molecular weight distribution with high polydispersity. Delignification proceeded mainly by the breaking of β-O-4′ linkages from arylglycerol β-aryl ether units. Other bonds were also affected but resisted better the solvolytic treatment.

Supplementary Materials: The following are available online at http://www.mdpi.com/2073-4395/10/8/1205/s1, Figure S1. Main substructures identified in the MWL and PFL of Paulownia. (**A**) Aryglycerol β-aryl ethers; (**A′**) γ-OH acylated aryglycerol β-aryl ethers; (**B**) resinols; (**C**) phenylcoumarans; (**D**) spirodienones; (**E**) dibenzodioxocins; (**I**) cinnamyl alcohol end-groups; (**J**) cinnamaldehyde end-groups; (**G**) guaiacyl units; (**S**) syringyl units; (**S′**), oxidized syringyl units bearing a carbonyl group at C_α; Figure S2. 1H NMR spectra of MWL and formosolv Paulownia lignin; Figure S3. TGA (dashed) and DTG (bold) curves of MWL and PFL; and

Table S1. Assignments of carbon chemical shifts (δ, ppm) in ^{13}C NMR spectrum of MWL and formosolv lignin; Table S2. Assignments of ^{13}C-^{1}H correlation signals in the HSQC NMR spectrum of the obtained lignin fractions.

Author Contributions: Conceptualization, E.D. and A.d.V.; methodology, A.d.V.; software, E.D. and A.d.V.; validation, A.d.V. and G.G.; formal analysis, E.D., A.d.V. and G.G.; investigation, E.D.; resources, A.d.V. and G.G.; data curation, E.D., P.G.d.R. and A.d.V.; writing—original draft preparation, E.D. and A.d.V.; writing—review and editing, P.G.d.R., G.G., A.R. and P.G.; visualization, E.D., P.G.d.R., A.d.V., G.G., A.R. and P.G.; supervision, A.d.V., G.G., A.R., P.G.; project administration, A.d.V., G.G.; funding acquisition, G.G. All authors have read and agreed to the published version of the manuscript.

Funding: This research was funded by MINECO (Spain) in the framework of the projects "Development of processes for the integral use of fast-growing biomass by means of the production of bioethanol and chemical products" with reference CTQ2012-30855 and "Multistage processes for the integral benefit of macroalgal and vegetal biomass" with reference CTM2015-68503-R, by Consellería de Cultura, Educación e Ordenación Universitaria (Xunta de Galicia) through the contract ED431C 2017/62-GRC to Competitive Reference Group BV1, and by the CITACA Strategic Partnership ED431E 2018/07, programs partially funded by European Regional Development Fund (FEDER).

Acknowledgments: Pablo G. del Río would like to express his gratitude to the Ministry of Science, Innovation and Universities of Spain for his FPU research grant (FPU16/04077). Authors are grateful to Alberto Núñez and Jorge Otero from the Research Support Services of the Univeridade da Coruña for conducting and advising on the interpretation of NMR, FTIR, and TGA techniques. Finally, the authors thank Jalel Labidi and his team, from the Universidad del País Vasco, for the technical support in the analysis of SEC samples.

Conflicts of Interest: The authors declare no conflict of interest.

References

1. Morales, A.; Gullón, B.; Dávila, I.; Eibes, G.; Labidi, J.; Gullón, P. Optimization of alkaline pretreatment for the co-production of biopolymer lignin and bioethanol from chestnut shells following a biorefinery approach. *Ind. Crops Prod.* **2018**, *124*, 582–592. [CrossRef]
2. Dávila, I.; Remón, J.; Gullón, P.; Labidi, J.; Budarin, V. Production and characterization of lignin and cellulose fractions obtained from pretreated vine shoots by microwave assisted alkali treatment. *Bioresour. Technol.* **2019**, *289*, 121726. [CrossRef] [PubMed]
3. Barakat, A.; de Vries, H.; Rouau, X. Dry fractionation process as an important step in current and future lignocellulose biorefineries: A review. *Bioresour. Technol.* **2013**, *134*, 362–373. [CrossRef] [PubMed]
4. Dagnino, E.P.; Felissia, F.E.; Chamorro, E.; Area, M.C. Optimization of the soda-ethanol delignification stage for a rice husk biorefinery. *Ind. Crops Prod.* **2017**, *97*, 156–165. [CrossRef]
5. Cheiwpanich, S.; Laosiripojana, N.; Champreda, V. Optimization of organosolv based fractionation process for separation of high purity lignin from bagasse. *Mater. Sci. Forum* **2017**, *883*, 92–96. [CrossRef]
6. Zhang, K.; Pei, Z.; Wang, D. Organic solvent pretreatment of lignocellulosic biomass for biofuels and biochemicals: A review. *Bioresour. Technol.* **2016**, *199*, 21–33. [CrossRef]
7. Li, M.F.; Yang, S.; Sun, R.C. Recent advances in alcohol and organic acid fractionation of lignocellulosic biomass. *Bioresour. Technol.* **2016**, *200*, 971–980. [CrossRef]
8. Wu, K.; Shi, Z.; Yang, H.; Liao, Z.; Yang, J. Effect of ethanol organosolv lignin from bamboo on enzymatic hydrolysis of avicel. *ACS Sustain. Chem. Eng.* **2017**, *5*, 1721–1729. [CrossRef]
9. Zhang, Y.; Hou, Q.; Xu, W.; Qin, M.; Fu, Y.; Wang, Z.; Willför, S.; Xu, C. Revealing the structure of bamboo lignin obtained by formic acid delignification at different pressure levels. *Ind. Crops Prod.* **2017**, *108*, 864–871. [CrossRef]
10. Dong, L.; Zhao, X.; Liu, D. Kinetic modeling of atmospheric formic acid pretreatment of wheat straw with "potential degree of reaction" models. *RSC Adv.* **2015**, *5*, 20992–21000. [CrossRef]
11. Li, M.F.; Sun, S.N.; Xu, F.; Sun, R.C. Formic acid based organosolv pulping of bamboo (Phyllostachys acuta): Comparative characterization of the dissolved lignins with milled wood lignin. *Chem. Eng. J.* **2012**, *179*, 80–89. [CrossRef]
12. García, J.C.; Zamudio, M.A.M.; Pérez, A.; López, F.; Colodette, J.L. Search for optimum conditions of Paulownia autohydrolysis process and influence in pulping process. *Environ. Prog. Sustain. Energy* **2011**, *30*, 92–101. [CrossRef]

13. Domínguez, E.; Romaní, A.; Domingues, L.; Garrote, G. Evaluation of strategies for second generation bioethanol production from fast growing biomass Paulownia within a biorefinery scheme. *Appl. Energy* **2017**, *187*, 777–789. [CrossRef]
14. Gong, C.; Bujanovic, B.M. Impact of hot-water extraction on acetone-water oxygen delignification of *Paulownia* spp. and lignin recovery. *Energies* **2014**, *7*, 857–873. [CrossRef]
15. del Río, J.C.; Rencoret, J.; Marques, G.; Gutiérrez, A.; Ibarra, D.; Santos, J.I.; Jiménez-Barbero, J.; Zhang, L.; Martínez, A.T. Highly acylated (acetylated and/or *p*-coumaroylated) native lignins from diverse herbaceous plants. *J. Agric. Food Chem.* **2008**, *56*, 9525–9534. [CrossRef]
16. Lourenço, A.; Rencoret, J.; Chemetova, C.; Gominho, J.; Gutiérrez, A.; Del Río, J.C.; Pereira, H. Lignin composition and structure differs between xylem, phloem and phellem in *Quercus suber* L. *Front. Plant Sci.* **2016**, *7*, 1–14. [CrossRef]
17. Blumenkrantz, N.; Asboe-Hansen, G. New Method for Quantitative Determination of Uranic Acids. *Anal. Biochem.* **1973**, *54*, 484–489. [CrossRef]
18. de Vega, A.; Ligero, P. Formosolv fractionation of hemp hurds. *Ind. Crops Prod.* **2017**, *97*, 252–259. [CrossRef]
19. Caridad, R.; Ligero, P.; Vega, A.; Bao, M. Formic acid delignification of *Miscanthus sinensis*. *Cellul. Chem. Technol.* **2004**, *39*, 235–244. [CrossRef]
20. Akhnazarova, S.; Kafarov, V. *Experiment Optimisation in Chemistry and Chemical Engineering*; MIR Publ.: Moscow, Russia, 1982.
21. Rencoret, J.; Marques, G.; Gutiérrez, A.; Nieto, L.; Jiménez-Barbero, J.; Martínez, Á.T.; José, C. Isolation and structural characterization of the milled-wood lignin from *Paulownia fortunei* wood. *Ind. Crops Prod.* **2009**, *30*, 137–143. [CrossRef]
22. Björkman, A. Studies on finely divided wood. I. Extraction of lignin with neutral solvents. *Sven. Papperstidn.* **1956**, *13*, 477–485.
23. Zhou, S.; Xue, Y.; Sharma, A.; Bai, X. Lignin valorization through thermochemical conversion: Comparison of hardwood, softwood and herbaceous lignin. *ACS Sustain. Chem. Eng.* **2016**, *4*, 6608–6617. [CrossRef]
24. Wen, J.L.; Sun, S.L.; Xue, B.L.; Sun, R.C. Quantitative structures and thermal properties of birch lignins after ionic liquid pretreatment. *J. Agric. Food Chem.* **2013**, *61*, 635–645. [CrossRef]
25. Constant, S.; Wienk, H.L.J.; Frissen, A.E.; De Peinder, P.; Boelens, R.; Van Es, D.S.; Grisel, R.J.; Weckhuysen, B.M.; Huijgen, W.J.; Gosselink, R.J.; et al. New insights into the structure and composition of technical lignins: A comparative characterisation study. *Green Chem.* **2016**, *18*, 2651–2665. [CrossRef]
26. Granata, A.; Argyropoulos, D.S. 2-Chloro-4,4,5,5-tetramethyl-1,3,2-dioxaphospholane, a reagent for the accurate determination of the uncondensed and condensed phenolic moieties in lignins. *J. Agric. Food Chem.* **1995**, *43*, 1538–1544. [CrossRef]
27. Balakshin, M.; Capanema, E. On the quantification of lignin hydroxyl groups with ^{31}P and ^{13}C NMR spectroscopy. *J. Wood Chem. Technol.* **2015**, *35*, 220–237. [CrossRef]
28. Wen, J.L.; Sun, S.L.; Xue, B.L.; Sun, R.C. Recent advances in characterization of lignin polymer by solution-state nuclear magnetic resonance (NMR) methodology. *Materials* **2013**, *6*, 359–391. [CrossRef]
29. Zhou, S.; Liu, L.; Wang, B.; Xu, F.; Sun, R. Microwave-enhanced extraction of lignin from birch in formic acid: Structural characterization and antioxidant activity study. *Process Biochem.* **2012**, *47*, 1799–1806. [CrossRef]
30. Sammons, R.J.; Harper, D.P.; Labbé, N.; Bozell, J.J.; Elder, T.; Rials, T.G. Characterization of organosolv lignins using thermal and FT-IR spectroscopic analysis. *BioResources* **2013**, *8*, 2752–2767. [CrossRef]
31. Babij, N.R.; McCusker, E.O.; Whiteker, G.T.; Canturk, B.; Choy, N.; Creemer, L.C.; Amicis, C.V.D.; Hewlett, N.M.; Johnson, P.L.; Knobelsdorf, J.A.; et al. NMR chemical shifts of trace impurities: Industrially preferred solvents used in process and green chemistry. *Org. Process Res. Dev.* **2016**, *20*, 661–667. [CrossRef]
32. Brandt, A.; Chen, L.; Van Dongen, B.E.; Welton, T.; Hallett, J.P. Structural changes in lignins isolated using an acidic ionic liquid water mixture. *Green Chem.* **2015**, *17*, 5019–5034. [CrossRef]
33. Sun, X.F.; Jing, Z.; Fowler, P.; Wu, Y.; Rajaratnam, M. Structural characterization and isolation of lignin and hemicelluloses from barley straw. *Ind. Crops Prod.* **2011**, *33*, 588–598. [CrossRef]
34. Xu, W.; Miller, S.J.; Agrawal, P.K.; Jones, C.W. Depolymerization and hydrodeoxygenation of switchgrass lignin with formic acid. *ChemSusChem* **2012**, *5*, 667–675. [CrossRef] [PubMed]
35. Nimz, H.; Robert, D.; Faix, O.; Nemr, M. Carbon-13 NMR spectra of lignins. Structural differences between lignins of hardwoods, softwoods, grasses and compression wood. *Holzforschung* **1981**, *35*, 16–26. [CrossRef]

36. Gordobil, O.; Egüés, I.; Labidi, J. Modification of Eucalyptus and Spruce organosolv lignins with fatty acids to use as filler in PLA. *React. Funct. Polym.* **2016**, *104*, 45–52. [CrossRef]
37. Nitsos, C.; Stoklosa, R.; Karnaouri, A.; Vörös, D.; Lange, H.; Hodge, D.; Crestini, C.; Rova, U.; Christakopoulos, P. Isolation and characterization of organosolv and alkaline lignins from hardwood and softwood biomass. *ACS Sustain. Chem. Eng.* **2016**, *4*, 5181–5193. [CrossRef]
38. Erdocia, X.; Prado, R.; Corcuera, M.Á.; Labidi, J. Effect of different organosolv treatments on the structure and properties of olive tree pruning lignin. *J. Ind. Eng. Chem.* **2014**, *20*, 1103–1108. [CrossRef]
39. Miyagawa, Y.; Kamitakahara, H.; Takano, T. Fractionation and characterization of lignin-carbohydrate complexes (LCCs) of *Eucalyptus globulus* in residues left after MWL isolation. Part II: Analyses of xylan-lignin fraction (X-L). *Holzforschung* **2013**, *67*, 629–642. [CrossRef]
40. Zhu, M.Q.; Wen, J.L.; Wang, Z.W.; Su, Y.Q.; Wei, Q.; Sun, R.C. Structural changes in lignin during integrated process of steam explosion followed by alkaline hydrogen peroxide of *Eucommia ulmoides* Oliver and its effect on enzymatic hydrolysis. *Appl. Energy* **2015**, *158*, 233–242. [CrossRef]
41. Villaverde, J.J.; Li, J.; Ek, M.; Ligero, P.; De Vega, A. Native lignin structure of *Miscanthus x giganteus* and its changes during acetic and formic acid fractionation. *J. Agric. Food Chem.* **2009**, *57*, 6262–6270. [CrossRef] [PubMed]
42. Ben, H.; Ferrell, J.R. In-depth investigation on quantitative characterization of pyrolysis oil by ^{31}P NMR. *RSC Adv.* **2016**, *6*, 17567–17573. [CrossRef]
43. Tejado, A.; Peña, C.; Labidi, J.; Echeverria, J.M.; Mondragon, I. Physico-chemical characterization of lignins from different sources for use in phenol-formaldehyde resin synthesis. *Bioresour. Technol.* **2007**, *98*, 1655–1663. [CrossRef] [PubMed]
44. Yang, H.; Yan, R.; Chen, H.; Lee, D.H.; Zheng, C. Characteristics of hemicellulose, cellulose and lignin pyrolysis. *Fuel* **2007**, *86*, 1781–1788. [CrossRef]
45. Wang, X.; Guo, Y.; Zhou, J.; Sun, G. Structural changes of poplar wood lignin after supercritical pretreatment using carbon dioxide and ethanol-water as co-solvents. *RSC Adv.* **2017**, *7*, 8314–8322. [CrossRef]
46. Nadji, H.; Diouf, P.N.; Benaboura, A.; Bedard, Y.; Riedl, B.; Stevanovic, T. Comparative study of lignins isolated from Alfa grass (*Stipa tenacissima* L.). *Bioresour. Technol.* **2009**, *100*, 3585–3592. [CrossRef]
47. Huang, D.; Li, R.; Xu, P.; Li, T.; Deng, R.; Chen, S.; Zhang, Q. The cornerstone of realizing lignin value-addition: Exploiting the native structure and properties of lignin by extraction methods. *Chem. Eng. J.* **2020**, *402*, 126237. [CrossRef]
48. Liao, J.J.; Latif, N.H.A.; Trache, D.; Brosse, N.; Hussin, M.H. Current advancement on the isolation, characterization and application of lignin. *Int. J. Biol. Macromol.* **2020**, *162*, 985–1024. [CrossRef]
49. Iravani, S.; Varma, R.S. Greener synthesis of lignin nanoparticles and their applications. *Green Chem.* **2020**, *22*, 612–636. [CrossRef]

© 2020 by the authors. Licensee MDPI, Basel, Switzerland. This article is an open access article distributed under the terms and conditions of the Creative Commons Attribution (CC BY) license (http://creativecommons.org/licenses/by/4.0/).

agronomy

MDPI

Article

Pretreatment of Hazelnut Shells as a Key Strategy for the Solubilization and Valorization of Hemicelluloses into Bioactive Compounds

Sandra Rivas *, Andrés Moure and Juan Carlos Parajó

Department of Chemical Engineering, University of Vigo (Campus Ourense), 32004 As Lagoas, Spain; amoure@uvigo.es (A.M.); jcparajo@uvigo.es (J.C.P.)

* Correspondence: sandrarivas@uvigo.es; Tel.: +34-988-387075

Received: 7 May 2020; Accepted: 25 May 2020; Published: 27 May 2020

Abstract: Hazelnut industries generate a large amount of byproducts. Among them, waste hazelnut shells (which account for about 50% of the nut weight), are potential raw materials to produce value added products. Hydrothermal pretreatment enables the solubilization of hemicelluloses, while cellulose and lignin remain in the solid phase almost unaltered, allowing their subsequent processing for an integral valorization of the feedstock. When the reaction was performed at the optimal temperature (210 °C), hemicelluloses were mainly converted into soluble substituted oligosaccharides (OS). Further membrane processing of the liquid phase from hydrothermal pretreatment enabled the refining of the OS, which accounted for up to 90.87 wt% of the nonvolatile solutes (NVC) in the refined solution, which also contained 5 g of natural bound phenolics/100 g NVC. The target products showed a dose-dependent antioxidant activity, conferred by the phenolic components. Substituted OS were made up of xylose backbones with a wide degree of polymerization distribution, and showed structures highly substituted by acetyl and uronic groups. The data included in this study provide the basis for assessing the large-scale manufacture of substituted oligosaccharides with bound phenolics as bioactive components of functional use in foods, cosmetics, or pharmaceuticals.

Keywords: biorefinery; hazelnut shells; hydrothermal pretreatment; hemicelluloses; oligosaccharides; antioxidant activity

1. Introduction

The current model of industrial development, based on an intensive usage of nonrenewable fossil resources, results in a number of negative effects, including health risks and global warming. Therefore, a gradual transition into sustainable alternatives based on cleaner and renewable raw materials is imperative [1]. In this sense, the biorefinery concept, based on the selective separation of the major feedstock components into "fractions" made up of compounds with similar properties, provides a framework for the integral conversion of biomass into a wide scope of fuels, chemicals, and materials. Lignocellulose biorefineries are expected to play a key role in the sustainable development of the industrial sector in the near future, and offer a sustainable path towards a bio-based economy through an efficient conversion of feedstocks from agriculture or forestry, including agroindustrial residues [2]. It can be noted that extracting value from wastes is compatible with the "circular economy" paradigm, which can be complemented with the utilization of green processing technologies.

Hazelnut (*Corylus avellana* L.) is a fruit made up of a shell with a kernel inside, which is used as a food. *Corylus avellana* originated from the Mediterranean region, and is one of the world's major commercial nut crops. Turkey is the main producer of *Corylus avellana*, covering approximately 70 percent of the world's production, followed by Italy, China, Spain, and USA. According to Food and Agriculture Organization (FAO) [3], more than one million tons of hazelnuts (with shell) were produced worldwide in 2017. Hazelnut shells (HS) account for more than 50% of the total nut weight, becoming an important byproduct in the hazelnut industry. HS are currently employed as low-value fuel [4,5]. However, HS is an abundant and low-cost raw material with a large potential as a raw material for biorefineries, where it can yield a number of value added products [6]. Some target products reported are xylooligosaccharides [7], phenolic antioxidants [4,5], bioethanol [8], materials for composites [9], levulinic acid, char [10], and furfural [11].

Nevertheless, the immense complexity of biomass together with the polymeric nature of its structural components makes fractionation almost inevitable. In this sense, decreasing the recalcitrance of the lignocellulosic matrix through cost effective and environmentally friendly technologies is a crucial objective [1]. Pretreatment is an essential processing step that entails the deconstruction of the lignocellulosic biomass by the disarray of lignin and the exposure of polysaccharides (cellulose and hemicelluloses) to hydrolytic reactions [12].

A number of biorefinery pretreatments are proposed in literature [2,13]. Some of them are used for obtaining products with industrial interest from HS as feedstock: hydrothermal [5,7,14], acid [15], alkaline [15,16] or the double step acid/alkaline pretreatments [17], oxidative (i.e., ozonolysis) [18], organosolv [19], or the combined effect of delignification pretreatments before acid hydrolysis [8].

Hydrothermal processing (also called autohydrolysis) is a pretreatment in which biomass is subjected to reaction with compressed hot water, enabling an extensive solubilization of hemicelluloses [20]. The reaction is catalyzed by hydronium ions resulting from water autoionization and in situ generated organic acids (mainly acetic acid coming from the hydrolysis of acetyl groups in hemicelluloses). As a result, hemicelluloses are depolymerized into low molecular weight polysaccharides, oligosaccharides (OS), monosaccharides, and minor amounts of other products. Additionally, the hydrothermal pretreatment increases the surface area and decreases the crystallinity of cellulose, resulting in an improved susceptibility toward the enzymatic hydrolysis. Additionally, lignin can be solubilized in subsequent processing stages.

In HS, hemicelluloses account for 24.6–30 wt% of the raw material. The major hemicellulose constituent is heteroxylan, made up of a backbone of linked xylose units substituted with acetyl and uronic groups, and bound phenolic compounds [5]. Consequently, optimal conditions of hydrothermal processing can result in relevant proportions of hemicellulose-derived xylooligosaccharides (XOS), which can be obtained at a reasonable yield and purity [21].

The implementation of an additional refining stage is necessary when high purity XOS are required, i.e., food-grade OS [22]. This point is important when OS are produced by hydrothermal pretreatment, because a variety of byproducts (monosaccharides, acetic acid, sugar degradation products, extractives, or acid soluble lignin) can be present in the reaction media. Membrane technology is an interesting alternative for XOS purification—the size-dependent separation achieved leads to concentrated solutions of purified XOS, while low molecular weight contaminants are removed in permeate. Some examples of XOS refining by membrane processing were reported for different agricultural residues, i.e., rice husks [23], peanut shells [24], and almond shells [25,26].

XOS are potential prebiotics, defined as nondigestible food ingredients that allow specific changes in both the composition and activity of the gastrointestinal microbiota, as stimulation of the growth of probiotic bifidobacteria and lactic acid bacteria, conferring benefits in the human health [27]. Some beneficial effects described in literature including the maintenance of the human health, the prevention of diseases, and the decreased risk of chronic diseases [22].

XOS containing esterified phenolic compounds are natural antioxidants, with potential applications in food, cosmetic, pharmacy, and nutraceutical industries. This type of XOS is considered as emerging prebiotics [7], and antioxidant activity [28] is derived from the presence of bound phenolics [27]. Therefore, this type of XOS could contribute to satisfying the increasing demand for ingredients of functional foods, whose demand is expected to reach more than 440 billion USD in 2022 [27].

In this work, hydrothermal pretreatment of waste HS was proposed as an initial step of a multistage process allowing the complete utilization of HS. This step aimed at the solubilization of hemicelluloses, leaving a treated solid made up of cellulose and lignin that could be valorized by further processing. HS were processed at different temperatures, and the solid and liquid phases were assayed for composition and yields to allow the formulation of material balances. The liquid phase from hydrothermal processing performed under optimal reaction was refined by membrane processing to yield a final product meeting the purity degree required for commercial food grade OS. The purified OS were assayed for monomeric constituents and structural features. Additionally, the total phenolic content and the in vitro antioxidant properties were considered to assess the potential of OS as ingredients for functional foods.

2. Materials and Methods

2.1. *Raw Material*

Hazelnuts were purchased in a local market (Ourense, Spain). HS were milled, sieved to obtain a particle size between 0.250–1 mm, homogenized, and stored.

2.2. *Hydrothermal Pretreatment (Autohydrolysis) of Hazelnut Shells*

HS samples were treated in a 600 mL stainless steel reactor (Parr Instrument Company, Moline, IL, USA). The raw material was mixed with distilled water at a liquid to solid ratio of 10 kg water/kg of dry HS, heated under nonisothermal conditions up to reach the target temperature (in the range 190–225 °C), and then cooled by circulating water through an internal loop. Figure 1 shows the heating and cooling profiles that provide the reaction time to reach the temperatures of the experiments performed in this study. At the end of the treatments, liquid and solid phases were separated by vacuum filtration. Solids were washed with distilled water, and air dried. The solids from hydrothermal pretreatment (autohydrolyzed solids, AS) and autohydrolysis liquors (AL) were characterized for moisture, solid yield, and composition, as described in Section 2.4.

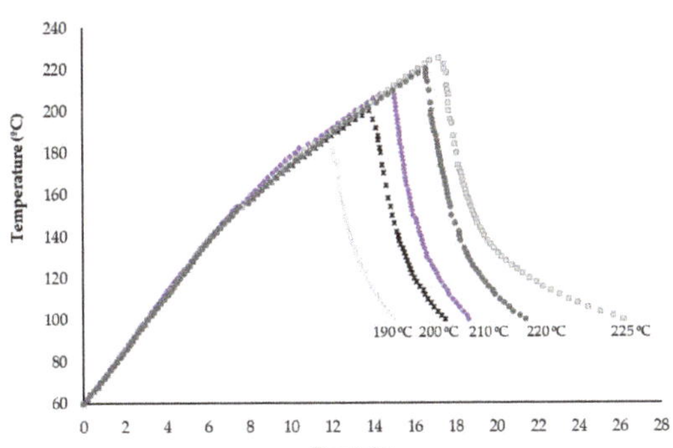

Figure 1. Heating and cooling profiles obtained for the autohydrolysis experiments performed at the assayed temperatures.

2.3. *Refining of Oligosaccharides by Membrane Processing*

AL obtained under optimal conditions for OS production was refined by discontinuous diafiltration (DD). Assays were performed in a stirred Amicon cell (Millipore) operating at a transmembrane pressure of 3.5 bar and room temperature. Water was added to AL at 2:1 volume ratio, diafiltered through a 0.3 kDa cutoff membrane (GE Osmonics Inc., Minnetonka, MN, USA) up to achieve a retentate volume equal to one of the feed. Samples from retentate and permeate were assayed for composition (using the same methodology as described in Section 2.4) and structural features.

2.4. *Analytical Procedures*

The moisture contents of HS and AS were assayed according to the T-264-cm-97 standard method [29]. The T-249-cm-85 method [30] was employed for measuring the contents of structural carbohydrates and Klason lignin. This method consisted of a two step quantitative acid hydrolysis performed with 72% and 4% H_2SO_4, respectively. The method led to an insoluble lignin residue (Klason lignin) and to a liquid phase. Samples from the liquid phase were filtered through 0.45 μm cellulose acetate membranes and analyzed by High Performance Liquid Chromatography (HPLC) using an Agilent 1200 series instrument (Agilent Technologies, Santa Clara, CA, USA), fitted with a refraction index detector (RID). Samples were assayed for monosaccharides (glucose, xylose, and arabinose), organic acids (acetic acid), and furans (furfural and hydroxymethylfurfural) using a 300 × 7.8 Aminex HPX-87H column (BioRad Life Science Group, Hercules, CA, USA) kept at 50 °C and eluted with 0.003 N H_2SO_4 at a 0.6 mL·min^{-1} flow rate. The ash content was assayed according to the T-211-om-02 method [31].

Samples of AL were filtered through 0.45 μm cellulose acetate membranes and assayed by HPLC as described above. Aliquots of AL were subjected to quantitative posthydrolysis (4% of H_2SO_4 at 121 °C for 20 min). The increases in the concentrations of monosaccharides and acetic acid caused by posthydrolysis provided the measure of oligomers concentration and their degree of substitution with acetyl groups. The content of total nonvolatile compounds (NVC) was measured by oven-drying samples at 105 °C until constant weight. All the analytical determinations were performed in triplicate.

Uronic acids were determined spectrophotometrically at 520 nm, using galacturonic acid as a standard [32].

High Performance Anion-Exchange Chromatography with Pulsed Amperometric Detection (HPAED–PAD) was used to assess the types of oligomers present in the reaction media. The analyses were performed using an ICS300 instrument (Dionex, Sunnyvale, CA, USA) equipped with a 250 × 2 mm CarboPac PA-1 column in combination with a 25 × 2 mm CarboPac PA guard column [33]. Matrix-Assisted Laser Desorption and Ionization Time-of-Flight Mass Spectrometry (MALDI TOF–MS) analyses were employed to allow a detailed structural characterization of the oligomers contained in AL. Assays were performed using an Autoflex III smartbeam instrument (Bruker Daltonics, Bremen, Germany) operating in linear positive ion mode. Spectra were acquired and treated using the Flex Control 3.0 and Flex-Analysis 3.0 software (Bruker Daltonics), respectively [34]. Commercial XOS (degree of polymerization, DP 2-6) from Megazyme (Wicklow, Ireland) were used as standards for HPAED–PAD and MALDI TOF–MS analyses.

The total phenolic content (TPC) was determined using the Folin–Ciocalteu assay [35], and the results are expressed as Gallic Acid Equivalents (GAE). The determination of antioxidant activities was carried out using the 2,2-diphenyl-1-picrylhydrazyl assay ((DPPH) [36], Trolox-Equivalent Antioxidant Capacity assay (TEAC) [37], and Ferric Reducing Ability of Plasma assay (FRAP) [38] methods. Identification of phenolic compounds was carried out by High Performance Liquid Chromatography with Diode Array Detection (HPLC–DAD) [39].

3. Results and Discussion

3.1. Composition of HS

The average composition of HS, listed in Table 1, was the same reported in a previous work performed with the same HS lot [5]. The major structural component of HS was lignin (40.08 wt% in oven-dry basis). Higher lignin contents, in the range 46–51.3 wt%, were reported for the same substrate [7,40,41]. Hemicelluloses accounted for 32.28% of the dry mass of HS, and were mainly constituted by xylan (22.69%), followed by uronic and acetyl substituents (overall content, 8.93%). Minor amounts of arabinan were also detected. Aydinli and Caglar (2012) [42] reported 28.9% of hemicelluloses, a value markedly higher that the 18.7% reported by Surek and Buyukkileci (2017) [7]. The glucan content of HS accounted 26.49% of the raw material, in the range reported by Aydinli and Caglar (2012) [42], and considerably higher than the 18.7% determined by Surek and Buyukkileci (2017) [7].

Table 1. Composition of HS expressed as g of component per 100 g of oven-dried raw material. The values are reported as the average of triplicate measurements ± standard deviation.

Component	Content (g/100 g of Dry HS)	
Glucan	26.49	±0.26
Xylan	22.69	±0.23
Arabinan	0.65	±0.01
Acetyl groups	4.37	±0.04
Uronic groups	4.56	±0.17
Klason Lignin	40.08	±0.21
Ash	0.78	±0.05
Total Identified (%)	98.85	

3.2. Hydrothermal Pretreatment

The hydrothermal pretreatments were carried out under nonisothermal conditions up to reach temperatures in the range 190–225 °C. Once the target temperatures were reached, the reaction media were cooled immediately and filtered. The aqueous phases contained hemicellulose-derived oligomers, monosaccharides, sugar decomposition products, and acetic acid. The generation rate of these compounds depended on the severity of the autohydrolysis conditions [20,43,44], here measured by the maximal temperature.

Table 2 shows data regarding the effects of the hydrothermal pretreatments performed at diverse temperatures on both the solid yield and the composition of liquid phases. The solid yield decreased steadily with temperature, a trend that was more marked up to 210 °C. This behavior can be explained because increasing temperatures promote the progressive solubilization of xylan and acetyl and uronic groups in hemicelluloses. Although arabinan also makes part of hemicelluloses, its practical importance is limited owing to the low contents. The progressive removal of hemicelluloses from the raw material led to the production of AS with increased glucan and lignin content. At the highest temperature assayed (225 °C), the content of both lignin and glucan accounted for 93.13% of AS. The glucan and lignin contents of AS varied in the ranges of 25.47%–38.70% and 41.08%–54.43%, respectively, corresponding to recovery yields of glucan and Klason lignin in solid phase higher than 90% and 85%, respectively. These results are favorable for the subsequent processing of AS by other methods, thus enabling the integral valorization of the raw material [2,13,20].

Table 2. Solid yields (measured as g of oven-dry AS/100 g of oven-dry HS) and composition of AS (measured as g component/100 g of oven-dry AS) obtained in experiments performed up to the desired temperatures. The results are reported as the average of triplicate measurements ± standard deviation.

Temperature (°C)	Solid Yield (g of AS/100 g of Dry HS)	Composition (g of Component/100 g of AS)						
		Glucan	Xylan	Arabinan	Acetyl Groups	Uronic Groups	Klason Lignin	Others (by Difference)
190	88.92	25.47 ± 0.16	20.7 ± 0.09	0.16 ± 0.02	3.96 ± 0.26	3.34 ± 0.03	41.08 ± 0.15	5.29
200	78.5	31.42 ± 0.19	15.44 ± 0.64	0.09 ± 0.00	3.01 ± 0.05	2.31 ± 0.13	43.90 ± 0.22	3.92
210	70.85	33.67 ± 0.21	9.70 ± 0.04	0.08 ± 0.00	1.76 ± 0.06	2.12 ± 0.22	49.65 ± 0.16	3.02
220	66.23	37.51 ± 0.19	6.66 ± 0.18	0.00	1.20 ± 0.10	0.64 ± 0.04	52.14 ± 0.21	1.86
225	64.84	38.70 ± 0.41	4.62 ± 0.33	0.00	0.71 ± 0.04	0.48 ± 0.02	54.43 ± 0.39	1.05

Table 3 lists the results determined for the composition of the reaction media in the same set of experiments. When the hydrothermal pretreatments were carried out at temperatures below 210 °C, the concentration of nonvolatile compounds (NVC) increased up to 26.46 g/L. This concentration remained fairly constant in the range 210–220 °C, and dropped at the highest temperature assayed. The data are expressed in terms of the identified NVC (INVC), calculated as the joint contributions of OS and monosaccharides. XOS, the most abundant components in AL, reached their highest concentration (16.24 g/L, accounting for 73.67% of the xylan present in the feedstock) at 210 °C. Harsher conditions resulted in decreased XOS concentrations, owing to the generation of xylose (and furfural under the most severe conditions assayed). Arabinooligosaccharides (ArOS) were only found (in little concentrations) in assays performed under mild conditions. The corresponding monomer (arabinose) reached its highest concentration (0.45 g/L) at 210 °C. Glucooligosaccharides (GOS) and glucose also reached limited concentrations, revealing the solubilization of a small fraction of glucan under the most severe conditions. Concerning the substituents, the concentration of acetyl groups (AG) bound to OS reached concentrations up to 3.55 g/L at 220 °C (corresponding to 77.05% of the amount present in the feedstock), whereas the maximal concentrations of uronic groups linked to OS (U) were found at a milder temperature (200 °C). From the results shown in Table 3, it can be calculated that the maximum concentration of substituted OS (including GOS, XOS, ArOS, AG, and U) was achieved at 210 °C, and reached 20.49 g/L. The concentrations of monosaccharides (maximum value, 2.68 g/L achieved at 220 °C) were comparatively low.

Table 3. Concentrations of products present in the liquid phase of hydrothermal treatments performed at temperatures ranging from 190 to 225 °C. Data are expressed in g/L.

		Temperature (°C)				
		190	200	210	220	225
INVC	Glucose	0.06	0.08	0.12	0.02	0.08
	Xylose	0.15	0.31	1.44	2.06	2.32
	Arabinose	0.17	0.25	0.45	0.34	0.28
	GOS	0.14	0.11	0.11	0.09	0.20
	XOS	3.18	9.35	16.24	15.71	14.33
	ArOS	0.22	0.17	0.01	0.00	0.00
	AG	0.74	2.24	3.09	3.55	3.32
	U	1.39	1.12	1.07	0.94	0.89
	Others *	2.48	3.15	3.92	3.75	4.41
	Total NVC	8.51	16.78	26.46	26.45	25.82
VC	Acetic acid	0.20	0.36	0.90	1.42	2.08
	Furfural	0.02	0.01	0.20	0.40	0.81
	HMF	0.00	0.00	0.01	0.02	0.03

GOS: glucooligosaccharides; XOS: xylooligosaccharides; ArOS: arabinooligosaccharides; AG, acetyl groups linked to oligosaccharides; U: uronic acid linked to oligosaccharides; HMF: hydroxymethylfurfural. NVC: total nonvolatile compounds; INVC: identified nonvolatile components; VC: volatile compounds. * Others: measured as the difference of the total NVC and the INVC.

The concentrations of volatile compounds (VC) increased smoothly up to 210 °C, and then increased markedly as a result of reactions taking place under severe conditions (for example, cleavage of AG into acetic acid and furfural generation from pentoses).

3.3. Refining of Oligosaccharides by Membrane Treatment

Based on the results discussed above, 210 °C was selected as the optimal temperature because this experiment (denoted AL210) led to the highest concentration of substituted OS. However, unwanted compounds (including monosaccharides or nonsaccharide compounds) were also present in the liquid phase, and could limit the application of the NVC fraction in the food, cosmetic, or pharmaceutical industries. OS refining of can be achieved using a number of separation techniques, including solvent extraction [21,45], adsorption [23,33], chromatography [45], and membrane filtration [21,23,25–27]. Membrane processing is currently seen as the most promising technology for the industrial manufacture of high purity, owing to the low energy requirements, easy manipulation of operational variables, and relatively easy scale up [25,45].

Figure 2 shows the scheme of the purification method used in this work, including data regarding the chemical characterization of streams and material balances. The solution AL210 (content of total OS, 20.49 g/L, accounting for 77.41% of the NVC) was employed as a feed. This stream also contained some unwanted components that should be removed, including monosaccharides (2.01 g/L), ONVC (3.97 g/L), and VC (1.11 g/L). The feed solution was refined by DD, which led to a retentate (containing 17.44 g NVC/L) and permeate (NVC concentration, 4.02 g/L). As expected, the retentate showed an increased proportion of nonvolatile solutes corresponding to OS (90.87 g/100 g of NVC, in comparison to 77.41 g OS/100 g of NVC determined for the feed solution AL210). This finding confirmed the suitability of membrane processing for OS purification, keeping a good balance between the concentrations of the target products in retentate (15.85 g/L) and in permeate (1.76 g/L). It can be noted that most unwanted compounds were present in permeate (i.e., 1.63 g ONVC/L and 0.63 g monosaccharides/L).

The available data allow the comparison of the molar ratios of oligomer components (XOS:AG:U) between the feed and the retentate (1:0.47:0.05 and 1:0.43:0.04, respectively). The molar ratio XOS:AG was slightly higher than the one reported in literature for autohydrolysis liquors from hazelnut shells [7]. A comparative molar ratio XOS:AG of 1:0.56 was also reported for membrane processing of autohydrolysis liquors from peanut shells [24].

AL210 also contained VC (1.11 g/L), a fraction mainly made up of acetic acid (0.90 g/L) and minor amounts of HMF and furfural. The VC concentration decreased considerably in the retentate (0.29 g/L) relative to the feed content.

From the above data, it can be concluded that DD of AL210 allowed a selective recovery of substituted OS in retentate, in which the target products accounted for 90.87% of the NVC fraction, while most ONVC, monosaccharides, and VC were removed in the permeate. This finding is in agreement with the results reported in literature [24], with an increasing purity of oligomers from autohydrolysis liquors of peanut shells from 55.70% up to 72.4% using DD. In a related study, Singh et al. (2019) [26] obtained XOS of low degree of polymerization from almond shells using enzymatic treatments and membrane assisted refining. To our knowledge, no previous studies reported on the membrane refining of HS-derived OS.

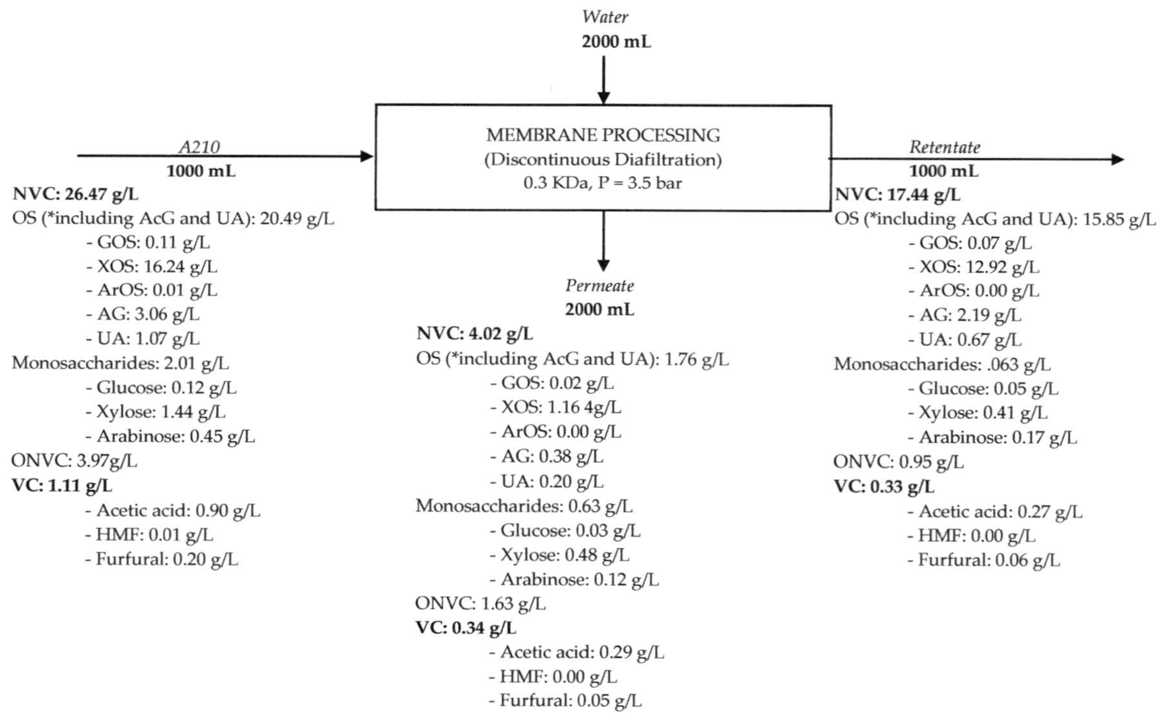

Figure 2. Diagram of membrane processing and composition of the involved streams: AL210, retentate, and permeate.

3.4. Structural Characterization of OS

Figure 3 shows the HPAEC–PAD elution profiles corresponding to AL210 and retentate. Data concerning commercial XOS in the range 2–6 are also included for comparison. It must be noted that the alkaline mobile phase used for HPAEC–PAD analysis caused the saponification of AG. Because of this, this technique provided useful information about the DP distribution of the oligomers, but not about the substitution pattern [33,43]. The elution profiles of AL210 and retentate showed similar patterns for compounds with DP > 3, peaks of oligomers with the same size. Oppositely, the peaks observed for DP2 and DP3 compounds were smaller in the case of retentate. This finding is in agreement the high recovery of XOS in the retentate, as discussed above.

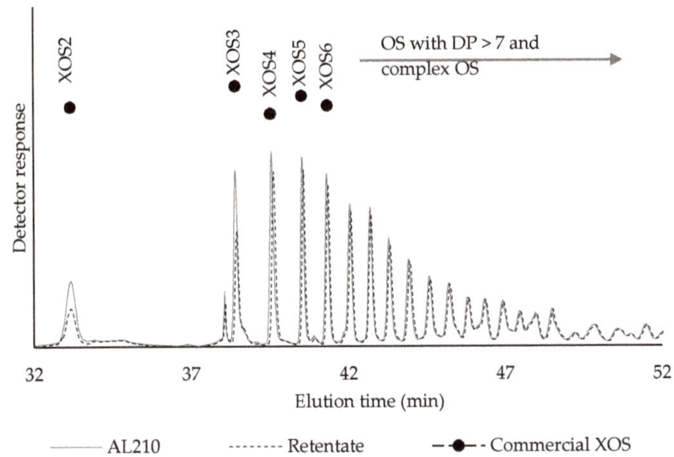

Figure 3. HPAEC–PAD elution profiles for the autohydrolysis liquors at 210 °C and retentate.

Additional information on the structures of oligomeric saccharides contained in retentate was obtained using MALDI–TOF MS (see Table 4). This technique is a powerful tool widely used for carbohydrate characterization, as oligomers can be investigated with minimal fragmentation [45,46]. The experimental data allowed the assessment of the pentose chain lengths, as well as their substitution pattern. The data reported correspond to sodium and potassium adducts. As 2,5 dihydroxybenzoic acid (DHB) was used in the analyses as a matrix, the components with $m/z < 500$ could not be identified with this method [23,24,34].

Based on the compositional information shown in Figure 2, the OS in the retentate were constituted mainly by pentose units. Based on the HPLC data obtained for arabinose and xylose, it can be concluded that the pentoses in OS backbones corresponded to xylose. The pentose chains were highly substituted by acetyl groups (AG) and/or *O*-methylglucuronic groups (U), following the patterns [mP nAG] or [mP nAG oU]. The spectra confirmed the presence of OS composed by pentose chains with DP in the range 3–16, containing up to 6 AG groups, and 1 or 2 U groups. These results are in agreement with the compositional data and with the molar ratios XOS:AG:U discussed above.

Table 4. OS structures in the retentate stream identified by MALDI–TOF MS.

m/z	Structure	*m/z*	Structure	*m/z*	Structure
537.14	3P 2AG+K^+	1133.37	7P 4AG+Na^+	1529.54	7P 4AG U+K^+/10P 4AG+Na^+
627.19	4P AG+K^+	1149.33	6P 3AG U+Na^+	1545.51	9P 3AG U+Na^+
653.23	4P 2AG+Na^+	1191.36	6P 4AG U+Na^+	1587.53	10P 5AG+K^+/9P 4AG U+Na^+
669.19	3P AG U+Na^+	1197.37	8P 2AG+K^+/7P AG U+Na^+	1629.55	10P 6AG+K^+
711.20	4P 3AG+K^+	1207.34	5P 3AG 2U+Na^+	1635.59	11P 3AG+K^+
801.23	5P 2AG+K^+/4P AG U+Na^+	1223.42	8P 3AG+Na^+	1645.55	9P 5AG U+K^+
843.24	5P 3AG+K^+/4P 2AG U+Na^+	1233.36	7P 6AG+K^+	1661.60	8P 4AG U+K^+
885.25	5P 4AG+K^+	1239.37	7P 2AG U+Na^+	1677.59	10P 3AG U+Na^+
901.25	4P 3AG U+K^+	1265.42	8P 4AG+Na^+	1719.59	10P 4AG U+Na^+
917.33	6P 2AG+Na^+	1281.39	7P 3AG U+Na^+	1809.65	11P 3AG U+Na^+
933.26	5P AG U+Na^+	1297.39	7P 3AG U+K^+	1851.63	11P 4AG U+Na^+
959.31	6P 3AG+Na^+	1323.41	8P 5AG+K^+	1941.70	13P 4AG+K^+
9752.70	5P 2AG U+Na^+	1339.41	7P 4AG U+Na^+	1983.67	12P 4AG U+Na^+
991.28	4P AG 2U+Na^+	1371.43	9P 3AG+K^+/8P 2AG U+Na^+	1993.65	11P 6AG+K^+
1017.29	6P 4AG+K^+/5P 3AG U+Na^+	1381.41	7P 5AG U+K^+	2067.69	13P 6AG+K^+
1033.29	5P 3AG U+K^+	1397.48	9P 4AG+Na^+	2115.74	14P 5AG+K^+
1059.31	5P 4AG U+Na^+	1413.45	8P 3AG U+Na^+	2199.70	14P 6AG+K^+
1065.31	7P 2AG+K^+/6P AG U+Na^+	1455.46	9P 5AG+K^+	2289.72	14P 5AG U+Na^+
1075.30	5P 4AG U+K^+	1471.46	8P 4AG U+K^+	2331.71	15P 6AG+K^+
1091.36	7P 3AG+Na^+	1503.50	10P 3AG+K^+/9P 2AG U+Na^+	2421.73	15P 5AG U+Na^+
1107.32	6P 2AG U+Na^+	1513.47	7P 4AG 2U+Na^+	2463.70	16P 6AG+K^+

P: pentoses; AG: acetyl groups; U: uronic groups. Na^+: sodium; K^+: potassium.

3.5. Total Phenolic Content and Antioxidant Activity

Pérez-Armada et al. (2019) [5] indicated that HS show a great potential as a source of natural antioxidants. These authors solubilized a part of the phenolics in HS by autohydrolysis, and recovered the target products using polymeric resins. In our work, the retentate from DD was subjected to acid hydrolysis to release a number of valuable compounds, including gallic, vanillic, and p-coumaric acids; aldehydes such as vanillin; and flavonoids such as catechin and (-) epicatechin.

Figure 4 shows the results achieved for the three streams involved in the membrane processing. The methods employed for this purpose included the total phenolic content (TPC, expressed as gallic acid equivalents/L or GAE/L), antioxidant activities determined using the methods TEAC (measured as g Trolox/L), FRAP (measured as g $FeSO_4{\cdot}7H_2O$/L), and DPPH (measured as g GAE/L needed for EC_{50}) [35–38].

The composition of phenolics (TPC) in AL210 was 1.63 g GAE/L, which is in the range reported for AL of other biomasses, i.e., eucalypt (1.64–1.98 g GAE/L) [47], vine shoots (1.33 g/L) [48], or peanut shells (1.58 g/L) [24]. The TPC of retentate decreased by 53.74% relative to the AL210 stream, and followed a behavior in agreement with reported literature concerning OS purification by membrane processes [23,24,27]. The TPC in the retentate accounted for 0.87 g/L, which implies a contribution of 5.00% relative to the NVC (17.44 g/L).

Despite the decreased phenolic content of retentate, the presence of phenolics is interesting due to their antioxidant activity, as these provide additional value to the target products as functional food ingredients [23,24,27]. According to the data in Figure 4, AL210 showed acceptable FRAP, TEAC, and DPPH activities. Interestingly, the DPPH radical scavenging capacity was not affected by DD, leading to the same EC_{50} value in feed and retentate. On the contrary, the ABTS radical scavenging activity determined by the TEAC assay dropped by 58.19% in retentate relative to the value measured for AL210. However, this antioxidant activity is still an interesting contribution to the functional properties of the retentate.

The assays based on radical scavenging reactions frequently show a dose-dependent response [23,49,50]. The data in Figure 5 show that the DPPH assay presented this type of behavior in the AL210 and retentate streams.

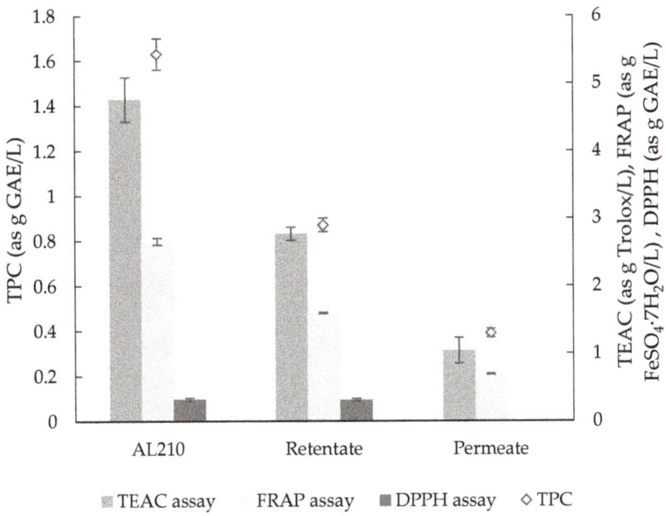

Figure 4. Total phenolic content (TPC) and antioxidant activity determined by TEAC, DPPH, and FRAP assays of AL210, retentate, and permeate. TPC, expressed as g GAE/L; TEAC, as g Trolox/L; FRAP, as g $FeSO_4{\cdot}7H_2O$/L; DPPH (EC_{50}), as g GAE/L.

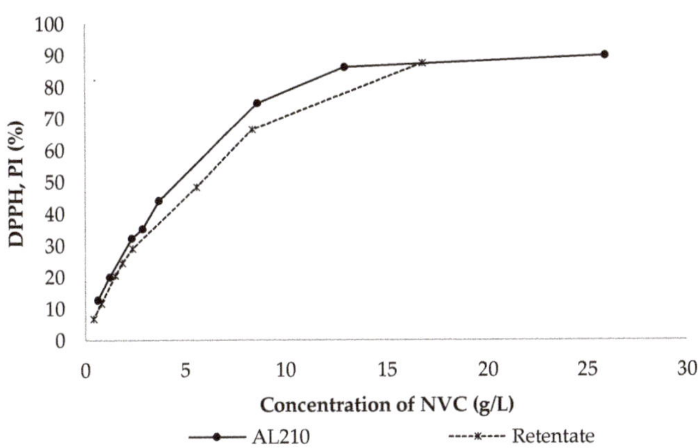

Figure 5. Effects of the concentration on the DPPH radical scavenging activity, expressed as percentage of inhibition, PI (%), for autohydrolysis liquors at 210 °C (AL210) and retentate.

In general terms, the data obtained in this work for antioxidant activities follow general trends similar to the ones reported in studies dealing with other lignocellulosic feedstocks (such as rice husks, pine and eucalypt woods, peanut shells or almond shells) [23,24,26].

4. Conclusions

This study deals with the hydrothermal pretreatment (autohydrolysis) of HS, conceived as the first step of an overall utilization of this feedstock in the scope of biorefineries. Operating at the temperature considered as optimal (210 °C), most hemicelluloses were broken down into soluble fragments, appearing in liquid phase as oligosaccharides. Unwanted, nonsaccharide products present in the liquid phase from autohydrolysis treatments were removed by DD. The target products (refined OS present in the retentate) accounted for 90.87% of the NVC fraction. OS presented a wide DP distribution, and the backbones were made up of xylose structural units substituted with acetyl and uronic groups. Phenolics bound to OS accounted for 5% of the total NVC, and showed dose-dependent antioxidant activity. The phenolics identified included gallic, vanillic, and *p*-coumaric acids; vanillin; and catechin. These results confirm the potential of the target products obtained in this study for applications in a number of fields, including the food, cosmetic, and pharmaceutical industries.

The autohydrolysis stage also produced solids containing up to 93.13% of glucan and Klason lignin, which represented recovery yields of these components above 90% and 85%, respectively. These data confirm the interest of hydrothermal processing as a first stage of an integrated process for the global valorization of HS.

Author Contributions: S.R. and A.M. designed the research. S.R. performed experiments and analyzed the data. S.R. and J.C.P. prepared the original draft. All authors discussed the data. All authors have read and agreed to the published version of the manuscript.

Funding: S.R. contracts were supported by Ministerio de Economía y Competitividad (FJCI-2015-23765) and Ministerio de Ciencia, Innovación y Universidades (IJC2018-037665-I). This research was also funded by "La Caixa" and Triptolemos Foundation.

Acknowledgments: S.R. thanks Ministerio de Economía y Competitividad and Ministerio de Ciencia, Innovación y Universidades for her "Juan de la Cierva" contracts (FJCI-2015-23765 and IJC2018-037665-I). She also thanks to "La Caixa" and Triptolemos Foundation for the financial support.

Conflicts of Interest: The authors declare no conflict of interest.

References

1. Mahmood, H.; Moniruzzaman, M.; Iqbal, T.; Khan, M.J. Recent advances in the pretreatment of lignocellulosic biomass for fuels and value-added products. *Curr. Opin. Green Sustain. Chem.* **2019**, *20*, 18–24. [CrossRef]
2. Jönson, L.J.; Martín, C. Pretreatment of lignocellulose: Formation of inhibitory by-products and strategies for minimizing their effects. *Bioresour. Technol.* **2016**, *199*, 103–112. [CrossRef]
3. FAO. Hazelnut Production. 2017. Available online: http://www.fao.org/faostat/en/#data/QC/visualize (accessed on 9 December 2019).
4. Yuan, B.; Lu, M.; Eskridge, K.M.; Isom, L.D.; Hanna, M.A. Extraction, identification and quantification of antioxidant phenolics from hazelnut (*Corylus avellana* L.) shells. *Food Chem.* **2018**, *244*, 7–15. [CrossRef]
5. Pérez-Armada, L.; Rivas, S.; González, B.; Moure, A. Extraction of phenolic compounds from hazelnut shells by green processes. *J. Food Eng.* **2019**, *225*, 1–8. [CrossRef]
6. Arslan, Y.; Takac, S.; Eken-Saracoğlu, N. Kinetic study of hemicellulosic sugar production from hazelnut shells. *Chem. Eng. J.* **2012**, *185–186*, 23–28. [CrossRef]
7. Surek, E.; Buyukkileci, A.O. Production of xylooligosaccharides by autohydrolysis of hazelnut (*Corylus avellana* L.) shell. *Carbohydr. Polym.* **2017**, *174*, 565–571. [CrossRef]
8. Arslan, Y.; Eken-Saracoğlu, N. Effects of pretreatment methods for hazelnut shell hydrolysate fermentation with *Pichia Stipitis* to ethanol. *Bioresour. Technol.* **2010**, *101*, 8664–8670. [CrossRef]
9. Balart, J.F.; García-Sanoguera, D.; Balart, R.; Boronat, T.; Sánchez-Nacher, L. Manufacturing and properties of biobased thermoplastic composites from poly(lactic) and hazelnut shell wastes. *Polym. Compos.* **2018**, *39*, 848–857. [CrossRef]

10. Licursi, D.; Antonetti, C.; Fulignati, S.; Vitolo, S.; Puccini, M.; Ribechini, E.; Bernazzani, L.; Raspolli Galletti, A.M. In-depth characterization of valuable char obtained from hydrothermal conversion of hazelnut shells to levulinic acid. *Bioresour. Technol.* **2017**, *244*, 880–888. [CrossRef]
11. Demirbas, A. Furfural production from fruit shells by acid-catalyzed hydrolysis. *Energy Source Part A* **2006**, *28*, 157–165. [CrossRef]
12. Haldar, D.; Pukait, M.K. Lignocellulosic conversion into value-added products: A review. *Process Biochem.* **2020**, *89*, 110–133. [CrossRef]
13. Carvalheiro, F.; Duarte, L.C.; Gírio, F. Hemicellulose biorefineries: A review on biomass pretreatments. *J. Sci. Ind. Res.* **2008**, *67*, 849–864.
14. Hosgun, E.H.; Bozan, B. Effect of temperature and time on the steam pretreatment of hazelnut shells for the enzymatic saccharification. *Chem. Eng. Trans.* **2014**, *37*, 379–384. [CrossRef]
15. Uzuner, S.; Cekmecilioglu, D. Hydrolysis of hazelnut shells as carbon source for bioprocessing applications and fermentation. *Int. J. Food Eng.* **2014**, *10*, 799–808. [CrossRef]
16. Uzuner, S.; Sharma-Shivappa, R.R.; Cekmecelioglu, D. Bioconversion of alkali pretreated hazelnut shells to fermentable sugars for generation of high value products. *Waste Biomass Valorization* **2017**, *8*, 407–416. [CrossRef]
17. Vadivel, V.; Moncalvo, A.; Dordoni, R.; Spigno, G. Effects of and acid/alkaline treatment on the release of antioxidants and cellulose from different agro-food wastes. *J. Waste Manag.* **2017**, *64*, 304–314. [CrossRef]
18. Uzuner, S.; Sharma-Shivappa, R.R.; Cekmecelioglu, D.; Kolar, P. A novel oxidative destruction of lignin and enzymatic digestibility of hazelnut shells. *Biocatal. Agric. Biotechnol.* **2018**, *13*, 110–115. [CrossRef]
19. Mancini, G.; Papirio, S.; Lens, P.N.; Esposito, G. Anaerobic digestion of lignocellulosic materials using ethanol-organosolv pretreatment. *Environ. Eng. Sci.* **2018**, *35*, 953–960. [CrossRef]
20. Garrote, G.; Domínguez, H.; Parajó, J.C. Autohydrolysis of corncob: Study of non isothermal xylooligosaccharide production. *J. Food Eng.* **2002**, *52*, 211–218. [CrossRef]
21. Moure, A.; Gullón, P.; Domínguez, H.; Parajó, J.C. Advances in the manufacture, purification and applications of xylo-oligosaccharides as food additives and nutraceuticals. *Process Biochem.* **2006**, *41*, 1913–1923. [CrossRef]
22. Aachary, A.A.; Prapulla, S.G. Xylooligosaccharides (XOS) as an emerging prebiotic: Microbial synthesis, utilization, structural characterization, bioactive properties and applications. *Compr. Rev. Food Sci. Food Saf.* **2011**, *10*, 2–16. [CrossRef]
23. Rivas, S.; Conde, E.; Moure, A.; Domínguez, H.; Parajó, J.C. Characterization, refining and antioxidant activity of saccharides derived from hemicelluloses of wood and rice husks. *Food Chem.* **2013**, *141*, 495–502. [CrossRef]
24. Rico, X.; Gullón, B.; Alonso, J.L.; Parajó, J.C.; Yáñez, R. Valorization of peanut shells: Manufacture of bioactive oligosaccharides. *Carbohydr. Polym.* **2018**, *183*, 21–28. [CrossRef]
25. Nabarlatz, D.; Torras, C.; García-Valls, R.; Montané, D. Purification of xylo-oligosaccharides from almond shells by ultrafiltration. *Sep. Purif. Technol.* **2007**, *53*, 238–243. [CrossRef]
26. Singh, R.D.; Nadar, C.G.; Muir, J.; Arora, A. Green and clean process to obtain low degree of polymerization xylooligosaccharides from almond shell. *J. Clean Prod.* **2019**, *241*, 118237. [CrossRef]
27. Bhatia, L.; Sharma, A.; Bachheti, R.K.; Chandel, A.K. Lignocellulose functional oligosaccharides: Production, properties and health benefits. *Prep. Biochem. Biotechnol.* **2019**, *49*, 744–758. [CrossRef]
28. Shahidi, F.; Alasalvar, C.; Kiyana-Pathirana, C.M. Antioxidant phytochemicals in hazelnut kernel (*Corylus avellana* L.) and hazelnut byproducts. *J. Food Chem.* **2007**, *55*, 1212–1220. [CrossRef]
29. Technical Association of the Pulp and Paper Induustry (TAPPI) Method T 264 cm-07. Preparation of wood for chemical analysis. In *TAPPI Test Methods 2002–2003*; TAPPI Press: Atlanta, GA, USA, 1997.
30. TAPPI Method T 249 cm-85. Carbohydrate composition of extractive free wood and wood pulp by gas–liquid chromatography. In *TAPPI Test Methods 2002–2003*; TAPPI Press: Atlanta, GA, USA, 1985.
31. TAPPI method T 211 om-0.2. Ash in wood, pulp, paper and paperboard: Combustion at 525 °C. In *TAPPI Test Methods 2002–2003*; TAPPI Press: Atlanta, GA, USA, 1993.
32. Blumenkrantz, N.; Asboe-Hansen, G. New method for quantitative determination of uronic acids. *Anal. Biochem.* **1973**, *54*, 484–489. [CrossRef]

33. Gullón, P.; Moura, P.; Esteves, M.P.; Girio, F.M.; Domínguez, H.; Parajó, J.C. Assessment on the fermentability of xylooligosaccharides from rice husks by probiotic bacteria. *J. Agric. Food Chem.* **2008**, *56*, 7482–7487. [CrossRef]
34. Rivas, S.; Gullón, B.; Gullón, P.; Alonso, J.L.; Parajó, J.C. Manufacture and properties of bifidogenic saccharides derived from wood mannan. *J. Agric. Food Chem.* **2012**, *60*, 4296–4305. [CrossRef]
35. Singleton, V.L.; Rossi, J.A. Colorimetry of total phenolics with phosphomolybdic phosphotungstic acid reagents. *Am. J. Enol. Vitic.* **1965**, *16*, 144–158.
36. Von Gadow, A.; Joubert, E.; Hansmann, C.F. Comparison of the antioxidant activity of aspalathin with that of other plant phenols of rooibos tea (*Aspalathus linearis*), a-tocopherol, BHT and BHA. *J. Agric. Food Chem.* **1997**, *45*, 632–637. [CrossRef]
37. Re, R.; Pellegrini, N.; Proteggente, A.; Pannala, A.; Yang, M.; Rice-Evans, C. Antioxidant activity applying an improved ABTS radical cation decolorization assay. *Free Radic. Biol. Med.* **1999**, *26*, 1231–1237. [CrossRef]
38. Benzie, I.F.F.; Strain, J.J. The ferric reducing ability of plasma (FRAP) as a measure of "antioxidant power": The FRAP assay. *Anal. Biochem.* **1996**, *239*, 70–76. [CrossRef]
39. Conde, E.; Moure, A.; Domínguez, H.; Parajó, J.C. Fractionation of antioxidants from autohydrolysis of barley husks. *J. Agric. Food Chem.* **2008**, *56*, 10651–10659. [CrossRef]
40. Demirbas, A. Oils from hazelnut shell and hazelnut kernel husk for biodiesel production. *Energy Sources Part A Recovery Util. Environ.* **2008**, *30*, 1870–1875. [CrossRef]
41. Hoşgün, E.Z.; Bozan, B. Effect of different types of thermochemical pretreatment on the enzymatic hydrolysis and the composition of hazelnut shells. *Waste Biomass Valorization* **2019**. [CrossRef]
42. Aydinli, B.; Caglar, A. The investigation of the effects of two different polymers and three catalysts on pyrolysis of hazelnut shell. *Fuel Process. Technol.* **2012**, *93*, 1–7. [CrossRef]
43. Rivas, S.; Santos, V.; Parajó, J.C. Aqueous fractionation of hardwood selective glucuronoxylan solubilisation and purification of the reaction products. *J. Chem. Technol. Biotechnol.* **2017**, *92*, 367–374. [CrossRef]
44. Santos, T.M.; Alonso, M.V.; Oliet, M.; Domínguez, J.C.; Rigual, V.; Rodriguez, F. Effect of autohydrolysis on *Pinus radiata* wood for hemicellulose extraction. *Carbohydr. Polym.* **2018**, *194*, 285–293. [CrossRef]
45. Qing, Q.; Li, H.; Kumar, R.; Wyman, C.E. Xylooligosaccharides production, quantification, and characterization in context of lignocellulosic biomass pretreatment. In *Aqueous Pretreatment of Plant Biomass for Biological and Chemical Conversion to Fuels and Chemicals*, 1st ed.; Wyman, C.E., Ed.; John Wiley & Sons, Ltd.: Hoboken, NJ, USA, 2013; pp. 391–415.
46. Nimptsch, K.; Süß, R.; Scnabelrauch, M.; Nimptsch, A.; Schiller, J. Positive ion MALDI-TOF mass spectra are more suitable than negative ion spectra to characterize sulphated glycosaminoglycan oligosaccharides. *Int. J. Mass Spectrom.* **2012**, *310*, 72–76. [CrossRef]
47. Conde, E.; Moure, A.; Domínguez, H.; Parajó, J.C. Production of antioxidants by non-isothermal autohydrolysis of lignocellulosic wastes. *LWT-Food Sci. Technol.* **2011**, *44*, 436–442. [CrossRef]
48. Gullón, B.; Eibes, G.; Moreira, M.T.; Dávila, I.; Labidi, J.; Gullón, P. Antioxidant and antimicrobial activities of extracts obtained from the refining of autohydrolysis liquors of vine shoots. *Ind. Crop. Prod.* **2017**, *107*, 105–113. [CrossRef]
49. Licursi, D.; Antonetti, C.; Mattonai, M.; Pérez-Armada, L.; Rivas, S.; Ribecchini, E.; Raspolli Galletti, A.M. Multi-valorisation of giant reed (*Arundo donax* L.) to give levulinic acid and valuable phenolic antioxidants. *Ind. Crop. Prod.* **2018**, *112*, 6–17. [CrossRef]
50. Huang, C.; Wang, X.; Liang, C.; Jiang, X.; Yang, G.; Xu, J.; Yong, Q. A sustainable process for procuring biological active fractions of high-purity xylooligosaccharides and water-soluble lignin from *Moso* bamboo prehydrolyzate. *Biotechnol. Biofuels* **2019**, *12*, 189. [CrossRef] [PubMed]

© 2020 by the authors. Licensee MDPI, Basel, Switzerland. This article is an open access article distributed under the terms and conditions of the Creative Commons Attribution (CC BY) license (http://creativecommons.org/licenses/by/4.0/).

agronomy

MDPI

Article

Development of Pretreatment Strategies for the Fractionation of Hazelnut Shells in the Scope of Biorefinery

Laura López, Sandra Rivas *, Andrés Moure, Carlos Vila and Juan Carlos Parajó

Department of Chemical Engineering, University of Vigo (Campus Ourense), 32004 As Lagoas, Ourense, Spain; laura.lopez.caamano@uvigo.es (L.L.); amoure@uvigo.es (A.M.); cvila@uvigo.es (C.V.); jcparajo@uvigo.es (J.C.P.)

* Correspondence: sandrarivas@uvigo.es; Tel.: +34-988-387075

Received: 22 September 2020; Accepted: 12 October 2020; Published: 14 October 2020

Abstract: Hazelnut shells are an important waste from the hazelnut processing industry that could be valorized in a multi-product biorefinery. Individual or combined pretreatments may be integrated in processes enabling the integral fractionation of biomass. In this study, fractionation methods based on alkaline, alkaline-organosolv, organosolv, or acid-catalyzed organosolv treatments were applied to raw or autohydrolyzed hazelnut shells. A comparative analysis of results confirmed that the highest lignin removal was achieved with the acid-catalyzed organosolv delignification, which also allowed limited cellulose losses. When this treatment was applied to raw hazelnut shells, 65.3% of the lignin was removed, valuable hemicellulose-derived products were obtained, and the cellulose content of the processed solids increased up to 54%. Autohydrolysis of hazelnut shells resulted in the partial solubilization of hemicelluloses (mainly in the form of soluble oligosaccharides). Consecutive stages of autohydrolysis and acid-catalyzed organosolv delignification resulted in 47.9% lignin removal, yielding solids of increased cellulose content (55.4%) and very low content of residual hemicelluloses. The suitability of selected delignified and autohydrolyzed-delignified hazelnut shells as substrates for enzymatic hydrolysis was assessed in additional experiments. The most susceptible substrates (from acid-catalyzed organosolv treatments) reached 74.2% cellulose conversion into glucose, with a concentration of 28.52 g glucose/L.

Keywords: biorefinery; hazelnut shells; hydrothermal pretreatment; delignification; enzymatic hydrolysis; fractionation; cellulose; hemicelluloses

1. Introduction

Lignocellulosic biomass (LB) is a key resource for the sustainable manufacture of bio-based chemicals and fuels. LB can be processed according to the biorefinery concept, which involves consecutive treatment stages to achieve an integral benefit of the feedstock, with minimal or no waste generation [1]. The success of biorefineries also depends on the right application of technology methodologies for obtaining multi-products [2]. The contribution of biorefineries to a future bio-economy inspires business opportunities based on product diversification while improving environmental performance [3].

In the field of biomass valorization, the biorefinery acts as a platform for chemicals and energy production through the inclusion of diverse conversion technologies [3,4].

The complexity of biomass utilization lies in both the polymeric nature of its main constituents (cellulose, hemicelluloses, and lignin) and their different chemical reactivities. In the last few decades, several fractionation methods have been developed, as summarized in recent literature reviews [2,5–8]. Galbe and Wallberg [6] classified the conventional methods according to their mode of action:

mechanical treatments (e.g., milling or grinding) reduce the particle size of biomass, increasing the surface area of the particles and improving the enzyme accessibility; dilute acid pretreatments (with H_2SO_4, H_3PO_4 or other strong acids), and hydrothermal pretreatments (with hot, compressed water or steam) promote the hydrolysis of hemicelluloses; whereas alkaline and organosolv processing enable the extraction of lignin.

Hydrothermal pretreatments (autohydrolysis, performed with hot, compressed water) causes the selective solubilization of hemicelluloses by depolymerization into soluble compounds of lower molecular weight. Soluble, low molecular weight polymers from hemicelluloses show potential for applications in the manufacture of barrier films or hydrogels [6]; oligosaccharides show prebiotic properties with potential in the pharmaceutical and food markets [9,10]; and monosaccharides can be transformed into value-added chemicals such as xylitol and furfural [11]. Additionally, autohydrolysis increases the biomass surface area and decreases the crystallinity of the cellulosic fraction remaining in the solid phase, facilitating its further hydrolysis.

Solutions of alkalis (such as NaOH, KOH, or ammonia) are suitable for removing lignin by saponification of intermolecular ester bonds, and for increasing the digestibility of cellulose [5]. Alternatively, delignification can be achieved by organosolv treatments with organic solvents (e.g., ethanol) or solvent-water mixtures, in the presence or absence of catalysts. From an environmental point of view, the separation of lignin from the organosolv media by precipitation with water facilitates the wastewater treatment. From an economic perspective, organosolv lignin is a valuable product with potential applications in diverse fields, including the manufacture of adhesives, fibers, films, biodegradable polymers [12], and natural antioxidants [13].

The solids from delignification treatments, with increased cellulose contents, may show increased susceptibility to enzymatic saccharification [1,14,15], or serve as substrates for manufacturing cellulose nanocrystals [16]. In the first case, glucose can be transformed in value-added compounds (e.g., bioethanol or lactic acid) [15,17]; whereas in the second one, cellulose nanocrystals are used in a number of fields, including biological applications.

"Conventional" pretreatments (or their combinations) are expected to play an important role in the future full-scale biorefineries [6]. Combined pretreatments may allow process designs enabling an improved fractionation and an efficient processing into a spectrum of multi-products [8]. For example, consecutive stages of autohydrolysis and delignification allow the separate recovery of hemicellulose-derived saccharides, lignin, and cellulose [18], enabling the integral fractionation of biomass, an aspect of crucial importance for a successful biorefinery approach [19].

Hazelnut (*Corylus avellana* L.) is a commercial crop, whose fruit is widely used in food industries [20]. Turkey is the main producer and exporter of hazelnut, followed by Italy, Azerbaijan, USA, China, Georgia, Iran, France, Chile, and Spain. According to the Food and Agriculture Organization (FAO), 863,888 tons of hazelnuts (with shell) were produced worldwide in 2018 [21]. Hazelnut shells (HS) are the most important byproduct of the hazelnut processing industry, representing more than half of the total nut weight. HS are a low–cost byproduct, usually burned, but could be valorized on the basis of their lignocellulosic nature. Recently, getting value from selected biochemicals in HS has become a motivation for research, due to the compelling economic benefits [3,22]. In terms of composition, HS are mainly made up of lignin: Perez-Armada et al. [23] reported 40.1 wt% of lignin, whereas Demirbas [24] and Surek and Buyukkileci [25] found around 46%, and Hoşgün and Bozan [20] near 50%. HS hemicelluloses are mainly made up of xylan, acetyl, and uronic moieties, which accounted jointly for 25–32.5% of the dry weight, whereas the reported cellulose contents are about 26–27% [10,23,26].

Based on the above ideas, this work deals with the development of HS fractionation methods suitable for a multi-product biorefinery. HS and solids resulting from HS autohydrolysis (denoted AS) were subjected to diverse delignification treatments to assess their efficiency. The soluble hemicellulose-derived products were identified, and the solids from selected delignification stages were assayed as substrates for enzymatic hydrolysis. In summary, this work provides an experimental assessment on biorefinery schemes enabling an integral valorization of HS.

2. Materials and Methods

2.1. Raw Material

Hazelnuts were purchased locally (Ourense, Spain). The shells (HS) were separated, milled to particle sizes within the range 0.250–1 mm, and subjected to analysis and chemical processing as described below.

2.2. Autohydrolysis of Hazelnut Shells

HS autohydrolysis was performed in a stirred 600 mL stainless steel reactor (Parr Instrument Company, Moline, IL, USA). HS were mixed with distilled water at a liquid to solid ratio (LSR) of 10 g/g of dry HS, heated up to 210 °C, and cooled immediately. Heating profiles previously reported for non-isothermal autohydrolysis shown the reaction time to reach the target temperature. Then, heating up from 60 °C to 210 °C lasted 16 min. These conditions were reported as optimal in literature [10,23]. The autohydrolyzed solids (AS) were separated by vacuum filtration, washed with distilled water, and air-dried before chemical characterization and solid yield determination. The autohydrolysis liquors (AL) were assayed for composition as described in Section 2.5.

2.3. Delignification Pretreatments

HS and AS were subjected to different delignification treatments. All experiments were performed at a LSR of 8 g/g, under the following conditions:

1. Alkaline delignification experiments of HS and AS were performed in an autoclave at 121 °C for 60 min using 2, 4, or 8 wt% NaOH solutions. The alkaline solutions and the operational conditions were selected according to the literature [20,27–29].
2. Alkaline-organosolv pretreatments of HS and AS were carried out in an autoclave at 121 °C or 135 °C for 60 min, in media with equal amounts of ethanol and alkaline solutions, containing 2–8 wt% NaOH with respect to the total solution.
3. Organosolv pretreatments were carried out in a stirred, 600 mL stainless steel reactor (Parr Instrument Company, Moline, IL, USA). HS and AS were treated with a mixture of ethanol/water (55/45 *w*/*w*) at 200 C for 60 or 120 min, according to the conditions reported in the literature [1,15].
4. Acid-catalyzed organosolv treatments of HS and AS were performed in the Parr reactor indicated above. The media contained a mixture of ethanol/aqueous H_2SO_4 (60/40 *w*/*w*), where the amount of H_2SO_4 corresponded 1 g H_2SO_4/100 g substrate. The reaction media were kept at 160–180 °C for 60–120 min.

Table 1 sumarizes the different delignification experiments performed in this work.

The solid and liquid phases from treatments were separated by vacuum filtration. The treated solids were washed first with solutions of the same composition as the ones used in the respective pretreatments, and then with distilled water. The solid phases from treatments were assayed for composition and solid yield. From the experimental data, the percentages of delignification, cellulose removal and hemicellulose removal were calculated as follows:

$$\%\ \text{Solid Yield (SY)} = 100 \times \frac{DS_{AD}}{DS_{BD}} \quad (1)$$

where DS_{AD} and DS_{BD} are the dry weights of the solid before and after delignification, respectively;

$$\%\ \text{delignification} = 100 \times \frac{TL_{BD} - TL_{AD} \times \frac{SY}{100}}{TL_{BD}} \quad (2)$$

where TL_{BD} and TL_{AD} are the percentages of total lignin present in the solids before and after delignification, respectively;

$$\% \text{ cellulose removal} = 100 \times \frac{C_{BD} - C_{AD} \times \frac{SY}{100}}{C_{BD}} \quad (3)$$

where C_{BD} and C_{AD} are the percentages of cellulose present in the solids before and after delignification, respectively;

$$\% \text{ hemicelluloses removal} = 100 \times \frac{H_{BD} - H_{AD} \times \frac{SY}{100}}{H_{BD}} \quad (4)$$

where H_{BD} and H_{AD} are the percentages of hemicelluloses (measured as the overall contributions of xylan, arabinan and acetyl groups) present in the solids before and after delignification, respectively.

The lignin present in the liquid phases from treatments performed under selected conditions was precipitated by adding HCl 5M to pH = 2, kept overnight at 4 °C, recovered by filtration, and dried in a vacuum oven at 40 °C. Aliquots of the liquid phases from treatments were analyzed for hemicellulose-derived compounds (oligosaccharides, monosaccharides, acetic acid, and furans) following the methods listed in Section 2.5.

Table 1. Set of experiments performed in this work for the delignification of HS and AS (LSR = 8 g/g).

Delignification Method	Experiment	Substrate and Operational Conditions
Alkaline (NaOH-water)	1	HS 121 °C 60 min, 2% NaOH
	2	AS 121 °C 60 min, 2% NaOH
	3	HS 121 °C 60 min, 4% NaOH
	4	AS 121 °C 60 min, 4% NaOH
	5	HS 121 °C 60 min, 8% NaOH
	6	AS 121 °C 60 min, 8% NaOH
Alkaline-organosolv (ethanol-aqueous NaOH)	7	HS 121 °C 60 min, 50/50 ethanol/aqueous NaOH, 2% total solution
	8	AS 121 °C 60 min, 50/50 ethanol/aqueous NaOH, 2% total solution
	9	HS 121 °C 60 min, 50/50 ethanol/aqueous NaOH, 4% total solution
	10	AS 121 °C 60 min, 50/50 ethanol/aqueous NaOH, 4% total solution
	11	HS 121 °C 60 min, 50/50 ethanol/aqueous NaOH, 8% total solution
	12	AS 121 °C 60 min, 50/50 ethanol/aqueous NaOH, 8% total solution
	13	HS 135 °C 60 min, 50/50 ethanol/ aqueous NaOH, 4% total solution
	14	AS 135 °C 60 min, 50/50 ethanol/aqueous NaOH, 4% total solution
Organosolv (ethanol-water)	15	HS 200 °C 60 min, 55/45 ethanol/water
	16	AS 200 °C 60 min, 55/45 ethanol/water
	17	AS 200 °C 120 min, 55/45 ethanol/water
Acid-catalyzed organosolv (ethanol/aqueous H_2SO_4)	18	HS 180 °C 60 min, 60/40 ethanol/aqueous H_2SO_4,
	19	AS 180 °C 60 min, 60/40 ethanol/aqueous H_2SO_4,
	20	AS 180 °C 120 min, 60/40 ethanol/aqueous H_2SO_4,
	21	HS 160 °C 120 min, 60/40 ethanol/aqueous H_2SO_4,
	22	AS 160 °C 120 min, 60/40 ethanol/aqueous H_2SO_4,

2.4. *Enzymatic Hydrolysis*

Solids resulting from consecutive stages of autohydrolysis (AS) and delignification under selected conditions were subjected to hydrolysis using the enzymatic complex Cellic CTec2 (Novozymes, Denmark). The enzymatic activity of Cellic CTec2 was 137 FPU (Filter Paper Units)/g of enzyme [30]. Hydrolysis assays were performed in Erlenmeyer flasks kept at 50 °C in an orbital incubator (150 rpm). The pH of media was adjusted to 4.8 by adding 50 mM citrate buffer. Experiments were performed for 0–96 h at LSR = 15 g/g using an enzyme to solid ratio (ESR) of 15 FPU/g substrate. All assays were performed by triplicate. Aliquots of samples were withdrawn at selected times, centrifuged, filtered, diluted and assayed for glucose by HPLC as described in Section 2.5. Cellulose conversion into glucose, and xylan conversion into xylose were calculated as follows:

$$\%\text{ cellulose conversion into glucose} = 100 \times \frac{[Glu]}{[Glu_{Pot}]} \tag{5}$$

where $[Glu]$ is the glucose concentration and $[Glu_{Pot}]$ is the potential glucose concentration (calculated assuming total conversion of the cellulose contained in the substrate);

$$\%\text{ xylan conversion to xylose} = 100 \times \frac{[Xyl]}{[Xyl_{Pot}]} \tag{6}$$

where $[Xyl]$ is the xylose concentration and $[Xyl_{Pot}]$ is the potential xylose concentration (calculated assuming total conversion of the xylan contained in the substrate).

2.5. Analytical Procedures

The chemical composition of HS, AS, and solids from delignification were analyzed using the following TAPPI standard methods: moisture: T-264-cm-97, ash: T-211-om-02, structural carbohydrates, and Klason lignin: T-249-cm-85 [31–33]. The latter method is based on a two-step quantitative acid hydrolysis (QAH) performed with 72 and 4% H_2SO_4, respectively. The insoluble residue from QAH was oven-dried and weighed for Klason lignin determination. The liquid phase from QAH was assayed for glucose, xylose, arabinose, acetic acid, furfural, and hydroxymethylfurfural by HPLC, using a 1200 series instrument (Agilent Technologies, Santa Clara, CA, USA) fitted with a refractive index detector) and a 300 × 7.8 Aminex HPX-87H column (BioRad Life Science Group Hercules, CA, USA). The instrument detector was kept at 50 °C. The mobile phase was 0.003 N H_2SO_4 eluted at 0.6 mL·min^{-1}. The results allowed the determination of cellulose, xylan, arabinan, and acetyl groups present in the solid substrates. The acid-soluble lignin (ASL) was quantified spectrophotometrically at 205 nm. The total lignin was calculated as the sum of Klason lignin and ASL. All the analyses were performed in triplicate.

Aliquots from autohydrolysis and liquid phases from delignification assays were subjected to quantitative posthydrolysis (4% of H_2SO_4 at 121 °C for 20 min), and the increase in the concentrations of monosaccharides and acetic acid caused by posthydrolysis measured the amounts oligomers and their degree of substitution with acetyl groups. All the analyses were performed in triplicate.

3. Results and Discussion

3.1. Autohydrolysis and Composition of HS and AS

In previous studies reported by our research group, 210 °C was identified as the optimal autohydrolysis temperature for producing soluble hemicellulosic oligosaccharides [10]. Under the same conditions, soluble antioxidant compounds were extracted from the substrate [23]. Both oligosaccharides and antioxidant compounds find applications in a number of fields, including the food, cosmetic and pharmaceutical industries. Based on this information, HS autohydrolysis assays were performed at 210 °C, and the corresponding AS were selected for assessing the separation of cellulose from lignin.

Upon autohydrolysis, 35.4% of the dry HS mass was dissolved, yielding an aqueous phase with the following composition (in g/L): xylooligosaccharides, 16.15; glucooligosaccharides, 0.07; acetyl groups, 3.65; xylose, 1.02; arabinose, 0.31; acetic acid, 0.86; and total phenolic content, 1.60 g equivalent gallic acid/L. The average compositions of HS and AS are presented in Table 2. As expected, autohydrolysis causes the selective separation of hemicelluloses resulted in AS with increased percentages of cellulose (38.7%) and total lignin (50.4%).

Table 2. HS and AS composition, expressed as g of component/100 g of dry HS or AS, respectively. Data reported as average values ± standard deviations.

Component	HS (g/100 g of dry HS)	AS (g/100 g of AS)
Cellulose	24.2 ± 0.1	38.7 ± 0.2
Xylan	23.2 ± 0.1	7.5 ± 0.2
Arabinan	0.3 ± 0.0	0.0 ± 0.0
Acetyl groups	4.6 ± 0.1	1.6 ± 0.1
Klason lignin	38.5 ± 0.6	49.7 ± 0.7
Acid Soluble Lignin (ASL)	1.2 ± 0.1	0.7 ± 0.1
Other components	8.0	1.80

3.2. Delignification Treatments

Table 3 lists data concerning the HS and AS delignification treatments, including SY and the removal percentages of the structural components (lignin, cellulose, and hemicelluloses). For comparative purposes, Table 4 lists results reported for the delignification of hazelnut shells and other biomasses. Little information has been reported on the delignification of native or pretreated HS, with alkaline delignification being the most studied processing method [20,27,28].

Table 3. Solid yield and removal percentages of lignin, cellulose, and hemicelluloses achieved in the experiments 1–22 in Table 1. The removal percentages were calculated from their source (HS and AS, respectively).

Experiment	Solid Yield (%)	Lignin Removal (%)	Cellulose Removal (%)	Hemicellulose Removal (%)
1	91.7	7.2	10.0	26.6
2	81.8	12.7	22.5	74.7
3	88.0	15.7	9.0	38.8
4	78.1	20.7	22.3	76.7
5	76.1	19.2	7.4	47.8
6	76.9	24.3	20.3	79.2
7	86.8	11.3	11.5	27.5
8	82.3	17.2	20.0	70.6
9	82.4	18.0	10.9	42.5
10	88.0	22.4	13.8	62.2
11	82.7	18.9	10.3	52.4
12	85.5	29.1	24.9	69.1
13	73.5	21.2	11.0	49.2
14	75.8	26.6	16.2	70.1
15	53.6	53.3	0.0	67.5
16	76.8	35.2	2.6	42.0
17	75.8	32.8	5.2	51.0
18	46.2	65.3	0.0	76.0
19	64.9	47.9	7.1	87.7
20	57.9	50.5	17.3	93.2
21	61.4	47.9	0.0	56.2
22	73.2	37.2	5.6	62.1

Table 4. Literature reported on the delignification of diverse types of biomass using operational conditions related to the ones used in this work.

Raw Material	Reagents	T (°C)	Time	LSR	% Deligni-fication	Pre-Processing	Reference
Hazelnut shell	NaOH 4% (*w*/*v*)	121	60 min	10/1 (*v*/*w*)	32	Steam explosion, 5 min, 198–200 °C	[27]
	NaOH 4% (*w*/*v*)	121	90 min	10/1 (*v*/*w*)	42.5	Steam explosion, 5 min, 198–200 °C	[27]
	H_2O_2 4% (*w*/*v*)	121	30 min	10/1 (*v*/*w*)	36	Steam explosion, 5 min, 198–200 °C	[27]
	$NaBH_4$ 4% (*w*/*v*)	121	60 min	10/1 (*v*/*w*)	48	Steam explosion, 5 min, 198–200 °C	[27]
	NaOH 5% (*w*/*v*)	121	60 min	10/1 (*v*/*w*)	19.7	-	[28]
	NaOH 2.25%	120	60 min	10/1 (*v*/*w*)	60	-	[20]
	NaOH 2.25%	200	60 min	10/1 (*v*/*w*)	73.28	-	[20]
Almond shell	Ethanol 70/30 (*v*/*v*)	200	90 min	6/1	10.8	Autohydrolysis, 180 °C, 30 min, LSR 8/1	[34]
	NaOH 7.5 wt.%	121	90 min	6/1	18.4	Autohydrolysis. 180 °C, 30 min, LSR 8/1	[34]
Rice husks	Ethanol 54/46, NaOH 8% (*w*/*w* on solid)	160	60 min	10/1	90.1	Acid, 0.3% H_2SO_4 (*w*/*v*), 152 °C, 33 min	[35]
	Ethanol 54/46, NaOH 8% (*w*/*w* on solid)	160	100 min	10/1	91.47	Acid, 0.3% H_2SO_4 (*w*/*v*), 152 °C, 33 min	[36]
Hazelnut tree prunings	NaOH 2%	121	60 min	10/1	30.7	-	[37]
	NaOH 2%	121	60 min	10/1	51.2	Hydrothermal, 190 °C, 45 min, LSR 10/1 (*v*/*w*)	[37]
Olive tree pruning	Ethanol 70/30 (*v*/*v*)	200	90 min	6/1	31.2	Autohydrolysis, 180 °C, 30 min, LSR 8/1	[34]
	NaOH 7.5 wt.%	121	90 min	6/1	14.6	Autohydrolysis, 180 °C, 30 min, LSR 8/1	[34]
Sugarcane bagasse	Ethanol 50/50 (*v*/*v*), NaOH 1.5% on dry fiber (*w*/*w*)	175	60 min	5/1 (*v*/*w*)	44.3	-	[38]
	Ethanol 30/70 (*v*/*v*), NaOH 3% on dry fiber (*w*/*w*)	195	60 min	7/1 (*w*/*w*)	17.1	Acid, 0.2 M H_2SO_4, LSR 5/1 (*w*/*w*), 40 min, 120 °C	[39]
Miscanthus biomass	Ethanol 80/20 (*v*/*v*), H_2SO_4 1% (*w*/*w*, based on solid)	170	60 min	8/1	84	-	[40]
	Ethanol 80/20 (*v*/*v*), H_2SO_4 1% (*w*/*w*, on solid)	170	60 min	8/1	88.5	Autohydrolysis. LSR 9/1. 150 °C, 8h	[40]
Eucalyptus globulus	Ethanol 60:40 (*w*/*w*)	180–200	60 min	8/1 (*w*/*w*)	81	Autohydrolysis. LSR 8/1 (*w*/*w*), Severity) 3.65–3.94	[1]
Eucalyptus nitens bark	Ethanol 52-65%	192–200	60–86 min	8/1 (*w*/*w*)	49–52	-	[15]
Wheat straw	Ethanol 60/40 (*w*/*w*)	200	60 min	10/1 (*v*/*w*)	67	-	[41]
	Ethanol 60/40 (*w*/*w*)	200	60 min	10/1 (*v*/*w*)	64.3	Acid, LSR 7.5/1 (*v*/*w*), 160 °C, 30 min	[41]

Alkaline treatments of HS performed with 2–8% NaOH solutions (experiments 1, 3, and 5) did not exceed 19.2% delignification. These results are in agreement with the data reported by Uzuner et al. [28], who assessed the effects of NaOH concentration on delignification, and achieved a limited delignification degree (20%). The data were justified on the basis of the high recalcitrance of HS lignin to depolymerization. Hoşgün and Bozan [20] reported 60% of lignin removal operating at 120 °C for 60 min in media containing 2.25% NaOH, conditions that allowed almost complete cellulose recovery.

In this study, as a general trend, the solid dissolution and delignification effects reached in experiments performed with AS (exp. 2, 4, and 6) increased with the NaOH concentration more than they did in assays using HS (exp. 1, 3, and 5). The highest lignin removal on the delignification stage (24.3%) was reached when AS was treated with 8% NaOH at 121 °C for 60 min (experiment 6). These results are in accordance with literature using steam-exploded substrates, taking into account both delignification and autohydrolysis (31.6% of lignin removal for the experiment 4) but are below for experiments containing NaOH during longer times or $NaBH_4$ (42.5 and 48% lignin removal, respectively) [27]. The alkaline delignification of hazelnut tree prunings (before and after hydrothermal processing) was also assessed in media containing 2% NaOH, with a substantially improved delignification when the pretreated biomass was used as a substrate [37].

In the set of experiments performed with aqueous alkaline solutions (exp 1–6), AS led to higher hemicellulose removal than the ones carried out with HS. For example, 79.2% hemicellulose removal was achieved when AS was treated with 8% of NaOH at 121 °C for 60 min (exp. 6), conditions under which more than 20% of cellulose was removed (revealing poor selectivity).

The comparison between experiments performed with aqueous alkaline solutions and alkaline-organosolv mixtures (7–12) showed that the presence of ethanol resulted in limited additional delignification (11.3–26.6% of lignin removal). Related results were observed for the solubilization of cellulose and hemicelluloses. The experiments performed at higher temperatures (exp. 13 and 14) led to improved delignification degrees. In related studies, alkaline-organosolv methods were successfully applied to rice husks [35,36]. Native and pretreated sugarcane bagasse were also delignified by alkaline-organosolv delignification, resulting in a significantly higher lignin removal in the native samples [38,39].

Fernández-Rodríguez et al. [34] compared organosolv and alkaline delignification treatments of pre-processed biomass, looking at the manufacture of lignin isolates suitable for valorization alkaline methods led to the best results for almond shell delignification, whereas the opposite behavior was observed for olive tree pruning. In our study, as a general trend, organosolv treatments (exp. 15–17) and acid-catalyzed organosolv assays (exp. 18–22) showed that the amount of cellulose remaining in the solid phase was hardly altered. In assays using HS as a substrate, the delignification increased with respect to the ones performed with AS. This conclusion can be confirmed by comparing the results obtained in assays free from the catalyst (for example, 53.3% lignin removal in exp. 15 in comparison with 35.2% in experiment 16), and also by data analysis of experiments performed in H_2SO_4 containing media (65.3% delignification in exp. 18 in comparison with 47.9% in exp. 19).

Coupling stages of autohydrolysis and organosolv delignification has been reported as a suitable strategy for the complete fractionation of biomass [1,15,37], providing an alternative to conventional, single-stage delignification. In a related study, Huijgen et al. [41] compared organosolv and prehydrolysis-organosolv treatments of wheat straw, concluding that the prehydrolysis before organosolv resulted in decreased lignin recovery yields, a fact ascribed to the formation of "pseudo-lignin" and to lignin recondensation during prehydrolysis. El Hage et al. [42] found that the severity of the autohydrolysis modifies the lignin structure, affecting the subsequent organosolv delignification. Obama et al. [40] reported lignin alteration upon autohydrolysis, with the participation of repolymerization reactions (C-C linkages) that negatively affect the further delignification.

In this study, enhanced removal of hemicelluloses from HS and AS was observed in experiments performed in acid-catalyzed organosolv assays. In runs with AS, high degrees of hemicellulose removal (87.7% in exp. 19, performed at 180 °C for 60 min; or 93.2% in exp. 20, which lasted 120 min) were achieved in acid-catalyzed organosolv treatments. These results are significantly higher than the ones observed for HS (76% hemicellulose removal in exp. 18). However, it can be noted that the AS hemicellulose content (9.1%) was significantly lower than the one of HS (28.1%). An opposite pattern was observed in experiments performed in the absence ofan acid catalyst, in which HS reached higher hemicellulose removal.

According to the above ideas, SY (which is affected by the contents of lignin and hemicelluloses) were lower for acid-catalyzed experiments. In experiment 18, the limited SY (46.2%) corresponded to 65.3% lignin removal and 76% hemicellulose removal from HS. Using AS as a substrate under the same conditions (exp. 19), the SY (64.9%) corresponded to 47.9% and 87.7% removal of lignin and hemicelluloses, respectively.

A comparative analysis of results confirmed that the conditions of experiments 18 and 19 (dealing with HS and AS, respectively) were the best ones identified in this study. The liquid phase from HS in exp. 18 contained the following hemicellulose-derived compounds: xylooligosaccharides, 10.01 g/L; arabinooligosaccharides, 0.28 g/L; glucooligosaccharides, 0.07 g/L; acetyl groups, 2.25 g/L; xylose, 10.74 g/L; glucose, 0.07 g/L; acetic acid, 2.29 g/L; furfural 1.38 g/L. In comparison, the composition of the liquid resulting from the acid-catalyzed delignification of AS (exp. 19) contained the following hemicellulose-derived compounds: glucooligosaccharides, 2.34 g/L; xylooligosaccharides, 1.25 g/L; acetyl groups, 0.28 g/L; xylose, 4.20 g/L; glucose, 1.98 g/L; acetic acid, 1.06 g/L; furfural 2.06 g/L; hydroxymethylfurfural, 0.24 g/L.

3.3. *Enzymatic Hydrolysis*

The results included in this section provide a quantitative assessment of the susceptibility of selected delignified and autohydrolyzed-delignified solids towards enzymatic hydrolysis. The composition of the solids is listed in Table 5. The experimental plan included acid-catalyzed organosolv treatments (assays 18–19), and experiments in ethanol/NaOH media (runs 13–14).

Table 5. Substrates and composition of solids employed as substrates for enzymatic hydrolysis. Data expressed as average values ± standard deviations.

Substrate for Enzymatic Hydrolysis	Delignified HS (from exp. 13)	Delignified AS (from exp. 14)	Delignified HS (from exp. 18)	Delignified AS (from exp. 19)
Solid composition (g/100 g of delignified solid)				
Cellulose	29.3 ± 0.5	42.8 ± 0.7	54.0 ± 1.2	55.4 ± 1.9
Xylan	19.3 ± 0.8	3.6 ± 0.1	11.9 ± 0.9	1.7 ± 0.3
Arabinan	0.1 ± 0.0	0.0 ± 0.0	0.0 ± 0.0	0.0 ± 0.0
Acetyl groups	0.0 ± 0.0	0.0 ± 0.0	2.0 ± 0.2	0.1 ± 0.1
Klason lignin	41.5 ± 3.5	48.3 ± 2.5	29.0 ± 0.7	40.1 ± 1.2
Acid soluble lignin (ASL)	1.1 ± 0.1	0.6 ± 0.1	0.6 ± 0.1	0.4 ± 0.0
Other	8.7	4.7	2.5	2.3

Both AS and the solids obtained under the conditions listed in Table 5 presented different behavior as hydrolysis substrates. Figure 1a shows the cellulose conversions into glucose achieved at selected reaction times. The highest cellulose conversion (74.2%, corresponding to a volumetric glucose concentration of 28.52 g/L) was reached when the solids from exp. 18 (HS subjected to acid organosolv delignification) were hydrolyzed for 96 h. In comparison, the solid from exp. 19 (AS delignified under the same conditions) was poorly hydrolyzed (conversion below 15%, or 5.62 g glucose/L). HS delignified with aqueous NaOH/ethanol under the conditions of exp. 13 showed higher hydrolysis conversion (52.5%) than the one achieved using AS delignified under the same conditions (exp. 14). Despite the low conversions reached in the experiments performed with autohydrolyzed or autohydrolyzed-delignified solids (obtained in exp. 14 and 19), the highest enzymatic susceptibility was observed for the substrate from exp. 14, carried out in aqueous NaOH/ethanol media. In comparative terms, the susceptibility of the diverse substrates to hydrolysis varied as follows: AS < acid-catalyzed-organosolv AS (exp. 19) <

aqueous NaOH-organosolv AS (exp. 14) < aqueous NaOH-organosolv HS (exp. 13) < acid-catalysed organosolv HS (exp. 18).

Literature data reports that the lignin content affects the enzymatic hydrolysis, since cellulases may bound to lignin irreversibly [40]. In our study, the solid from exp. 18 (which showed the highest conversion into glucose) presented the highest cellulose/lignin ratio (1.86). In comparison, the solid from exp. 19 (cellulose/lignin ratio, 1.36) reached considerably lower conversion into glucose than the solids from exp. 13 and 14 (which presented cellulose/lignin ratios of 0.68 and 0.87, respectively). Çöpür et al. [27] reported the behavior of different pretreatment techniques on regard to their efficiencies for the subsequent enzymatic hydrolysis of glucan into glucose from hazelnut husks. The highest glucan to lignin ratio of pretreated solids was used as criteria for the subsequent saccharification. Then, they reported a glucan to lignin ratio of 0.99 when HS were treated with combined steam explosion-NaOH delignification, and a high conversion of glucan into glucose (74.7%) in enzymatic hydrolysis performed at LSR 20. Hoşgün and Bozan [20] compared the enzymatic susceptibility of HS treated by different methods (acid, alkali, and steam) and found that the maximum glucose recovery (58.7%) corresponded to samples pretreated with NaOH. Similar results were reported with alkali delignified HS when solids were enzymatically hydrolyzed at higher LSR and ESR than were used in this work [28].

Hallac et al. [43] reported that the relationship between the lignin content of pretreated *Buddleja davidii* samples and their enzymatic susceptibility was not proportional. Huijgen et al. [41] compared the digestibility of delignified and prehydrolyzed-delignified wheat straw, and concluded that prehydrolysis prior to organosolv improved the enzymatic cellulose digestibility, despite the low lignin removal in delignification. Yang and Pan [44] discussed some aspects of the complexity of lignin effects on the enzymatic hydrolysis: a) high lignin content does not necessarily predict poor enzymatic digestibility; b) not all lignins have the same inhibitory effect on enzymatic hydrolysis of cellulose, since lignin from different sources had varying inhibitory effects, greatly related to lignin structures and properties; c) not all the lignin in the same substrate has the same inhibitory effect. On the basis of these observations, they studied the inhibitory effect of lignin with varied physicochemical properties from different biomass sources on the enzymatic hydrolysis of cellulose, and concluded that the lignin inhibition to enzymatic hydrolysis of cellulose was related to the hydrophobicity or the phenolic hydroxyl groups of lignin.

The data in Figure 1b show that the residual xylan in the hydrolysis substrates was extensively hydrolyzed into xylose. For example, in exp. 18, 80% of xylan was converted into xylose, which reached 7.03 g/L. Sun et al. [30] reported that some accessory enzymes in the enzymatic hydrolysis of natural lignocellulosic substrates favored an efficient conversion, and underlined the role of the high specific xylanase activity of the complex Cellic CTec2.

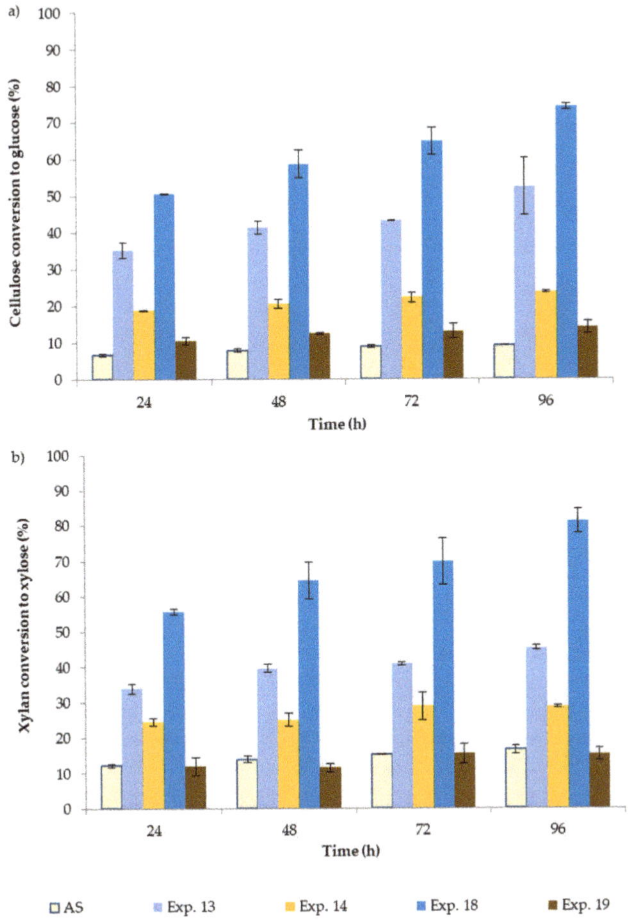

Figure 1. Effect of the enzymatic hydrolysis in the conversion of: (**a**) cellulose to glucose; (**b**) xylan to xylose, of spent solids from HS and AS processed under selected conditions of alkaline-organosolv, acid-catalyzed organosolv, and AS.

4. Conclusions

This study deals with the integral fractionation of HS, based on selected delignification methods (in alkaline, alkaline-organosolv, organosolv or acid-catalyzed organosolv media). Both HS and autohydrolyzed HS (AS) were used as substrates for delignification.

HS treated in acid-catalyzed organosolv media reached the highest delignification degree (65.3%), enabling the production of valuable hemicellulose-derived products (about half of them in the form of oligosaccharides), with limited cellulose losses from the solid phase. Alternatively, autohydrolysis led to the partial solubilization of hemicelluloses (mainly as oligosaccharides), and the subsequent acid-catalyzed organosolv of AS resulted in the simultaneous solubilization of the remaining hemicelluloses (87.7%) and lignin (47.9%).

The enzymatic hydrolysis of solids from delignification or autohydrolysis-delignification treatments confirmed that acid-catalyzed organosolv of HS provided the best substrate for enzymatic hydrolysis (74.2% cellulose conversion into glucose, with a volumetric concentration of 28.52 g glucose/L).

Then, the strategy approaches reported in this work can be considered as "conventional" but promising alternatives for the integral fractionation of HS in the scope of biorefinery.

Author Contributions: S.R. designed the research and discussed the data. S.R., L.L., A.M. and C.V. performed experiments and analyzed the data. S.R. and J.C.P. prepared the original draft. All authors discussed the data. All authors have read and agreed to the published version of the manuscript.

Funding: S.R. contract was supported by Ministerio de Ciencia, Innovación y Universidades (IJC2018-037665-I).

Acknowledgments: S.R. thanks Ministerio de Ciencia, Innovación y Universidades for her "Juan de la Cierva" contract (IJC2018-037665-I).

Conflicts of Interest: The authors declare no conflict of interest.

References

1. Romaní, A.; Garrote, G.; López, F.; Parajó, J.C. *Eucalyptus globulus* Wood fractionation by autohydrolysis and organosolv delignification. *Bioresour. Technol.* **2011**, *102*, 5896–5904. [CrossRef]
2. Mahmood, H.; Moniruzzaman, M.; Iqbal, T.; Khan, M.J. Recent advances in the pretreatment of lignocellulosic biomass for fuels and value-added products. *Curr. Opin. Green Sustain. Chem.* **2019**, *20*, 18–24. [CrossRef]
3. Ubando, A.T.; Feliz, C.B.; Chen, W.-H. Biorefineries in circular bioeconomy: A comprehensive review. *Bioresour. Technol.* **2020**, *299*, 122585. [CrossRef] [PubMed]
4. Kumar, B.; Bhardwaj, N.; Agrawal, K.; Chaturvedi, V.; Verma, P. Current perspective on pretreatment technologies using lignocellulosic biomass: An emerging biorefinery concept. *Fuel Process. Technol.* **2020**, *199*, 106244. [CrossRef]
5. Jönson, L.J.; Martín, C. Pretreatment of lignocellulose: Formation of inhibitory by-products and strategies for minimizing their effects. *Bioresour. Technol.* **2016**, *199*, 103–112. [CrossRef] [PubMed]
6. Galbe, M.; Wallberg, O. Pretreatment for biorefineries: A review of common methods for the efficient utilisation of lignocellulosic materials. *Biotechnol. Biofuels* **2019**, *12*, 294. [CrossRef]
7. Bhatia, S.K.; Jagtap, S.S.; Bedekar, A.A.; Bhatia, R.K.; Patel, A.K.; Pant, D.; Banu, J.R.; Rao, C.V.; Kim, Y.-G.; Yang, Y.-H. Recent developments in pretreatment technologies on lignocellulosic biomass: Effect of key parameters, technological improvements, and challenges. *Bioresour. Technol.* **2020**, *300*, 122774. [CrossRef]
8. Pachapur, V.L.; Kaur Brar, S.; Le Bihan, Y. Integrated wood biorefinery: Improvements and tailor-made two-step strategies on hydrolysis techniques. *Bioresour. Technol.* **2020**, *299*, 122632. [CrossRef]
9. Aachary, A.A.; Prapulla, S.G. Xylooligosaccharides (XOS) as an emerging prebiotic: Microbial synthesis, utilization, structural characterization, bioactive properties and applications. *Compr. Rev. Food Sci. Food Saf.* **2011**, *10*, 2–16. [CrossRef]
10. Rivas, S.; Moure, A.; Parajó, J.C. Pretreatment of hazelnut shells as a key strategy for the solubilization and valorization of hemicelluloses into bioactive compounds. *Agronomy* **2020**, *10*, 760. [CrossRef]
11. Kumar, V.; Bionod, P.; Sindhu, R.; Gnansounou, E.; Ahuwalia, V. Bioconversion of pentose sugars to value added chemicals and fuels: Recent trends, challenges and possibilities. *Bioresour. Technol.* **2018**, *269*, 443–451. [CrossRef] [PubMed]
12. Quesada-Medina, J.; López-Cremades, F.J.; Olivares-Carrillo, P. Organosolv extraction of lignin from hydrolyzed almond shells and application of the δ-value theory. *Bioresour. Technol.* **2010**, *101*, 8252–8260. [CrossRef] [PubMed]
13. Michelin, M.; Liebentritt, S.; Vicente, A.A.; Teixeira, J.A. Lignin from an integrated process consisting of liquid hot water and ethanol organosolv: Physicochemical and antioxidant properties. *Int. J. Biol. Macromol.* **2018**, *120*, 159–169. [CrossRef] [PubMed]
14. Moniz, P.; Lino, J.; Duarte, L.C.; Roseiro, L.B.; Boeriu, C.G.; Pereira, H.; Carvalheiro, F. Fractionation of hemicelluloses and lignin from rice straw by combining autohydrolysis and optimised mild organosolv delignification. *Bioresources* **2015**, *10*, 2626–2641. [CrossRef]
15. Romaní, A.; Larramendi, A.; Yáñez, R.; Cancela, A.; Sánchez, A.; Teixeira, J.A.; Domingues, L. Valorization of *Eucalyptus nitens* bark by organosolv pretreatment for the production of advanced biofuels. *Ind. Crop. Prod.* **2019**, *132*, 327–335. [CrossRef]

16. Robles, E.; Fernández-Rodríguez, J.; Barbosa, A.M.; Gordobil, O.; Carreño, N.L.V.; Labidi, J. Production of cellulose nanoparticles from blue agave waste treated with environmentally friendly processes. *Carbohydr. Polym.* **2018**, *183*, 294–302. [CrossRef] [PubMed]
17. Karnaouri, A.; Asimakopoulou, G.; Kalogiannis, K.G.; Lappas, A.; Topakas, E. Efficient D-lactic acid production by *Lactobacillus delbrueckii* subsp. *Bulgaricus through conversion of organosolv pretreated lignocellulosic biomass. Biomass Bioenergy* **2020**, *140*, 105672. [CrossRef]
18. Chen, X.; Li, H.; Sun, S.; Cao, X.; Sun, R. Effect of hydrothermal pretreatment on the structural changes of alkaline ethanol lignin from wheat straw. *Sci. Rep.* **2016**, *6*, 39354. [CrossRef]
19. Rossberg, C.; Bremer, M.; Machill, S.; Koenig, S.; Kerns, G.; Boeriu, C.; Windeisen, E.; Fischer, S. Separation and characterisation of sulphur-free lignin form different agricultural residues. *Ind. Crop. Prod.* **2015**, *73*, 81–89. [CrossRef]
20. Hoşgün, E.Z.; Bozan, B. Effect of different types of thermochemical pretreatment on the enzymatic hydrolysis and the composition of hazelnut shells. *Waste Biomass Valorization* **2020**, *11*, 3739–3748. [CrossRef]
21. FAO. Hazelnut Production. 2018. Available online: http://www.fao.org/faostat/en/#data/QC/visualize (accessed on 9 October 2020).
22. Yuan, B.; Lu, M.; Eskridge, K.M.; Isom, L.D.; Hanna, M.A. Extraction, identification and quantification of antioxidant phenolics from hazelnut (*Corylus avellana* L.) shells. *Food Chem.* **2018**, *244*, 7–15. [CrossRef] [PubMed]
23. Pérez-Armada, L.; Rivas, S.; González, B.; Moure, A. Extraction of phenolic compounds from hazelnut shells by green processes. *J. Food Eng.* **2019**, *255*, 1–8. [CrossRef]
24. Demirbas, A. Oils from hazelnut shell and hazelnut kernel husk for biodiesel production. *Energy Sources Part A Recovery Util. Environ.* **2008**, *30*, 1870–1875. [CrossRef]
25. Surek, E.; Buyukkileci, A.O. Production of xylooligosaccharides by autohydrolysis of hazelnut (*Corylus avellana* L.) shell. *Carbohyd. Polym.* **2017**, *174*, 565–571. [CrossRef] [PubMed]
26. Aydinli, B.; Caglar, A. The investigation of the effects of two different polymers and three catalysts on pyrolysis of hazelnut shell. *Fuel Process. Technol.* **2012**, *93*, 1–7. [CrossRef]
27. Çöpür, Y.; Tozluoglu, A.; Özkan, M. Evaluating pretreatment techniques for converting hazelnut husks to bioethanol. *Bioresour. Technol.* **2013**, *129*, 182–190. [CrossRef]
28. Uzuner, S.; Sharma-Shivappa, R.R.; Cekmecelioglu, D. Bioconversion of alkali pretreated hazelnut shells to fermentable sugars for generation of high value products. *Waste Biomass Valorization* **2017**, *8*, 407–416. [CrossRef]
29. Hoşgün, E.Z.; Berikten, D.; Kivanç, M.; Bozan, B. Ethanol production from hazelnut shells through enzymatic saccharification and fermentation by low-temperature alkali pretreatment. *Fuel* **2017**, *196*, 280–287. [CrossRef]
30. Sun, F.F.; Hong, J.; Hu, J.; Saddler, J.N.; Fang, X.; Zhang, Z.; Shen, S. Accesory enzymes influence cellulase hydrolysis of the model substrate and the realistic lignocellulosic biomass. *Enzym. Microb. Technol.* **2015**, *79*, 42–48. [CrossRef]
31. TAPPI Method T 264 cm-07. Preparation of wood for chemical analysis. Available online: https://imisrise.tappi.org/TAPPI/Products/01/T/0104T264.aspx (accessed on 21 January 2020).
32. TAPPI Method T 211 om-07. Ash in wood, pulp, paper and paperboard: Combustion at 525 °C. Available online: https://imisrise.tappi.org/TAPPI/Products/01/T/0104T211.aspx (accessed on 21 January 2020).
33. TAPPI Method T 249 cm-00. Carbohydrate composition of extractive free wood and wood pulp by gas–liquid chromatography. Available online: https://imisrise.tappi.org/TAPPI/Products/01/T/0104T249.aspx (accessed on 21 January 2020).
34. Fernández-Rodríguez, J.; Erdocia, X.; Sánchez, C.; González Alriols, M.; Labidi, J. Lignin depolymerization for phenolic monomers production by sustainable processes. *J. Energy Chem.* **2017**, *26*, 622–631. [CrossRef]
35. Dagnino, E.P.; Felissia, F.E.; Chamorro, E.; Area, M.C. Optimization of the soda-ethanol delignification stage for a rice husk biorefinery. *Ind. Crop. Prod.* **2017**, *97*, 156–165. [CrossRef]
36. Dagnino, E.P.; Felissia, F.E.; Chamorro, E.; Area, M.C. Studies on lignin extraction from rice husk by a soda-ethanol treatment: Kinetics, separation, and characterization of products. *Chem. Eng. Res. Des.* **2018**, *129*, 209–216. [CrossRef]
37. Sabanci, K.; Buyukkileci, A.O. Comparision of liquid hot water, very dilute acid and alkali treatments for enhancing enzymatic digestibility of hazelnut tree pruning residues. *Bioresour. Technol.* **2018**, *261*, 158–165. [CrossRef] [PubMed]

38. Mesa, L.; González, E.; Ruiz, E.; Romero, I.; Cara, C.; Felissia, F.; Castro, E. Preliminary evaluation of organosolv pre-treatment of sugar cane bagasse for glucose production: Application of 2^3 experimental design. *Appl. Energy* **2010**, *87*, 109–114. [CrossRef]
39. Mesa, L.; González, E.; Cara, C.; González, M.; Castro, E.; Mussatto, S.I. The effect of organosolv pretreatment variables on enzymatic hydrolysis of sugarcane bagasse. *Chem. Eng. J.* **2011**, *168*, 1157–1162. [CrossRef]
40. Obama, P.; Ricochon, G.; Muniglia, L.; Brosse, N. Combination of enzymatic hydrolysis and ethanol organosolv pretreatments: Effect on lignin structures, delignification yields and cellulose-to-glucose conversion. *Bioresour. Technol.* **2012**, *112*, 156–163. [CrossRef] [PubMed]
41. Huijgen, W.J.J.; Smit, A.T.; de Wild, P.J.; den Uil, H. Fractionation of wheat straw by prehydrolysis, organosolv delignification and enzymatic hydrolysis for production of sugars and lignin. *Bioresour. Technol.* **2012**, *114*, 389–398. [CrossRef]
42. El Hage, R.; Chrusciel, L.; Desharnais, L.; Brosse, N. Effect of autohydrolysis of *Miscanthus x giganteus* on lignin structure and organosolv delignification. *Bioresour. Technol.* **2010**, *101*, 9321–9329. [CrossRef]
43. Hallac, B.B.; Sannigrahi, P.; Pu, Y.; Ray, M.; Murphy, R.J.; Ragauskas, A.J. Effect of ethanol organosolv pretreatment on enzymatic hydrolysis of *Buddleja davidii* Stem Biomass. *Ind. Eng. Chem. Res.* **2010**, *49*, 1467–1472. [CrossRef]
44. Yang, Q.; Pan, X. Correlation between lignin physicochemical properties and inhibition to enzymatic hydrolysis of cellulose. *Biotechnol. Bioeng.* **2016**, *113*, 1213–1224. [CrossRef]

Publisher's Note: MDPI stays neutral with regard to jurisdictional claims in published maps and institutional affiliations.

© 2020 by the authors. Licensee MDPI, Basel, Switzerland. This article is an open access article distributed under the terms and conditions of the Creative Commons Attribution (CC BY) license (http://creativecommons.org/licenses/by/4.0/).

agronomy

MDPI

Article

Production of Cellulose Nanofibers from Olive Tree Harvest—A Residue with Wide Applications

Mónica Sánchez-Gutiérrez [1,2], Eduardo Espinosa [1,*], Isabel Bascón-Villegas [1,2], Fernando Pérez-Rodríguez [2], Elena Carrasco [2] and Alejandro Rodríguez [1,*]

1 Bioagres Group, Chemical Engineering Department, Universidad de Córdoba, Marie-Curie Building, 14014 Córdoba, Spain; v02sagum@uco.es (M.S.-G.); q12bavii@uco.es (I.B.-V.)

2 Food Science and Technology Department, Universidad de Córdoba, Darwin Building, 14014 Córdoba, Spain; b42perof@uco.es (F.P.-R.); bt2cajie@uco.es (E.C.)

* Correspondence: eduardo.espinosa@uco.es (E.E.); a.rodriguez@uco.es (A.R.)

Received: 20 April 2020; Accepted: 11 May 2020; Published: 13 May 2020

check for updates

Abstract: With the aim of identifying new sources to produce cellulose nanofibers, olive tree pruning biomass (OTPB) was proposed for valorization as a sustainable source of cellulose. OTPB was subjected to a soda pulping process for cellulose purification and to facilitate the delamination of the fiber in the nanofibrillation process. Unbleached and bleached pulp were used to study the effect of lignin in the production of cellulose nanofibers through different pretreatments (mechanical and TEMPO-mediated oxidation). High-pressure homogenization was used as the nanofibrillation treatment. It was observed that for mechanical pretreatment, the presence of lignin in the fiber produces a greater fibrillation, resulting in a smaller width than that achieved with bleached fiber. In the case of TEMPO-mediated oxidation, the cellulose nanofiber characteristics show that the presence of lignin has an adverse effect on fiber oxidation, resulting in lower nanofibrillation. It was observed that the crystallinity of the nanofibers is lower than that of the original fiber, especially for unbleached nanofibers. The residual lignin content resulted in a greater thermal stability of the cellulose nanofibers, especially for those obtained by TEMPO-mediated oxidation. The characteristics of the cellulose nanofibers obtained in this work identify a gateway to many possibilities for reinforcement agents in paper suspension and polymeric matrices.

Keywords: olive tree harvest; lignocellulose nanofibers; circular economy; valorization; pretreatments; high-pressure homogenization

---

1. Introduction

The concept of the circular economy—a system where waste generation is minimized by reintroducing residues and by-products into the production cycle—can be applied, to a large extent, to production processes that use natural resources. This is one of the bases that support the so-called bioeconomy, the need for the integral valorization of natural resources. In Europe, agriculture constitutes approximately 63% of the total biomass supply; forestry, about 36%; and fisheries, less than 1% [1]. It is therefore essential to focus on the recovery of waste generated by the agricultural sector in order to guide the economic strategy towards a circular economy and bioeconomy.

Spain is the leading country in olive and olive oil production with an average annual output of 9.8 million tons of olives, more than five times that of the second largest producer, Italy, with 1.8 million tons per year [2]. Spain represents 47% of worldwide olive production and 72% of European production. As consequence of this production, after harvest, a large number of different types of lignocellulosic materials are generated (pruning, leaves, stones, pomace, etc.), which generally have no industrial application and must be discarded. It is estimated that for the production of one kg of fruit, more than 0.8 kg of waste is generated, meaning more than 7.5 million tons of olive harvest waste per year, waste that could be valorized in Spain. Olive tree pruning biomass (OTPB), in common with any lignocellulosic material, mainly consists of cellulose, lignin and hemicellulose; and other non-structural minority compounds such as pigments, proteins, ashes, etc. This biomass can be fractionated into its various components by means of biorefinery processes. This fractionation of the OTPB into its lignocellulosic components has been widely studied by the scientific community, testing its application as a source of sugar [3], substrate for ethanol production [4], lignin [5], energy [6], building materials [7] and cellulose fibers for paper and cardboard production [8].

One of the most interesting avenues for the valorization of the agricultural residues is the production of nanocellulose as an alternative to wood sources [9]. Nanocellulose presents unique properties such as a high surface area, unique optical properties, lightweightness, stiffness and a high strength, in addition to its inherent properties in common with cellulose (renewable, biodegradable and sustainable) [10]. These properties allow the possibility of using this nanomaterial in many industrial sectors, expecting to reach a global turnover around 10,000 M€ in 2020 [11]. The wide range of applications of nanocellulose-based materials include the paper and cardboard industry [12], electronic devices [13], energy [14], cosmetics [15], composites [16], wastewater treatment [17], catalysts [18], construction [19], drug carriers [20] and biomedicine [21]. The use of agricultural residues, such as OTPB, as a source for the local, renewable and sustainable production of nanocellulose will allow countries with insufficient forest resources to produce these high value-added products.

Cellulose nanofibers (CNFs), also known as nanofibrillated cellulose, are one of the existing types of nanocellulose (along with cellulose nanocrystals and bacterial cellulose). CNFs are long (several microns), flexible (presenting both types of cellulose region, crystalline and amorphous), nanometric (1–100 nm in width) and are extracted from cellulosic fibers by mechanical methods [22]. The mechanical treatment aims at the isolation of the cellulose nanofibers by the delamination of the fibers. Several mechanical treatments have been studied, including high pressure homogenization [23], twin-screw extrusion [24], micro-fluidization [25] and ultrafine-friction grinding [26], the most commonly used. One of the great disadvantages of these treatments is the large number of passes that the fibers have to undergo and the long time required to produce delamination. Therefore, to facilitate and increase the effectiveness of the treatment, fibers are subjected to a previous process, known as pretreatment. Likewise, there are many pretreatments, but the most widely used and most effective are mechanical pretreatment [27], enzymatical pretreatment [28], TEMPO-mediated oxidation [29] and surface functionalization [30]. To study the effectiveness of the different treatments, it is crucial to determine the chemical composition of the source, to optimize the process of fiber obtention and to adequately characterize the final product.

In this work, olive tree pruning biomass has been valorized as a lignocellulosic source for the obtention of cellulose nanofibers as a high value-added product. The suitability of the chemical composition of the raw material and the fiber in cellulose nanofiber production has been studied. In order to study the effect of lignin on the effectiveness of nanofibrillation and its properties, the cellulose fiber was subjected to a bleaching process. Both types of fiber, bleached and unbleached, were subjected to two independent pretreatments, mechanical pretreatment and TEMPO-mediated oxidation, followed by high pressure homogenization treatment. The cellulose nanofibers obtained were widely characterized in terms of their chemical composition, morphology, thermal stability and crystallinity.

2. Materials and Methods

2.1. Sample Preparation

In this study, the raw material was obtained after the annual pruning of an olive tree plantation in the province of Córdoba (Spain), following olive harvest. The olive tree prunings were air-dried at room temperature until their moisture content was below 8% and stored until use. Before the raw material was subjected to the pulping process, it was chipped in an automatic grinder to obtain chips of 4–5 cm length to facilitate the fractioning of the lignocellulosic components.

2.2. Pulping Process

The olive chips were subjected to a pulping process in a 15 L capacity reactor, heated by an external heating jacket and rotated by means of a horizontal axis. The process carried out was a soda pulping process using 16% NaOH (on dry matter) as a reaction agent, at 170 °C for 60 min and a liquid/solid ratio of 8:1. The conditions were selected according to previous studies and the experience of the research group for the production of cellulose pulp for paper production [31,32] After pulping, the treated chips were dispersed in a pulp disintegrator for 30 min at 1200 rpm. Once the chips were disintegrated, the fiber was passed through a Sprout-Bauer beater and separated by sieving through a netting of 0.14 mm mesh size. The cellulosic pulp was centrifuged to remove excess water and left to dry at room temperature until use. Afterwards, the unbleached pulp was subjected to a bleaching process. For this purpose, 0.3 g of sodium chlorite per gram of pulp was incubated in a 0.3% pulp suspension in water at 80 °C for 3 h. After cooling, the pulp was filtered and washed with acetone and several cycles of distilled water (Figure S1). This bleaching process allows the removal of practically all the lignin present in the fiber, maintaining the entire carbohydrate composition [33]. This makes it possible to study the effect of lignin on the production of cellulose nanofibers and their characteristics.

2.3. Raw Material and Cellulosic Pulp Characterization

The olive tree pruning biomass and the cellulosic pulp obtained were characterized in terms of the chemical composition of the lignocellulose matrix. Both were characterized according to their content of ethanol extractables, hot water extractables, ashes, lignin, hemicelluloses and α-cellulose according to the TAPPI standards T-204, T-435, T-211, T-222, T-9m-54 and T-203 cm-09, respectively. The determination of each component of the chemical characterization was performed in triplicate and the means and standard deviations were calculated.

2.4. Cellulose Nanofiber Production

To obtain cellulose nanofibers (CNFs), two independent pretreatments were used, mechanical beating and TEMPO-mediated oxidation, both followed by a high-pressure homogenization treatment.

The mechanical pretreatment consisted of a mechanical refining (PFI beater) according to the ISO 5264-2:2002 standard, during 40,000 revolutions, to reach a Schopper–Riegler Degree (°SR) of 90 [26]. This pretreatment allows the fibrillation of the cellulose fibers by shear forces to facilitate nanofibrillation in the subsequent treatment. The TEMPO-mediated oxidation was carried out following the methodology described by Saito et al. [29]. The reaction was carried out at pH 10 and started with the addition of a specific amount of NaClO solution in order to obtain an oxidative power of 5 mmol per g of fiber. Once the addition of NaClO was complete, the pH was maintained by adding a 0.5 M NaOH solution. The reaction was finished when the pH remained stable.

A 1% pretreated fiber suspension was subjected to a nanofibrillation process in a high-pressure homogenizer (PandaPlus 2000, GEA Niro, Düsseldorf, Germany) in order to isolate the nanofibers that form the cellulose fibers. To avoid the occlusion of the homogenizer, gradual fibrillation was performed in the following sequence: 4 passes at 300 bars, 3 passes at 600 bars and 3 passes at 900 bars. This treatment has been demonstrated as an effective way of obtaining cellulose nanofibers from different raw materials and pretreatments [34].

By means of the mechanical and TEMPO-mediated oxidation pretreatments, CNFs were obtained, although in the case of unbleached pulp, residual lignin content remained in the final product (lignocellulose nanofibers; LCNF).

2.5. Cellulose Nanofiber Characterization

In order to evaluate the suitability of the different pretreatments and the effect of the residual lignin in the final products, the CNFs/LCNFs obtained were deeply characterized. The nanofibrillation yield, which determines the nanometric fraction of the CNF suspension by the separation of the non-nanometric material by centrifugation, was determined according to the methodology described by Besbes et al. [35]. For this, a 0.1% cellulose nanofiber suspension was centrifuged at 11,000 × *g* for 12 min. The dry weight of the non-nanometric material precipitated during centrifugation compared to the dry weight of the initial suspension was used to inversely determine the nanofibrillation yield. The optical transmittance at 800 nm of the 0.1% cellulose nanofiber suspension was measured using a Lambda 25 UV-Spectrometer. The carboxyl content (CC) was determined using conductimetric titration as described by Besbes et al. [35]. The cationic demand (CD) was determined using a particle charge detector Mütek PCD 05 following the protocol described by Espinosa et al. [23]. The values of cationic demand and carboxyl content are used for the theoretical calculation of the specific surface area of the cellulose nanofibers assuming a simultaneous interaction between the hydroxyl and carboxyl groups of the cellulose nanofiber surface and PolyDADMAC in the monolayer coating [23]. Assuming the cylindrical geometry of the cellulose nanofiber and using the specific surface, it is possible to determine the width of the nanofibers. This method has been evaluated in previous publications, and the theoretical values are very good approximations to the values observed by electron microscopy [23].

2.6. Viscosity, Degree of Polymerization and Length

The intrinsic viscosity (η_s) of the cellulose nanofibers was determined according to the ISO 5351:2010 standard. The degree of polymerization is related to the intrinsic viscosity (in $mL \cdot g^{-1}$) using the empirical relationship suggested by Marx-Figini [36]:

$$DP\ (<950):\ DP = (\eta s/0.42) \tag{1}$$

$$DP\ (>950):\ DP^{0.76} = (\eta s/2.28) \tag{2}$$

The length of the cellulose nanofiber was estimated from the degree of polymerization values using the equation proposed by Shinoda et al. [37]:

$$\text{Length (nm)} = 4.286 \times DP\text{-}757 \tag{3}$$

The measurements were made in triplicate, and the mean value and standard deviation were calculated.

2.7. Fourier-Transform Infrared Spectroscopy (FTIR) Analysis

The chemical structure of the cellulose nanofiber was analyzed by FTIR analysis. A FTIR-ATR Perkin-Elmer Spectrum Two was used to collect 20 infrared spectra in the range of 450–4000 cm^{-1} with a resolution of 4 cm^{-1}. The analysis was performed on a CNF film prepared by hot-drying the cellulose nanofiber suspension.

2.8. X-Ray Diffraction (XRD) Analysis

The X-ray diffraction patterns of the cellulosic pulp and CNFs were obtained using a Bruker D8 Discover with a monochromatic source CuKα1 with an angular range of 5°–50° at a 1.56°/min scan speed. The crystallinity index (CI) was calculated following the equation proposed by Segal et al. [38].

2.9. Thermogravimetric Analysis (TGA)

The thermal stability of the cellulosic fiber and cellulose nanofibers were measured using a Mettler Toledo Thermogravimetric analyzer (TGA/DSC 1). The measurements were performed by heating the samples (10.0 ± 1.0 mg) from room temperature to 600 °C at a heating rate of 10 °C/min under a nitrogen atmosphere with a nitrogen gas flow of 50 mL·min^{-1}. The temperature at which the degradation rate was at its maximum (T_{max}) was evaluated analyzing the TGA equivalent derivate (DTG).

3. Results and Discussion

3.1. Cellulosic Fiber Production and Characterization

The chemical composition of the OTPB and the cellulosic pulp obtained is shown in Table 1. This composition is similar to that reported in previous work [39]. OTPB was subjected to a soda pulping process to facilitate the deconstruction of the cellulose fiber and the purification of the lignocellulosic components. The soda pulping process showed a yield of 32.0%, similar to other more polluting processes such as kraft pulping (33%) [40].

Table 1. Chemical composition of olive tree pruning biomass and cellulosic pulp.

	Ext. EtOH (%)	Ext. AQ (%)	Ashes (%)	Lignin (%)	Hemicellulose (%)	α-Cellulose (%)
Olive tree pruning biomass	10.11 ± 0.74	6.2 ± 0.46	1.20 ± 0.04	21.80 ± 1.10	25.70 ± 0.47	41.41 ± 0.76
Olive cellulosic pulp	1.18 ± 0.02	0.02 ± 0.01	2.20 ± 0.01	14.60 ± 0.52	25.68 ± 0.08	59.67 ± 0.02

As can be observed, the non-structural elements (Ext. EtOH and Ext. AQ) were drastically reduced after the pulping process. In addition, the lignin content in the fiber was reduced to 14.6%. On the other hand, the cellulosic fraction was purified and concentrated to almost 60% (similar to the value achieved by the kraft process) [40]. The hemicellulose content is a key parameter in the effectiveness of the nanofibrillation process. This component acts as a hydrated steric barrier to microfibril aggregation, preventing the re-agglomeration of the delaminated fiber. Chacker et al. [41] analyzed the role of hemicelluloses in the nanofibrillation process, determining that a hemicellulose content of about 25% in fiber is the ideal value to obtain the maximum efficiency in nanofibrillation. In pulps with a 12% of hemicellulose content, the fibrillation yield decreases by half in comparison with higher hemicellulose content pulps. The OTPB pulp obtained in this work retains most of the hemicelluloses present in the initial raw material, showing a content of 25.68%, higher than that in the OTBP kraft pulp studied in previous work [37]. Compared to other cellulosic pulps successfully used in the production of cellulose nanofibers, OTPB showed a higher hemicellulose content than *Eucalyptus* kraft pulp (19.40%), kraft pine pulp (14%) and other agricultural residues such as corn (20%), wheat (23.30%), barley (18.30%), oat (16.40%), banana leaves (20.28%), tomato (11%) and lime residues (10%), oil palm empty fruit bunches (22%) and Brazilian satintail plants (9%) [40,42–46]. It is therefore concluded that the pulping process carried out produces cellulose pulp with an optimum chemical composition for the production of cellulose nanofibers.

The cellulosic pulp obtained was characterized in terms of thermal stability and crystallinity. Figure 1a shows the thermal degradation behavior of the OTBP cellulosic pulp. The OTBP pulp showed a multi-step degradation process by the presence of several components such as lignin, hemicellulose and cellulose that are degraded at different temperatures in the range studied. The initial weight loss step in the region of 30–120 °C is associated with the evaporation of the absorbed and bound water in the fiber. The thermal degradation in the temperature range 120–350 °C is related to the breaking of glycosidic bonds, the pyrolysis of polysaccharides and the depolymerization of lignin, hemicellulose and cellulose. In the last region at 350–600 °C, the weight loss is due to the pyrolysis of cellulose fibers and the remaining carbonaceous residue [47]. The DTG peak shows that the temperature of the maximum degradation (T_{max}) of fiber is observed at 348 °C. Figure 1b shows the X-ray diffraction patterns of the fiber structure. It is possible to observe that it presents two major diffractions peaks at 2θ = 16.1° and 22.5° corresponding to the 110 and 200 reflection planes of cellulose I's structure.

The crystallinity index of the cellulosic fiber can be calculated by comparing the reflection intensity of the peak at 22.5° (crystalline region) and the valley region between the two peaks associated with the amorphous region [38]. The CI observed for the OTPB pup was 60.26%. Considering that the only lignocellulosic element that can present crystallinity is α-cellulose, it is deduced that all of the cellulose present in the fiber (59.67 ± 0.02) shows a crystalline disposition, compared to the amorphous elements, hemicellulose and lignin, which do not provide crystallinity to the sample.

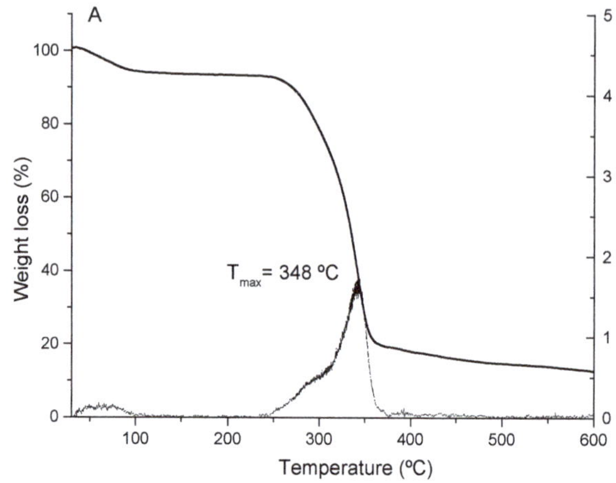

(**a**)

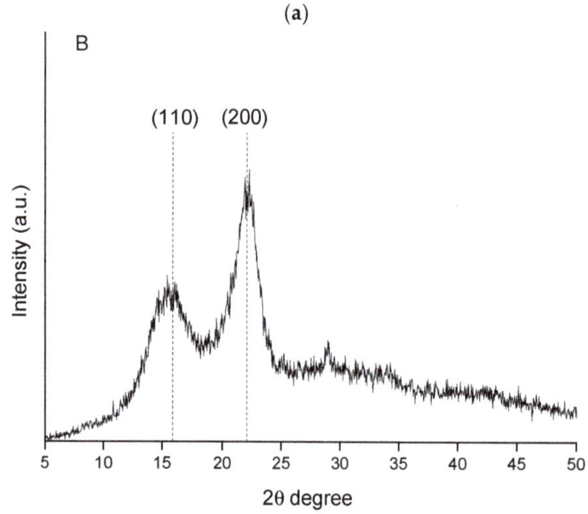

(**b**)

Figure 1. (**a**) Thermogravimetric analysis (TGA) and TGA equivalent derivate (DTG) curves and (**b**) the XRD pattern of OTBP pulp.

3.2. Cellulose Nanofiber Isolation and Characterization

The OTPB pulp was bleached to eliminate the lignin content while maintaining the carbohydrates in the fiber (hemicellulose and cellulose) and thus study the effect of lignin on the production of cellulose nanofibers through different pretreatments. Unbleached and bleached pulp were used for the production of lignocellulose nanofibers (LCNF) and cellulose nanofibers (CNF), respectively, through two different pretreatments, mechanical (Mec) and TEMPO-mediated oxidation (TO). The characterization of the different cellulose nanofibers in terms of nanofibrillation yield, transmittance, cationic demand, carboxyl content and morphology is shown in Table 2.

Table 2. Lignocellulose nanofiber and cellulose nanofiber characterization.

Sample	η (%)	T_{800} (%)	CD (μeq/g)	CC (μmols/g)	σ (m²/g)	Width (nm)	Length (nm)
LCNF-Mec	15.33 ± 0.47	9.12	253.33 ± 18.64	150.72 ± 15.17	49.97	50	4671
LCNF-TO	17.98 ± 0.89	13.74	223.85 ± 18.62	152.43 ± 6.63	34.78	71	1478
CNF-Mec	13.34 ± 0.02	18.27	240.06 ± 18.86	147.83 ± 3.63	44.78	55	3331
CNF-TO	26.44 ± 4.15	50.59	521.27 ± 9.33	311.95 ± 19.02	101.93	24	705

η: nanofibrillation yield; T_{800}: optical transmittance; CD: cationic demand; CC: carboxyl content; σ: specific surface area.

The nanofibrillation yield (η) for the LCNF and CNF ranges from 13.34% to 26.44%. These low yields in comparison with those for cellulose nanofibers obtained by enzymatic hydrolysis or the TEMPO-mediated oxidation of CNF from fully bleached wood pulp show that the obtained suspension is composed of cellulose nanofibers with large widths and cellulose microfibers [48,49]. The optical transmittance (T800) of the cellulose nanofiber suspension is an indirect indicator of the nanofibrillation yield. The cellulose microfibers contained in the suspension produce a higher light scattering compared to the nanofibers, so this parameter is highly related to yield and nanometric width. As with nanofibrillation yield, only slight differences in T800 are observed between the various nanofibers, except for CNF-TO. Since the chemical compositions of LCNF and CNF are different, the transmittance of the suspensions should not be considered as a key parameter in the characterization of the suspensions since lignin affects the refractive index. It is observed that CNF-TO presents a higher transmittance due to the fact that it presents a significantly higher nanofibrillation yield than the rest of nanofibers, and in addition, it does not contain lignin in its composition.

Cationic demand (CD) refers to the ability of the anionic surface of nanofibers to capture and interact with cationic substances. This value is highly related to the specific surface of the nanofiber; the larger the surface, the greater the capacity for interaction and the carboxyl content on that surface. The values of both parameters for LCNF-Mec, LCNF-TO and CNF-Mec are similar or even higher than what has been reported in the literature for CNF obtained by mechanical pretreatment or TEMPO-mediated oxidation from fibers with high lignin content [34,50–53]. It is observed again that there are not great differences in the cationic demand and carboxyl content, except for CNF-TO. In CNF-TO, TEMPO-mediated oxidation is much more effective that when it is performed on LCNF, as revealed by the increase in carboxyl content. It is observed that CNF-TO increases the carboxyl content by more than double compared to CNF-Mec. This increase is produced by the conversion of hydroxyl groups at the C6 positions on the surface of the cellulose fibers into carboxyl groups, enabling the delamination of the fiber by the electrostatic repulsion of the charged fiber surface [54]. On the other hand, in the case of LCNF, differences between both pretreatments are negligible. The presence of lignin in the fiber can affect the effectiveness of the oxidation reaction because the reaction activator, NaClO, is also consumed as the bleaching agent, producing the oxidation and dissolution of the lignin, thus preventing the selective activation of the catalyst. In fibers previously reported in literature with a lignin content lower than 10%, a partial oxidation of the -OH groups of the cellulose is produced, reaching maximum values of 300 μmol/g, higher than those reached for LCNF-TO described in this work (152.34 μmols/g), but not as high as those obtained for bleached wood pulps that can reach 1000 μmol/g [34,49,52].

The specific surface values, again, show differences in CNF-TO, which shows a considerably higher result than the other cellulose nanofibers. This is a very important parameter when using cellulose nanofibers as a reinforcing agent in materials produced from lignocellulosic materials such as paper, cardboard or fiberboards [12,55]. A larger specific surface area allows for a higher bonding capacity with adjacent fibers, thus improving the mechanical properties of the final product. Cellulose nanofibers with similar specific surface areas produce an increase about 100% in the mechanical properties of paper and carboard with low amounts of LCNF addition (3%) [53].

Nanofiber width, despite being within the nanometric range (24–71 nm), presents some differences that are discussed. For mechanical pretreatment, the presence of lignin in the fiber (LCNF-Mec) produces greater fibrillation in the fiber, producing a smaller width than CNF-Mec. This could be due to the lignin antioxidant action that prevents the re-bonding of the covalent bonds broken during the mechanical treatments [56]. Regarding TEMPO-mediated oxidation, differences are shown with the presence of lignin, being adverse because of the effect explained above. The length of nanofibers is an important parameter when analyzing the suitability of the application of cellulose nanofibers. The lignin content can affect the effectiveness of the method used for length determination through intrinsic viscosity. However, this method allows an estimation of the effect of the different pretreatments on the length parameter. In a generalized way, decreases in the lengths were observed when the fiber was subjected to TEMPO-mediated oxidation of 68.4% and 78.8% for LCNF-TO and CNF-TO, respectively, with respect to those following mechanical pretreatment. It is caused by the degradation of the cellulose amorphous regions into gluconic acid or cellulose-derived small fragments by depolymerization and β-elimination [57]. The length of the nanofibers is strongly related to the mechanical properties of the final composites made of cellulose nanofibers. It is therefore necessary to achieve a balance between the nanometric size reached during the nanofibrillation process and the shortening of the fiber due to its degradation. The aspect ratio (L/D) is a parameter that shows the relationship between length and width. It is observed that the different cellulose nanofibers showed aspect ratios of 93.44, 20.82, 60.56 and 29.38 for LCNF-Mec, LCNF-TO, CNF-Mec and CNF-TO, respectively. It is shown that although the mechanically pretreated nanofibers present a higher width than CNF-TO, they had a higher aspect ratio due to the low degradation that they underwent in the production process. Therefore, even though CNF-TO has a larger specific surface area and is thus more suitable for application in products made from lignocellulosic material (paper, cardboard, etc.), LCNF-Mec and CNF-Mec would show better behavior when added as a reinforcing agent on polymeric matrices [58].

The chemical composition of the different cellulose nanofibers was analyzed by a FTIR technique (Figure 2). All analyzed samples, as expected, show a spectrum typical of lignocellulosic materials. The peaks at 3300 and 2900 cm^{-1} are associated with the stretching vibration of the OH and CH groups present in the cellulose chains. The peaks in the range of 1350–1250 cm^{-1} are attributed to the presence of chemical groups of the hemicelluloses. The peaks at 1190, 1070 and 890 cm^{-1} are associated with the stretching and rocking vibrations of the C-O, C-H and CH_2 groups of cellulose [52]. However, there are some differences between the various cellulose nanofibers. It is observed that cellulose nanofibers obtained from OTPB bleached pulp (CNF-Mec and CNF-TO) do not show the peak at 1510 cm^{-1} that is observed in lignocellulose nanofibers (LCNF). This peak is related to the C=C symmetrical stretching of the aromatic rings, characteristic of the lignin. As expected, due to the nearly total elimination of lignin content in the bleached pulp, this peak is not observed in CNF. Another difference is observed in the peak at 1610 cm^{-1}, corresponding to the C = O stretching vibration in carboxyl groups. An important increase in the intensity of the peak is observed in the CNF-TO due to the regioselective conversion of C6 primary hydroxyl groups to carboxyl groups by the TEMPO-mediated oxidation.

The effects of the different pretreatments on the crystallinity of the cellulose nanofibers are shown in Figure 3. It is observed that the same peaks related to the 110 and 200 reflection planes of cellulose I are observed again, implying that the crystalline structure of the original fiber is maintained. The crystallinity index (CI) was calculated in the same way as for OTPB pulp. It shows that cellulose nanofibers present a lower CI (24%–49%) than the original fiber (60.26%). With regards to nanofibers obtained by mechanical pretreatment, they are produced by the disordering of the crystalline regions of the cellulose chain by the shear forces produced in the high-pressure homogenization process and during mechanical pretreatment. For TEMPO-mediated oxidized nanofibers, they are produced by the conversion of ordered cellulose structures into disordered structures by the sodium glucuronosyl units during the oxidation reaction [59]. CNF is observed to have a greater crystallinity than LCNF. This is due to the lignin elimination during the bleaching process and the consequent elimination of the amorphous component of the lignocellulose matrix, increasing the total crystallinity of the fiber. In addition, it is observed that mechanical pretreatment produced a greater disordering in the cellulose chain than the TEMPO-mediated oxidation.

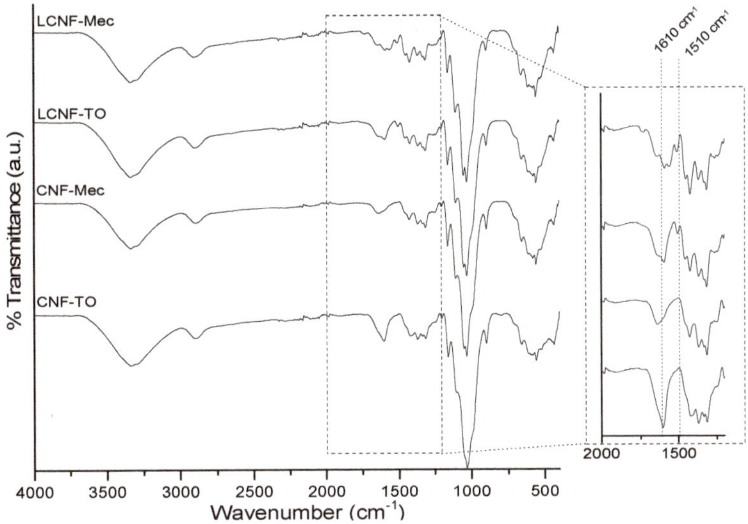

Figure 2. FTIR spectra of the different cellulose nanofibers.

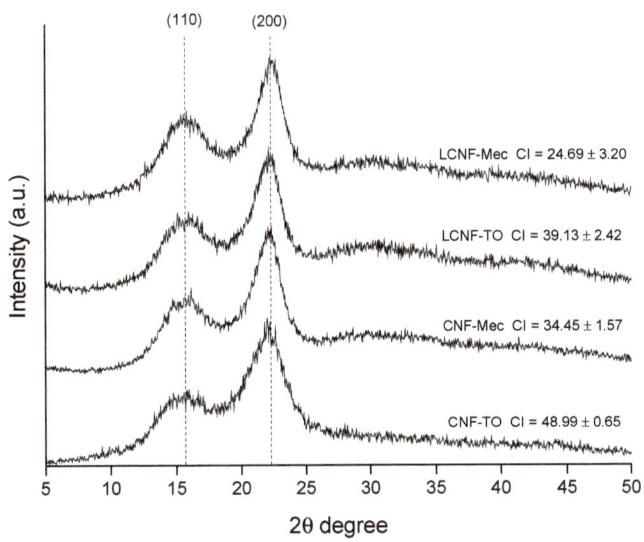

Figure 3. XRD diffraction patterns and crystallinity indices of the cellulose nanofibers.

The thermal stability of the different cellulose nanofibers was studied through the analysis of the TGA and DTG curves (Figure 4). The thermal degradation behavior shows three degradation stages observed in the initial fiber: (i) moisture loss, (ii) glycosidic bond degradation and (iii) cellulose pyrolysis. It is observed that LCNF (Figure 4a) and CNF (Figure 4b) present lower values for maximum thermal degradation, i.e., a lower T_{max}, than that obtained for OTPB pulp (348 °C). This is due to the larger specific surface of the nanometric-size fibers, which means that they are more exposed to heat, and degradation occurs more quickly than in the original fiber. It can be seen that for cellulose nanofibers obtained by mechanical pretreatment, there are no differences according to the presence or absence of lignin, both showing a T_{max} = 343 °C. However, analyzing the total mass loss, it is observed that a residual mass at 600 °C of 15.14% remains for LCNF-Mec compared to 8.95% for the CNF-Mec. This fact is not indicative of a higher thermal stability, but it indicates that a greater carbonaceous residue is produced after the pyrolysis of the lignocellulosic components due to the aromatic structure of lignin. Regarding the cellulose nanofibers obtained by TEMPO-mediated oxidation, noticeable differences are observed, showing maximum degradations at 325 °C and 298 °C for LCNF-TO and CNF-TO, respectively. CNF-TO presents worse thermal stability than LCNF-TO and the products obtained by mechanical pretreatment, since in addition to the nanometric size, it has a greater number of free ends (higher cationic demand and carboxyl content), which favors thermal degradation [60]. In addition, contrarily to what has been observed in the nanofibers obtained by mechanical pretreatment, a large increase in the residual mass was produced in CNF-TO (29.76%) in comparison with the values obtained for LCNF-TO (16.54%). This fact is produced by the introduction of carboxyl groups on the surface of the fiber during TEMPO-mediated oxidation, increasing the carboxyl content, especially for CNF-TO (311.95 μmol/g) as observed through its characterization. It is therefore concluded that CNF-Mec and LCNF-Mec, in addition to presenting higher aspect ratios that can result in greater reinforcement effects in polymeric matrices, can be used in polymers with higher transition temperatures compared to CNF-TO and LCNF-TO due to their greater thermal stability.

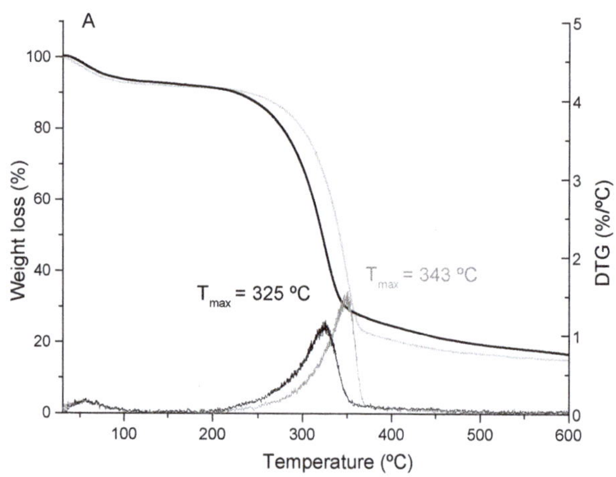

(a)

Figure 4. *Cont.*

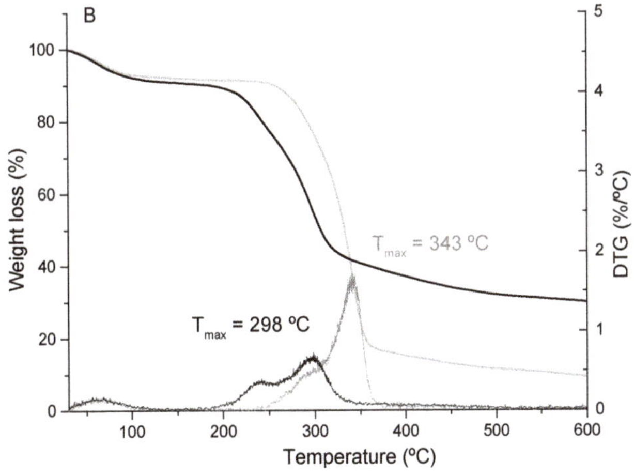

(b)

Figure 4. TGA and DTG curves for the different cellulose nanofibers: (**a**) lignocellulose nanofibers (LCNF) and (**b**) cellulose nanofibers (CNF). Black curves for those obtained by TEMPO-mediated oxidation and grey from mechanical pretreatment.

4. Conclusions

Olive tree pruning biomass (OTPB) was identified as a lignocellulosic source for the production of cellulose nanofibers from cellulosic pulps obtained by a sustainable pulping process. The cellulose nanofibers were produced by two different pretreatments, mechanical and TEMPO-mediated oxidation, followed by high-pressure homogenization. The influence of the residual lignin content on the effectiveness of the different pretreatments was analyzed by the thorough characterization of the cellulose nanofibers produced. All the cellulose nanofibers produced in this work were in the nanometric range; however, important differences were observed. TEMPO-mediated oxidation is more effective with bleached pulp; however, mechanical pretreatment was favored by the presence of lignin. The presence of lignin results in cellulose nanofibers with low crystallinity indices following mechanical pretreatment and TEMPO-mediated oxidation (24.69% and 39.13%) in comparison with those produced from bleached nanofibers (39.13% and 48.99%). The thermal stability of the cellulose nanofibers produced by mechanical treatment shows similar values regardless of the presence of lignin (343 °C); however, in the TEMPO-mediated oxidation, the lignin content produces a greater thermal stability (325 °C) in comparison with that obtained with bleached nanofibers (298 °C). The characteristics of the cellulose nanofibers obtained are of great interest for their application in different sectors.

Supplementary Materials: The following are available online at http://www.mdpi.com/2073-4395/10/5/696/s1, Figure S1: Cellulose pulp images.

Author Contributions: The article was conceptualized by A.R. and F.P.-R. The experimental work was realized by M.S.-G., I.B.-V. and E.C. The article was written by E.E. The article was reviewer by A.R. All authors have read and agreed to the published version of the manuscript.

Funding: The authors are grateful to Spain's DGICyT, MICINN for funding this research within the framework of the Project CTQ2016-78729-R.

Acknowledgments: The authors wish to acknowledge the support of the Central Service for Research Support (SCAI) at the university of Córdoba.

Conflicts of Interest: The authors declare no conflict of interest

References

1. Gurría, P.; Ronzon, T.; Tamosiunas, S.; López, R.; García Condado, S.; Guillén, J.; Cazzaniga, N.E.; Jonsson, R.; Banja, M.; Fiore, G.; et al. *Biomass Flows in the European Union*; Publications Office of the European Union: Luxembourg, Luxembourg, 2017.
2. Food and Agriculture Organization of the United Nations (FAO). 2018 FAOSTAT. Available online: http://www.fao.org/faostat/en/#data (accessed on 2 April 2020).
3. Miranda, I.; Simões, R.; Medeiros, B.; Nampoothiri, K.M.; Sukumaran, R.K.; Rajan, D.; Pereira, H.; Ferreira-Dias, S. Valorization of lignocellulosic residues from the olive oil industry by production of lignin, glucose and functional sugars. *Bioresource Technol.* **2019**, *292*, 121936. [CrossRef] [PubMed]
4. Fernandes-Klajn, F.; Romero-García, J.M.; Díaz, M.J.; Castro, E. Comparison of fermentation strategies for ethanol production from olive tree pruning biomass. *Ind. Crop. Prod.* **2018**, *122*, 98–106. [CrossRef]
5. Santos, J.I.; Fillat, Ú.; Martín-Sampedro, R.; Eugenio, M.E.; Negro, M.J.; Ballesteros, I.; Rodríguez, A.; Ibarra, D. Evaluation of lignins from side-streams generated in an olive tree pruning-based biorefinery: Bioethanol production and alkaline pulping. *Int. J. Biol. Macromol.* **2017**, *105*, 238–251. [CrossRef]
6. Suardi, A.; Latterini, F.; Alfano, V.; Palmieri, N.; Bergonzoli, S.; Pari, L. Analysis of the Work Productivity and Costs of a Stationary Chipper Applied to the Harvesting of Olive Tree Pruning for Bio-Energy Production. *Energies* **2020**, *13*, 1359. [CrossRef]
7. Liuzzi, S.; Rubino, C.; Martellotta, F.; Stefanizzi, P.; Casavola, C.; Pappalettera, G. Characterization of biomass-based materials for building applications: The case of straw and olive tree waste. *Ind. Crop. Prod.* **2020**, *147*, 112229. [CrossRef]
8. Mutjé, P.; Pèlach, M.A.; Vilaseca, F.; García, J.C.; Jiménez, L. A comparative study of the effect of refining on organosolv pulp from olive trimmings and kraft pulp from eucalyptus wood. *Bioresource Technol.* **2005**, *96*, 1125–1129. [CrossRef]
9. García, A.; Gandini, A.; Labidi, J.; Belgacem, N.; Bras, J. Industrial and crop wastes: A new source for nanocellulose biorefinery. *Ind. Crop. Prod.* **2016**, *93*, 26–38. [CrossRef]
10. Balea, A.; Fuente, E.; Monte, C.M.; Merayo, N.; Campano, C.; Negro, C.; Blanco, A. Industrial Application of Nanocelluloses in Papermaking: A Review of Challenges, Technical Solutions, and Market Perspectives. *Molecules* **2020**, *25*, 526. [CrossRef]
11. Inshakova, E.; Inshakov, O. World market for nanomaterials: Structure and trends. *MATEC Web Conf.* **2017**, *129*, 02013. [CrossRef]
12. Boufi, S.; González, I.; Delgado-Aguilar, M.; Tarrès, Q.; Pèlach, M.À.; Mutjé, P. Nanofibrillated cellulose as an additive in papermaking process: A review. *Carbohyd. Polym.* **2016**, *154*, 151–166. [CrossRef]
13. Tayeb, P.H.; Tayeb, A. Nanocellulose applications in sustainable electrochemical and piezoelectric systems: A review. *Carbohyd. Polym.* **2019**, *224*, 115149. [CrossRef] [PubMed]
14. Hsu, H.H.; Zhong, W. Nanocellulose-Based Conductive Membranes for Free-Standing Supercapacitors: A Review. *Membranes* **2019**, *9*, 74. [CrossRef] [PubMed]
15. Bongao, H.C.; Gabatino, R.R.A.; Arias, C.F.H.; Magdaluyo, E.R. Micro/nanocellulose from waste Pili (Canarium ovatum) pulp as a potential anti-ageing ingredient for cosmetic formulations. *Mater. Today-Proc.* **2020**, *22*, 275–280. [CrossRef]
16. Klemm, D.; Cranston, E.D.; Fischer, D.; Gama, M.; Kedzior, S.A.; Kralisch, D.; Kramer, F.; Kondo, T.; Lindström, T.; Nietzsche, S.; et al. Nanocellulose as a natural source for groundbreaking applications in materials science: Today's state. *Mater. Today* **2018**, *21*, 720–748. [CrossRef]
17. Putro, J.N.; Kurniawan, A.; Ismadji, S.; Ju, Y.-H. Nanocellulose based biosorbents for wastewater treatment: Study of isotherm, kinetic, thermodynamic and reusability. *Environ. Nanotechnol. Monit. Manage.* **2017**, *8*, 134–149. [CrossRef]
18. Zhang, C.; Feng, X.; Wang, B.; Mao, Z.; Xu, H.; Zhong, Y.; Zhang, L.; Chen, X.; Sui, X. Nanocellulose sponges as efficient continuous flow reactors. *Carbohyd. Polym.* **2019**, *224*, 115184. [CrossRef]
19. Barnat-Hunek, D.; Grzegorczyk-Frańczak, M.; Szymańska-Chargot, M.; Łagód, G. Effect of Eco-Friendly Cellulose Nanocrystals on Physical Properties of Cement Mortars. *Polymers* **2019**, *11*, 2088. [CrossRef]
20. Liu, Y.; Sui, Y.; Liu, C.; Liu, C.; Wu, M.; Li, B.; Li, Y. A physically crosslinked polydopamine/nanocellulose hydrogel as potential versatile vehicles for drug delivery and wound healing. *Carbohyd. Polym.* **2018**, *188*, 27–36. [CrossRef]

21. Moohan, J.; Stewart, A.S.; Espinosa, E.; Rosal, A.; Rodríguez, A.; Larrañeta, E.; Donnelly, F.R.; Domínguez-Robles, J. Cellulose Nanofibers and Other Biopolymers for Biomedical Applications. A Review. *Appl. Sci.* **2019**, *10*, 65. [CrossRef]
22. Dufresne, A. *Nanocellulose, from Nature to High Performance Tailored Materials;* Walter de Gruyter GmbH.: Berlin, Germany, 2012.
23. Espinosa, E.; Tarres, Q.; Delgado-Aguilar, M.; Gonzalez, I.; Mutje, P.; Rodriguez, A. Suitability of wheat straw semichemical pulp for the fabrication of lignocellulosic nanofibres and their application to papermaking slurries. *Cellulose* **2016**, *23*, 837–852. [CrossRef]
24. Rol, F.; Karakashov, B.; Nechyporchuk, O.; Terrien, M.; Meyer, V.; Dufresne, A.; Belgacem, M.N.; Bras, J. Pilot-Scale Twin Screw Extrusion and Chemical Pretreatment as an Energy-Efficient Method for the Production of Nanofibrillated Cellulose at High Solid Content. *ACS Sustain. Chem. Eng.* **2017**, *5*, 6524–6531. [CrossRef]
25. Wang, W.; Mozuch, M.D.; Sabo, R.C.; Kersten, P.; Zhu, J.Y.; Jin, Y. Production of cellulose nanofibrils from bleached eucalyptus fibers by hyperthermostable endoglucanase treatment and subsequent microfluidization. *Cellulose* **2015**, *22*, 351–361. [CrossRef]
26. Espinosa, E.; Rol, F.; Bras, J.; Rodríguez, A. Production of lignocellulose nanofibers from wheat straw by different fibrillation methods. Comparison of its viability in cardboard recycling process. *J. Clean. Prod.* **2019**, *239*, 118083. [CrossRef]
27. Espinosa, E.; Dominguez-Robles, J.; Sanchez, R.; Tarres, Q.; Rodriguez, A. The effect of pre-treatment on the production of lignocellulosic nanofibers and their application as a reinforcing agent in paper. *Cellulose* **2017**, *24*, 2605–2618. [CrossRef]
28. Tarres, Q.; Saguer, E.; Pelach, M.A.; Alcala, M.; Delgado-Aguilar, M.; Mutje, P. The feasibility of incorporating cellulose micro/nanofibers in papermaking processes: The relevance of enzymatic hydrolysis. *Cellulose* **2016**, *23*, 1433–1445. [CrossRef]
29. Saito, T.; Kimura, S.; Nishiyama, Y.; Isogai, A. Cellulose Nanofibers Prepared by TEMPO-Mediated Oxidation of Native Cellulose. *Biomacromolecules* **2007**, *8*, 2485–2491. [CrossRef]
30. Rol, F.; Saini, S.; Meyer, V.; Petit-Conil, M.; Bras, J. Production of cationic nanofibrils of cellulose by twin-screw extrusion. *Ind. Crop. Prod.* **2019**, *137*, 81–88. [CrossRef]
31. Jiménez, L.; Pérez, I.; de la Torre, J.; García, J.C. The effect of processing variables on the soda pulping of olive tree wood. *Bioresource Technol.* **1999**, *69*, 95–102. [CrossRef]
32. Martin-Sampedro, R.; Rodríguez, A.; Requejo, A.; Eugenio, M.E. Improvement of TCF bleaching of olive tree pruning residue pulp by addition of a laccase and/or xylanase pre-treatment. *Bioresources* **2012**, *7*, 1488–1505. [CrossRef]
33. Espinosa, E.; Bascón-Villegas, I.; Rosal, A.; Pérez-Rodríguez, F.; Chinga-Carrasco, G.; Rodríguez, A. PVA/(ligno)nanocellulose biocomposite films. Effect of residual lingin content on the structural, mechanical, barrier and antioxidant properties. *Int. J. Biol. Macromol.* **2019**, *141*, 197–206. [CrossRef]
34. Espinosa, E.; Sánchez, R.; González, Z.; Domínguez-Robles, J.; Ferrari, B.; Rodríguez, A. Rapidly growing vegetables as new sources for lignocellulose nanofibre isolation: Physicochemical, thermal and rheological characterisation. *Carbohyd. Polym.* **2017**, *175*, 27–37. [CrossRef] [PubMed]
35. Besbes, I.; Alila, S.; Boufi, S. Nanofibrillated cellulose from TEMPO-oxidized eucalyptus fibres: Effect of the carboxyl content. *Carbohyd. Polym.* **2011**, *84*, 975–983. [CrossRef]
36. Marx-Figini, M. The acid-catalyzed degradation of cellulose linters in distinct ranges of degree of polymerization. *J. Appl. Polym. Sci.* **1987**, *33*, 2097–2105. [CrossRef]
37. Shinoda, R.; Saito, T.; Okita, Y.; Isogai, A. Relationship between Length and Degree of Polymerization of TEMPO-Oxidized Cellulose Nanofibrils. *Biomacromolecules* **2012**, *13*, 842–849. [CrossRef] [PubMed]
38. Segal, L.; Creely, J.J.; Martin, A.E.; Conrad, C.M. An empirical method for estimating the degree of crystallintiy of native cellulose using X-ray diffractometer. *Text. Res. J.* **1959**, *29*, 786–974. [CrossRef]
39. Requejo, A.; Rodríguez, A.; Colodette, J.L.; Gomide, J.L.; Jiménez, L. Optimization of ECF bleaching and refining of kraft pulping from olive tree pruning. *BioResources* **2012**, *7*, 4046–4055.
40. Fillat, Ú.; Wicklein, B.; Martín-Sampedro, R.; Ibarra, D.; Ruiz-Hitzky, E.; Valencia, C.; Sarrión, A.; Castro, E.; Eugenio, M.E. Assessing cellulose nanofiber production from olive tree pruning residue. *Carbohyd. Polym.* **2018**, *179*, 252–261. [CrossRef]
41. Chaker, A.; Alila, S.; Mutje, P.; Vilar, M.R.; Boufi, S. Key role of the hemicellulose content and the cell morphology on the nanofibrillation effectiveness of cellulose pulps. *Cellulose* **2013**, *20*, 2863–2875. [CrossRef]

42. Rajan, K.; Djioleu, A.; Kandhola, G.; Labbé, N.; Sakon, J.; Carrier, D.J.; Kim, J.-W. Investigating the effects of hemicellulose pre-extraction on the production and characterization of loblolly pine nanocellulose. *Cellulose* **2020**. [CrossRef]
43. Kassab, Z.; Kassem, I.; Hannache, H.; Bouhfid, R.; Qaiss, A.E.K.; El Achaby, M. Tomato plant residue as new renewable source for cellulose production: Extraction of cellulose nanocrystals with different surface functionalities. *Cellulose* **2020**. [CrossRef]
44. Jongaroontaprangsee, S.; Chiewchan, N.; Devahastin, S. Production of nanocellulose from lime residues using chemical-free technology. *Mater. Today-Proc.* **2018**, *5*, 11095–11100. [CrossRef]
45. Septevani, A.A.; Rifathin, A.; Sari, A.A.; Sampora, Y.; Ariani, G.N.; Sudiyarmanto; Sondari, D. Oil palm empty fruit bunch-based nanocellulose as a super-adsorbent for water remediation. *Carbohyd. Polym.* **2020**, *229*, 115433. [CrossRef] [PubMed]
46. Benini, K.C.C.d.C.; Voorwald, H.J.C.; Cioffi, M.O.H.; Rezende, M.C.; Arantes, V. Preparation of nanocellulose from Imperata brasiliensis grass using Taguchi method. *Carbohyd. Polym.* **2018**, *192*, 337–346. [CrossRef] [PubMed]
47. Alemdar, A.; Sain, M. Isolation and characterization of nanofibers from agricultural residues - Wheat straw and soy hulls. *Bioresource Technol.* **2008**, *99*, 1664–1671. [CrossRef]
48. Delgado-Aguilar, M.; González Tovar, I.; Tarrés, Q.; Alcalá, M.; Pèlach, M.À.; Mutjé, P. Approaching a Low-Cost Production of Cellulose Nanofibers for Papermaking Applications. *BioResources* **2015**, *10*, 5435–5455. [CrossRef]
49. González, I.; Boufi, S.; Pèlach, M.A.; Alcalà, M.; Vilaseca, F.; Mutjé, P. Nanofibrillated cellulose as papper additive in eucalyptus pulp. *BioResources* **2012**, *7*, 5167–5180. [CrossRef]
50. Vallejos, M.E.; Felissia, F.E.; Area, M.C.; Ehman, N.V.; Tarrés, Q.; Mutjé, P. Nanofibrillated cellulose (CNF) from eucalyptus sawdust as a dry strength agent of unrefined eucalyptus handsheets. *Carbohyd. Polym.* **2016**, *139*, 99–105. [CrossRef]
51. Tarrés, Q.; Ehman, N.V.; Vallejos, M.E.; Area, M.C.; Delgado-Aguilar, M.; Mutjé, P. Lignocellulosic nanofibers from triticale straw: The influence of hemicelluloses and lignin in their production and properties. *Carbohyd. Polym.* **2017**, *163*, 20–27. [CrossRef]
52. Espinosa, E.; Sánchez, R.; Otero, R.; Domínguez-Robles, J.; Rodríguez, A. A comparative study of the suitability of different cereal straws for lignocellulose nanofibers isolation. *Int. J. Biol. Macromol.* **2017**, *103*, 990–999. [CrossRef]
53. Tarrés, Q.; Espinosa, E.; Domínguez-Robles, J.; Rodríguez, A.; Mutjé, P.; Delgado-Aguilar, M. The suitability of banana leaf residue as raw material for the production of high lignin content micro/nano fibers: From residue to value-added products. *Ind. Crop. Prod.* **2017**, *99*, 27–33. [CrossRef]
54. Saito, T.; Shibata, I.; Isogai, A.; Suguri, N.; Sumikawa, N. Distribution of carboxylate groups introduced into cotton linters by the TEMPO-mediated oxidation. *Carbohyd. Polym.* **2005**, *61*, 414–419. [CrossRef]
55. Theng, D.; Arbat, G.; Delgado-Aguilar, M.; Vilaseca, F.; Ngo, B.; Mutjé, P. All-lignocellulosic fiberboard from corn biomass and cellulose nanofibers. *Ind. Crop. Prod.* **2015**, *76*, 166–173. [CrossRef]
56. Solala, I.; Iglesias, M.C.; Peresin, M.S. On the potential of lignin-containing cellulose nanofibrils (LCNFs): A review on properties and applications. *Cellulose* **2019**. [CrossRef]
57. Sang, X.K.; Qin, C.R.; Tong, Z.F.; Kong, S.; Jia, Z.; Wan, G.C.; Liu, X.L. Mechanism and kinetics studies of carboxyl group formation on the surface of cellulose fiber in a TEMPO-mediated system. *Cellulose* **2017**, *24*, 2415–2425. [CrossRef]
58. Zimmermann, T.; Bordeanu, N.; Strub, E. Properties of nanofibrillated cellulose from different raw materials and its reinforcement potential. *Carbohyd. Polym.* **2010**, *79*, 1086–1093. [CrossRef]
59. Puangsin, B.; Yang, Q.L.; Saito, T.; Isogai, A. Comparative characterization of TEMPO-oxidized cellulose nanofibril films prepared from non-wood resources. *Int. J. Biol. Macromol.* **2013**, *59*, 208–213. [CrossRef]
60. Sharma, P.R.; Varma, A.J. Thermal stability of cellulose and their nanoparticles: Effect of incremental increases in carboxyl and aldehyde groups. *Carbohyd. Polym.* **2014**, *114*, 339–343. [CrossRef]

© 2020 by the authors. Licensee MDPI, Basel, Switzerland. This article is an open access article distributed under the terms and conditions of the Creative Commons Attribution (CC BY) license (http://creativecommons.org/licenses/by/4.0/).

agronomy

MDPI

Article

Techno-Economic Study of Castor Oil Crop Biorefinery: Production of Biodiesel without Fossil-Based Methanol and Lignoethanol Improved by Alkali Pretreatment

Vajiheh Rahimi [1], Marzieh Shafiei [2] and Keikhosro Karimi [1,3,*]

1 Department of Chemical Engineering, Isfahan University of Technology, Isfahan 84156-83111, Iran; v.rahimi@ce.iut.ac.ir

2 Department of Chemical Engineering, Vrije Universiteit Brussel, 1050 Brussels, Belgium; marzieh.shafiei@vub.be

3 Industrial Biotechnology Group, Research Institute for Biotechnology and Bioengineering, Isfahan University of Technology, Isfahan 84156-83111, Iran

* Correspondence: karimi@iut.ac.ir; Tel.: +98-3133935623

Received: 27 August 2020; Accepted: 1 October 2020; Published: 10 October 2020

check for updates

Abstract: Castor, a non-edible oil crop that flourishes even under extreme cultivation conditions, can be cultivated in wastewater with a lower cultivation cost than similar plants, e.g., rapeseed and soybean. This plant, containing seeds and lignocellulosic residues, has a promising perspective for biofuel production. The oil extracted from the seeds is inexpensive and can be efficiently converted to biodiesel, while the lignocellulosic parts are suitable for ethanol production after pretreatment with NaOH. Biodiesel typically produced from the fossil-based methanol; however, it can also be produced from the ethanol. In this study, ethanol used for biodiesel production is produced from the lignocellulosic residues (scenario 1), which are more sustainable and environmentally friendly; the process was compared with that of the methanol (scenario 2). In this study, techno-economic analyses were used to compare the technical and economic aspects of producing biodiesel from methanol and the produced ethanol. Simulations of the processes were carried out by Aspen plus software, and economic studies were conducted by Aspen Economic Analyzer. The prices of produced ethanol as a byproduct in scenarios 1 and 2 were USD 0.701 and 0.693 per liter, respectively, which are greater than that of gasoline. The prices of biodiesel produced as a primary product for scenarios 1 and 2 are USD 0.410 and 0.323/L, lower than the price of diesel in the Middle East region. The profitability indices for scenarios 1 and 2 are 1.29 and 1.41, respectively. Therefore, despite environmental benefits, the biorefinery based on producing biodiesel from methanol is more economically feasible than that produced from ethanol.

Keywords: biorefinery; techno-economic study; castor plant; biodiesel; bioethanol; alkali pretreatment

1. Introduction

To prevent environmental issues caused by the use of fossil fuels, research on renewable sources of energy has increased. In this regard, biodiesel, bioethanol, and biogas are among the promising alternatives [1]. Biodiesel—the monoalkyl ester of vegetable, algal, and animal oils—is nowadays employed as a replacement to diesel. It is produced from edible oils and non-edible oils, while non-edible oils are more sustainable and promising resources [2–5]. Among the renewable liquid biofuels, ethanol currently plays a major role as a blend for the improvement of gasoline properties. Ethanol production from lignocelluloses has received significant attention due to its low price and availability in large quantities without food and energy conflict [6,7].

Energy crops that are specifically cultivated for energy production are among the most important sources of energy. These plants are divided into three groups of oily, cellulosic, and sugar-rich plants. These plants, e.g., castor plant, are suggested to supply a major part of future energy in the form of liquid and gaseous biofuels [8].

The castor plant is among the oily crops, containing seeds with 47–49% (*w*/*w*) non-edible oil, mainly include ricinoleic acid [9]. The castor plant can be cultivated under different climate conditions and in wastewater. The cultivation cost of this plant is lower than the other oily plants, such as jatropha and rapeseed. Moreover, castor seeds and seed cake are restricted to be used as human and animal feeds, as they are highly poisonous [10]. The castor plant's residues, including stem, leaves, and seed processing residues (seed cake), are potential sources for bioethanol and biogas production and its oil is suitable for biodiesel production with high efficiency [9].

Biofuel production from lignocellulosic feedstocks and energy crops has different technical and economic bottlenecks. The biorefinery concept is suggested to address the process's drawbacks and make the bioenergy from lignocelluloses competitive with the current forms of energy. Biorefineries are referred to as facilities to produce biofuels, energy, and biochemicals from renewable feedstocks. Recently, the biorefinery based on using energy crops as feedstock received significant attention. This has been predicted to have a significant role in addressing climate change and decreasing dependence on fossil fuels in economically feasible pathways [11,12].

An essential factor in biorefinery development is the consumption versus produced energy. The energy produced in the form of heat, power, and liquid or gaseous fuels through the biorefinery should be higher than the energy consumption in different units of processes. Therefore, the energy analysis for these processes is very important. To predict the feasibility of the biorefinery, technical and economic studies are necessary. The technical and economic analyses are methods that identify the strengths and weaknesses of each process and show future plans and perspectives of biorefinery development [13].

In recent years, techno-economic analyses were used as a tool to evaluate the feasibility of biodiesel, bioethanol, and biogas production from lignocellulosic materials. The economics of biodiesel production from edible, non-edible, and waste oil by different plants showed that the most effective factors on biodiesel price and economic parameters are raw oil cost and plant scale [14–22]. The techno-economic analyses showed that the raw material cost, type of byproducts, plant scales, and tax policies in the varied area have the greatest effect on economic factors of ethanol production from lignocelluloses. In addition, these studies showed that bioethanol production in a biorefinery in large scales is a profitable state [13,23–25]. In another study, we evaluated the potential of biodiesel, biogas, and heat production from the castor plant. The results showed that biogas production from the lignocellulosic part is not economically feasible [26]. The experimental results also showed that ethanol with high efficiency could be obtained from the lignocellulosic residues of castor plant. This lignoethanol was then successfully used for the transesterification process for biodiesel production [9,27]. This biodiesel is produced from two renewable feedstocks, i.e., castor oil and bioethanol, and thus is much more eco-friendly compared with the biodiesel produced from the fossil-based methanol [28,29]. To our knowledge, no references were detected for techno-economic study for the biorefinery based on castor plant for the production of bioethanol and biodiesel from lignoethanol.

The potential of castor plant for biofuel production is approved experimentally [9,27]. Based on experimental data, the economy of biorefinery for biodiesel and bioethanol production based on castor plant was investigated for the first time in this study. Two different processes, i.e., biodiesel production using lignoethanol (scenario 1) and methanol (scenario 2), were studied. Aspen Plus was used for the simulation of processes, Aspen Process Economic Analyzer (APEA) was employed to evaluate the economic parameters, and sensitivity analyses were conducted to determine the effective factors.

2. Methods

Two scenarios for biodiesel and bioethanol production in a biorefinery based on castor plant using NaOH pretreatment were developed. These processes were designed based on the promising experimental data previously obtained [9,27]. Aspen Plus simulated all the processes, and then APEA evaluated the economic features.

2.1. Process Development

In the first scenario, castor oil is used to produce biodiesel by transesterification with ethanol that is produced from the lignocellulosic materials of castor plant, including leaves, stem, and seed cake. In the second scenario, castor oil is used to produce biodiesel by transesterification with methanol and the lignocellulosic materials used for bioethanol production similar to scenario 1.

2.1.1. Scenario 1: The Biorefinery for Biodiesel (from Oil and Ethanol), Bioethanol, and Heat Production

This process includes feed handling, transesterification, biodiesel purification, pretreatment, ethanol production via simultaneous saccharification and fermentation (SSF), purification of bioethanol by distillation, and fuel grade production of ethanol by dehydration, heat production, wastewater treatment (WWT), and utility services units. The block flow diagram (BFD) of this scenario is shown in Figure 1.

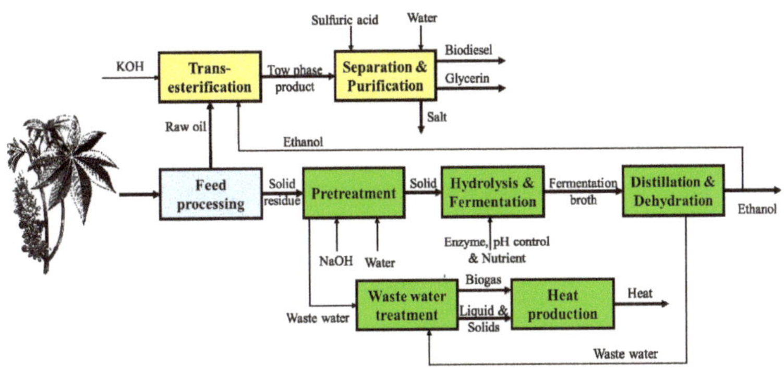

Figure 1. The block flow diagram (BFD) of biorefining of an environmentally friendly process for biodiesel and ethanol production (scenario 1).

In the harvest time, the seeds are separated, and seeds and residues were then sent to plant with trucks. In the feed handling unit, oil is extracted from the seeds by a cold mechanical press, filtered, kept for 24 h at 80 °C in the tank storage to remove extra moisture, and then sent to the transesterification reactor. Seed cake, stem, and leaves are crushed and dumped into storage piles until transfer to the pretreatment unit. Similar units for feed handling are considered for both scenarios [30].

Biodiesel is produced in a 25 m^3 ransesterification reactor at 62.5 °C for 3.46 h by using 1 g potassium hydroxide (KOH) per kg oil and 0.29:1 ethanol to oil mass ratio to obtain 85% biodiesel efficiency. The two-phase stream, i.e., fatty acid esters (biodiesel) rich and glycerol rich streams, is removed from the reactor [1,9,20,27,31].

In the separation unit, the excess alcohol (here ethanol) contents of both biodiesel and glycerin are first recovered. The ethanol recovery is carried out by using a three-phase distillation tower, containing 22 sieve trays with 80% efficiency and operates under vacuum pressure of 0.2 bar (absolute). The product is finally sent to the dehydration unit for water separation, and the bottom stream of the tower sends to the biodiesel purification unit [31].

To help the phase separation, water is added to the process in the centrifuge. The products of transesterification are contaminated by the remaining raw material (i.e., oil), ethanol, catalyst (KOH), and soap. Thus, according to ASTM D6751 or EN 14214, the produced biodiesel is purified to contain less than 0.05% *v/v* impurities. A distillation tower containing 22 trays (sieve trays, 80% efficiency, operates at 0.1 bar pressure) is used for biodiesel purification. The purified biodiesel (99.95% *v/v*) obtained from the distillate is sent to the storage tanks, and the unreacted oil that exits from the bottom is sent back to the transesterification reactor. The glycerin-rich stream is first neutralized by the sulfuric acid and then purified in a distillation column with 4 sieve trays (80% efficiency, 0.4 bar pressure). The process flow diagram of the biodiesel production unit is presented in Figure 2 [1,24].

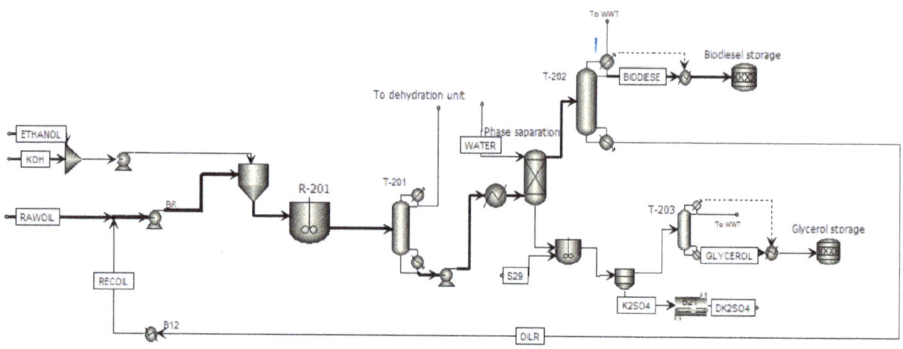

Figure 2. Process flow diagram of biodiesel production (transesterification and biodiesel purification units) (cf. Table 1 for process conditions).

Table 1. The process conditions for equipment for biodiesel production in scenario 1 in Figure 2.

Equipment/Conditions	Transesterification Reactor (R-201)	Distillation Tower 1 (T-201)	Distillation Tower 2 (T-202)	Distillation Tower 3 (T-203)
Input T (°C)	62.5	62.5	50	25
Upper output T (°C)	62.5	216.7	517	244
Lower output T (°C)	-	42.1	148.3	75
Input P (bar)	1	1	1	1
Upper output P (bar)	-	0.2	0.1	0.4
Lower output P (bar)	1	0.2	0.2	0.5
Number of theoretical stages	-	17	17	3

The pretreatment of castor residues is conducted with 8% *w/v* NaOH (100 °C, 60 min, 22% solid loading). The pretreated residues are then filtered and washed in a 9-stage counter current solid washer to minimize the freshwater consumption. Water containing the impurities is then sent to WWT unit.

SSF is used for ethanol production by hydrolysis of the pretreated solids (13% solid loading) with Cellic CTec3 enzyme (1.8% *w/w* of cellulose, Novozymes, Denmark [32]) and fermentation with *Saccharomyces cerevisiae* at 37 °C for 72 h under anaerobic conditions [25.33]. The nine main fermenters are designed with a volume of 350 m^3. The seed yeast is prepared in a series of fermenters (stainless steel 304), starting from 5 L to 300 m^3 final volume with 1:10 scale-up ratio. The beer (fermentation broth) stream, leaving the fermenters, contains 3.86% *v/v* ethanol, unfermented hexoses and pentoses, and lignin. The beer is stored in a 150 m^3 tank and then sent to distillation step.

The purification of ethanol is conducted using stripper, rectifier, and scrubber columns (Table 2). The produced carbon dioxide is separated in a flash drum stage, and degassed beer is preheated to enter the second stage of stripper (24 sieve trays, 80% efficiency, 0.2 absolute bar). The vapor side-draw from the third stage, comprising 40.17% *v/v* ethanol, is fed to the rectifier column (32 sieve trays, 80% efficiency, 1.8 absolute bar). This column produced ethanol in 94% *v/v* concentration in vapor form. The vent from the top of the beer column, as well as the vents from beer stream storage and fermenters, is sent to the third column that is a water scrubber. The scrubber is a simple packed column that recovered 99% of vented ethanol by washing with water. The ethanol is exiting from the bottom and returned to the beer column [33]. The solid containing the unconverted residues of the substrate, e.g., lignin, is presented in the wastewater that leaves the distillation. It is sent to the wastewater treatment and used for heat production.

Saturated vapor from the rectifier column, along with the recovered ethanol from biodiesel separation unit, is superheated and fed to the molecular sieve unit for purification to 99.9% *v/v* concentration of ethanol. This unit includes two adsorption columns. They are used alternatively in adsorbing and regenerating operation. The regeneration of the adsorption columns is performed by pure hot ethanol vapor. The final product, i.e., pure ethanol, is cooled and a part of it (40%) is sent to the biodiesel production unit, and the remaining ethanol pumped to storage to be sold as a fuel [33].

Table 2. The process conditions for the equipment of distillation and dehydration in Figure 3.

Equipment/Conditions	Beer Column	Rectifier Column	Scrubber Column	Dehydration Section
Input temperature (°C)	100	120	38.4	92.1
Highest output T (°C)	122.2	119.7	36.2	25
Lowest output T (°C)	115.7	92.1	36	-
Input P (bar)	4.8	2	1	1.7
Highest output P (bar)	1.9	1.7	0.9	1
Lowest output P (bar)	2.1	2	0.9	-
Number of theoretical stages	19	25	9	-

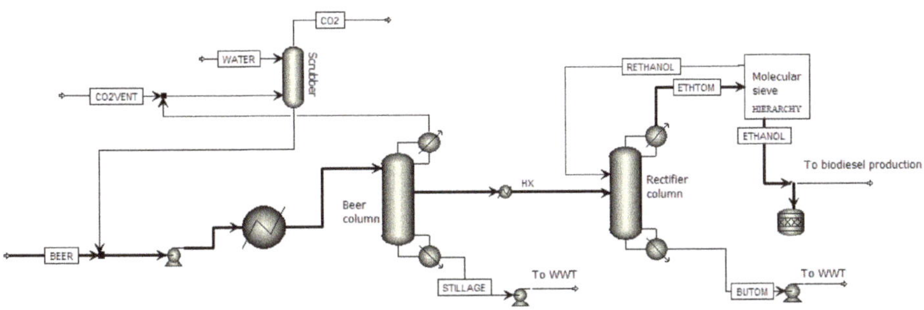

Figure 3. Process flow diagram of distillation and dehydration unit.

The bottom product of the rectifier column is pure water (>99% *w/w*) that is added to clean process water. The bottom product of the beer column, i.e., stillage, mainly contains non-fermentable sugars and lignin. Lignin is separated from this stream by filtration and is sent to the WWT unit. The liquor that is removed from the lignin separation step and the liquid effluent from the pretreatment unit is sent to the filtration and electrodialysis for the removal of dissolved compounds and NaOH that remained in the liquid. To reduce the COD and produce biogas, an anaerobic digester is used. The produced biogas consists of 44% carbon dioxide and 54% *v/v* biomethane. The rest of the organic materials are converted to CO_2, water, and sludge, using aerobic digestion. A clarifier is used to settle the sludge, and the water is treated and added to clean process water. Biogas, lignin, and dewatered sludge are combusted, and heat is produced as a valuable byproduct in the heat production unit.

2.1.2. Scenario 2: The Biorefinery for Biodiesel (from Oil and Methanol), Bioethanol, and Heat Production

The overview of the process used in scenario 2 is presented in Figure 4. The castor plant seed and residual handling, oil extraction, and ethanol production units are similar to scenario 1. However, the transesterification of oil and process parameters are different. In this process, methanol is used for the transesterification in 10 m^3 reactor (40 °C, 1.5 h, 1.5 g KOH/kg oil catalyst, 0.4:1 methanol to oil ratio). The excess alcohol is separated in the methanol recovery and recycled.

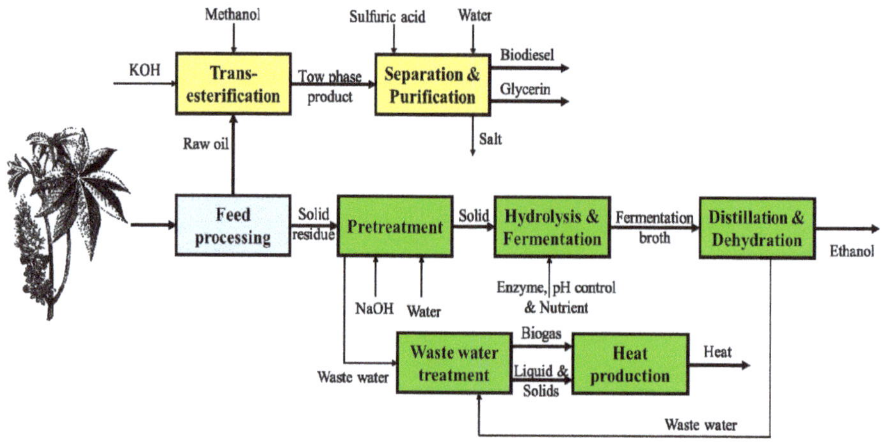

Figure 4. The BFD of biorefining of the fossil-based process for biodiesel and ethanol production (scenario 2).

2.2. Plant Location and Capacity

In this study, Iran is selected as a place for the biorefinery, where fossil fuel resources and methanol are widely available. However, the area has water shortages and is dealing with severe environmental impacts. Therefore, castor, as a low water demanding plant that can be cultivated in wastewater, is used. Considering the total world production of castor [9,27], a plant with the capacity of handling 120,000 tons of castor plant per year (based on the dry weight) is used.

2.3. Process Simulation and Economic Calculations

The main processes of the biorefineries are simulated by Aspen Plus to obtain the composition and properties of streams. The physical properties databank of NREL (National Renewable Energy Laboratory, Golden, CO, USA) is used for the definition of lignocellulosic components [34]. The sizing and cost of all equipment are estimated by Aspen PEA. The assumptions are: plant capacity, 120,000 tons/year castor plant (dry basis); plant location, Asia (Iran); costs index, 2019; depreciation, straight-line for 7 years with zero salvage value; contingency factor, 18%; construction period, two years with a 6-month start-up; working capital, 5% of the total investment cost; project life, 30 years; tax rate, 0% based on Iran regulation for biofuel production.

The sensitivity analysis to identify the factors affecting the profitability and final price of biodiesel is carried out.

3. Results

Experimental data [9,27] showed promising results for the production of bioenergy from castor plant oil and residues. The residues are used for ethanol production, and ethanol is used as an alternative to methol for biodiesel production. The economic feasibility of this process is evaluated through the simulation by Aspen plus and analysis by Aspen PEA.

3.1. Material and Energy Balance

The amounts of feedstocks, products, and utilities are calculated and shown in Table 3. The primary raw materials used in the two scenarios (i.e., NaOH, nutrients, pH control agents, and enzymes) are equal. However, in the biodiesel production unit of scenario 2, based on experimental data, KOH consumption is more than scenario 1, and therefore, the amount of H_2SO_4 that is used for catalyst neutralization in scenario 2 is more than that in scenario 1. In scenario 1, a part of ethanol produced from the residues is used in the transesterification unit. Based on the experimental data, the conversion of this reaction in scenario 1 is 85.0%, while it is 88.2% in scenario 2. In scenario 1, 15% of raw oil is unreacted and recycled to the transesterification reactor, resulting in higher biodiesel production than that in scenario 2. Moreover, the biodiesel composition in each scenario is different. It is in the form of fatty acid ethyl ester in scenario 1, while it is fatty acid methyl ester in scenario 2. Consequently, the energy content of biodiesel in each scenario is different from another.

Table 3. Energy and material balances for each scenario based on simulation results.

Components	Scenario 1	Scenario 2	Price (USD/kg)
Raw materials (ton/year)			
Solid residuals	63,840	63,840	0.02
Seed	56,160	56,160	0.04
NaOH [1]	6530	6530	0.19
Nutrient [2]	4460	4460	0.31
Methanol	-	2130	0.25
H_2SO_4	170	240	0.02
KOH [1]	180	270	0.84
pH control [3]	170	170	0.18
Enzyme	7	7	7.5
Products (ton/year)			
Biodiesel	10,490	7400	
Ethanol	7950	16,140	0.94 (USD/L)
Glycerol	2200	1850	0.85
Heat	30	30	5.7 (USD/MJ)
CO_2 [4]	6,292,000	6,292,000	0.06 (USD/m^3)
Salts	150	340	0.59
Utilities			
Make up water	1,171,440	1,240,043	0.03 (USD/m^3)
High pressure steam	69,305	77,706	13.07 (USD/ton)
Low pressure steam	87,611	87,611	13.07 (USD/ton)
Fuel	6.32	5.41	11.86 (USD/MWh)
Power [5]	5295	5289	6.38×10^{-6} (USD/kWh)

[1] The purity of NaOH is 60% weight per volume and used KOH is solid. [2] The nutrients are a mixture of 60% (*w/w*) of super triple phosphate and 40% (*w/w*) of urea. [3] The pH controls with Na_2CO_3. [4] m^3 per year. [5] Megawatt per year.

The main difference in the scenarios is the production of net ethanol. In scenario 1, about 40% of produced ethanol is used for biodiesel production, resulting in a much lower net ethanol production yield than that in scenario 2.

The prices of all components are shown in Table 3. Based on the price and amount of each component, seeds and solid residuals are the primary feeds of biorefinery. The price of NaOH (due to high consumption) and nutrients, which are expensive, is in the next position.

The distribution of raw material cost and product sales is shown in Figure 5. In scenario 1, the highest income is from biodiesel sale; in scenario 2, the transesterification reaction is conducted with methanol, and the highest revenue is from bioethanol sale.

The type and amount of utilities are shown in Table 3 for the two scenarios. The water is mainly used as makeup in the cooling system, in the pretreatment unit for washing the solid material, and to regulate NaOH concentration and in the fermentation step for adjusting the total solid in the fermenters. High pressure steam is used in distillation towers and also for heating. Low pressure steam is used to regulate the temperature and preheat the feeds. Fuel and hot oil are used for heating the oil streams in the biodiesel production unit in each scenario [33].

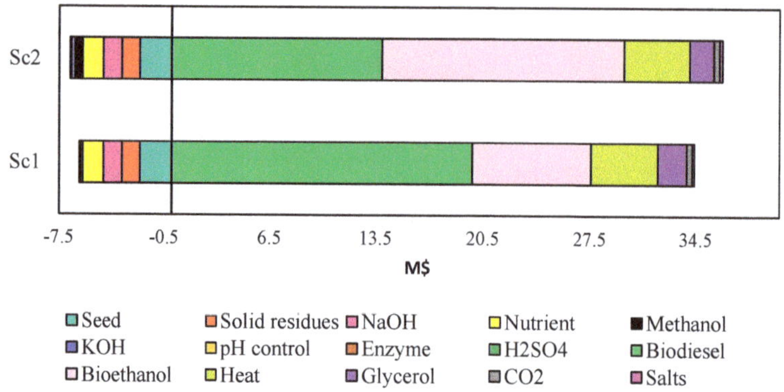

Figure 5. The distribution of raw material cost and product sales.

The breakdown of energy consumption in different units of each scenario is shown in Figure 6. Distillation towers in ethanol purification and energy consumption in biodiesel production are the most energy demanding units. Pretreatment of lignocellulosic materials at 100 °C and the anaerobic digestion at 55 °C, as well as the filtration system in the wastewater treatment unit, also need considerable amounts of energy.

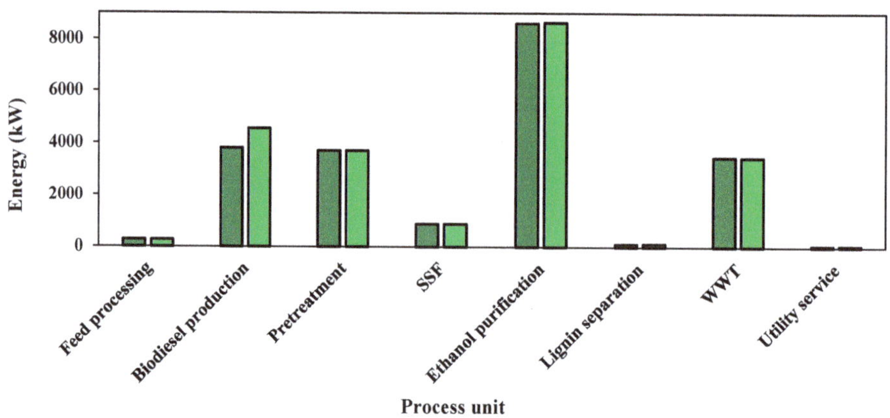

Figure 6. Breakdown of energy consumption in units of each scenario. Dark and light green represent scenarios 1 and 2, respectively.

3.2. Energy Efficiency

The ratio of energy outputs to inputs in the process is defined as energy efficiency [32]. The energy outputs involve the energy available in biodiesel, bioethanol, and glycerol as well as produced heat. The energy inputs comprise the energy available in raw materials as well as steam, electricity, and fuel.

The energy consumptions of the scenarios are presented in Figure 6. The lower heating values are used to calculate the energy content of raw materials and products. The energy efficiency of scenarios 1 and 2 are 64.8% and 64.1%, respectively. In scenario 2, the energy consumption of the biodiesel production unit is more than that in scenario 1. On the other hand, the energy content of biodiesel is higher than that of ethanol, and in scenario 1, the amount of produced biodiesel is considerably higher than scenario 2. Thus, for the production of a higher amount of energy, transesterification with ethanol is preferred.

3.3. Total Capital Cost

Based on the simulation results, economic analysis was conducted by Aspen PEA for each scenario. Total project investments are USD 51.69 million and USD 48.90 million for scenarios 1 and 2, respectively. The retention times for transesterification reactions in scenarios 1 and 2 are 3.64 and 1.5 h, respectively. The volume of equipment in scenario 1 is much higher than those in scenario 2, which results in the higher total capital cost of the plant in scenario 1 [31].

The most expensive units of plants are SSF and wastewater treatment (WWT) (Figure 7). SSF needs a long retention time (24 h) and a relatively high number of equipment, and the WWT unit needs filtration together with large anaerobic digesters with a long retention time [13].

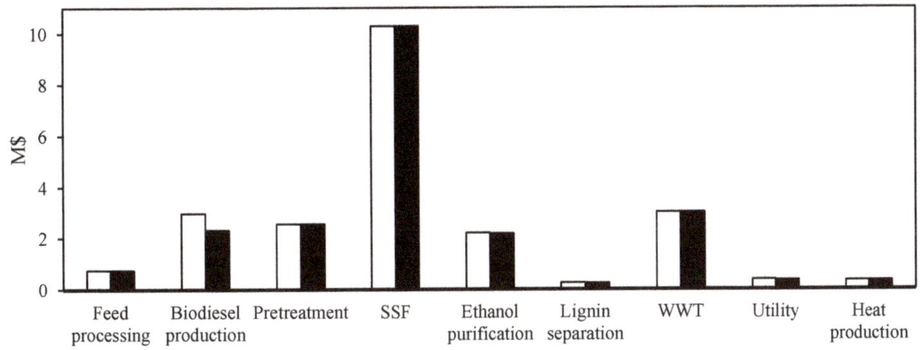

Figure 7. Breakdown of capital investment in units. The empty and filled columns represent scenarios 1 and 2, respectively.

3.4. Operating Cost

The breakdown of the operating costs is summarised in Figure 8. The direct manufacturing costs include the expenses for labor, supervisory, raw materials, utilities, and equipment maintenance and repairs. Tax and overhead are the fixed charges, while the expenditures for research and development, financing, and administrative form the general expenses [33]. The coziest raw materials are castor seed, plant residues, and sodium hydroxide. The byproducts of plants are bioethanol, glycerin, potassium phosphate, heat, and carbon dioxide. The highest revenue from the byproducts is obtained from bioethanol.

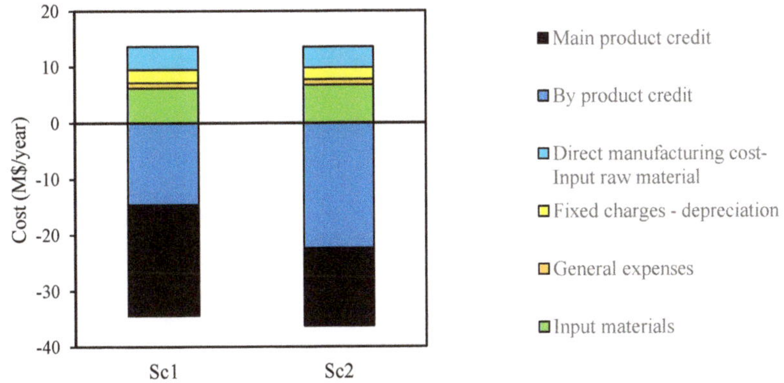

Figure 8. The breakdown of the operating costs.

3.5. *Manufacturing and Equivalent Costs*

The manufacturing costs of biodiesel and bioethanol are calculated by considering the operating cost for all scenarios, and the breakdown is shown in Figure 8. The manufacturing costs of biodiesel for scenarios 1 and 2 are 0.328 and 0.271 USD/L, respectively. The final price of biodiesel is 0.336 and 0.279 USD/L, respectively. In spite of more biodiesel production in scenario 1, the final price is higher than that in scenario 2, which is due to the higher operating and total investment costs. The final price of biodiesel in the two scenarios is lower than the diesel fuel price (Table 4). The diesel equivalent prices of biodiesel, as the main product in each scenario, are presented Table 4. This parameter in scenario 2 is lower than the diesel price; thus, this process is more economically feasible and can compete with fossil-fuel-based diesel.

Table 4. The manufacturing cost and final price of biofuels.

Product	Manufacturing Cost (USD/L)	Final Price (USD/L)	Final Price (Gasoline Equivalent) (USD/L)
Biodiesel			
Scenario 1	0.328	0.336	0.410
Scenario 2	0.271	0.279	0.323
Bioethanol			
Scenario 1	0.474	0.482	0.701
Scenario 2	0.468	0.476	0.693
Gasoline *	-	0.400	0.400
Diesel fuel *	-	0.380	0.380
Medical Ethanol	-	0.940	1.370

* The price of fossil fuels is based on the price offered by Organization of the Petroleum Exporting Countries (OPEC) for the Middle East region on December 2016.

The manufacturing cost, final price, and gasoline-equivalent price were calculated for bioethanol in each scenario (Table 4). In the two scenarios, the gasoline equivalent price of bioethanol is more than the gasoline price. Therefore, bioethanol in Iran could not be an appropriate substitute for gasoline. The amount of ethanol produced in scenario 2 is higher than that in scenario 2.

3.6. *Profitability Parameters*

Considering the interest rate and money time value in the operating costs and total capital investment, discounted cash flow and the profitability parameters, i.e., profitability index, net return rate on investment, and payout period are calculated (Table 5) [33].

Table 5. The profitability parameters of four scenarios.

Scenario	1	2
Payout period (year)	5.21	4.64
Net return rate (NRR)	29.21	41.69
Profitability index (PI)	1.29	1.41

The payout period is the minimum time to get back the total capital investment. The minimum time is calculated for scenario 2 because of the lowest capital investment requirement and higher revenue. Net return rate is the ratio of the net present value and the present value (PV) of cumulative outflows. NRR of the two scenarios is positive, indicating the profitability of the two processes.

The relative profitability of processes was evaluated by the profitability index (PI), which is calculated by dividing the income value by the costs. The results (Table 5) show that both scenarios are profitable since the project is profitable when PI is greater than one. However, scenario 2 is more profitable, which is due to the production of higher amounts of ethanol.

3.7. Sensitivity Analysis

The effects of changes in the price of the most expensive raw materials (Figure 5), i.e., seed, castor plant residues, methanol, and NaOH, on bioethanol and biodiesel price are presented in Figure 9. The results indicate that seed oil and methanol prices significantly affect the manufacturing price of biodiesel. For example, 13% and 23% increases in the biodiesel manufacturing cost for scenarios 1 and 2 are observable after a 50% increase in seed oil price, respectively. The next effective parameter is the methanol price. Increasing the methanol price by 50% leads to a 5% increase in the manufacturing cost of biodiesel in scenario 2.

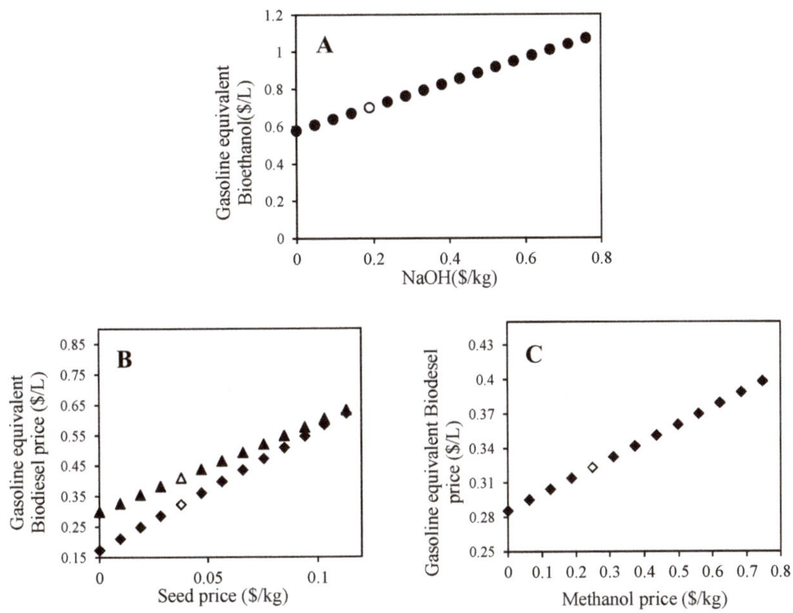

Figure 9. The effect of NaOH price on bioethanol price in scenario 1 (•) (**A**), effect of the price of plant residuals on the bioethanol manufacturing cost in scenarios 1 (▲) and 2 (♦) (**B**), and effect of methanol price on biodiesel price in scenario 2 (♦) (**C**). The empty sign represents the base case values of the profitability index.

The price of plant residuals and NaOH had a direct effect on the bioethanol cost. As shown in Figure 9A,B by increasing the price of plant residuals and NaOH, the bioethanol manufacturing cost is increased and has a secure margin compared with the ethanol in the Iranian market. In both scenarios, bioethanol is the main byproduct. The biodiesel price is calculated by dividing the net yearly production expenses into the annual biodiesel production rate. The net yearly production expenses are calculated by subtraction of ethanol revenues from total production costs of the plant. Thus, the biodiesel manufacturing cost is decreased by increasing the ethanol production price.

The effects of bioethanol and heat prices, the main byproducts, on the biodiesel manufacturing cost are presented in Figure 10. The results indicate that there is no revenue from the byproduct by 50% increase in the byproduct price in the base cases. The price and amount of glycerol, CO_2, and salts have minimal effects and do not significantly affect the process economy. Moreover, the effects of bioethanol and heat price on biodiesel price are also investigated. For both scenarios, the credit from bioethanol significantly affected the manufacturing cost of biodiesel.

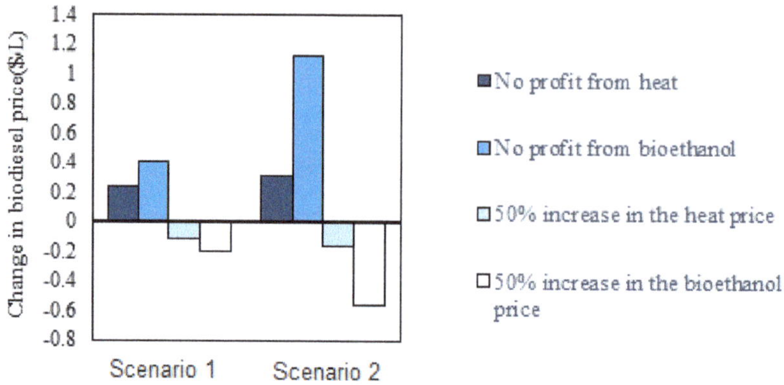

Figure 10. The effects of bioethanol and heat prices on the biodiesel manufacturing cost.

The effects of the primary feedstock (castor seeds) price on the profitability index of each scenario are investigated (Figure 11). Increasing the seed oil price leads to decreasing the profitability index. By changing the price of seed oil in the range of 0–400%, the profitability index is larger than one for both scenarios, indicating a wide safe margin for PI.

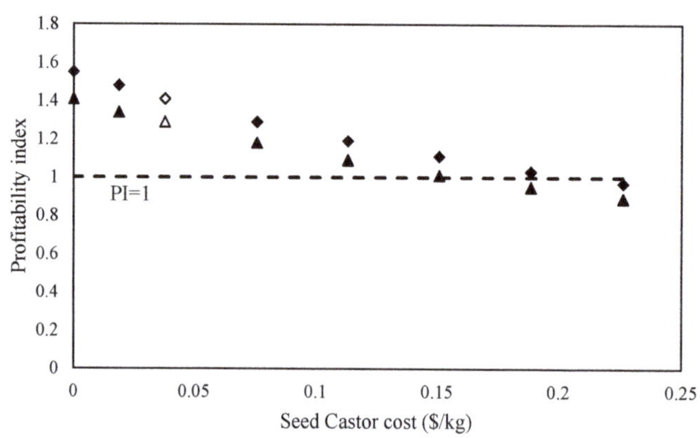

Figure 11. The effects of seed oil cost on the profitability index in scenarios 1 (▲) and 2 (♦). The empty sign represents the base case values of the profitability index.

4. Conclusions

Using ethanol produced from the lignocellulosic waste for biodiesel production from castor oil has excellent environmental benefits. However, the results of this study showed that the production of biodiesel from the castor oil by transesterification with fossil-based methanol is more economically profitable. In other words, there is a trade-off between economic aspects of the process and environmental issues. Moreover, the sensitivity analysis performed on the price of the main product showed that the price of the primary seed oil is the most important determinant of the price of the final product. This analysis also showed that the production of ethanol has an enormous effect on the price of the final product, i.e., biodiesel.

Author Contributions: Conceptualization, M.S. and K.K.; methodology, M.S.; software, V.R.; validation, M.S. and K.K.; formal analysis, V.R. and M.S.; investigation, V.R. and M.S.; resources, M.S.; data curation, V.R. and M.S.; writing—original draft preparation, V.R.; writing—review and editing, M.S. and K.K.; visualization, V.R. and M.S.; supervision, M.S. and K.K.; project administration, M.S. and K.K.; funding acquisition, K.K. All authors have read and agreed to the published version of the manuscript.

Funding: This research received no external funding.

Conflicts of Interest: The authors declare no conflict of interest.

References

1. Nigam, P.S.; Singh, A. Production of liquid biofuels from renewable resources. *Prog. Energy Combust. Sci.* **2011**, *37*, 52–68. [CrossRef]
2. Demirbaş, A. Progress and recent trends in biodiesel fuels. *Energy Convers. Manag.* **2009**, *50*, 14–34. [CrossRef]
3. Berman, P.; Nizri, S.; Wiesman, Z. Castor oil biodiesel and its blends as alternative fuel. *Biomass-Bioenergy* **2011**, *35*, 2861–2866. [CrossRef]
4. Chakrabarti, M.; Ali, M.; Baroutian, S.; Saleem, M. Techno-economic comparison between B10 of Eruca sativa L. and other indigenous seed oils in Pakistan. *Process. Saf. Environ. Prot.* **2011**, *89*, 165–171. [CrossRef]
5. Tabatabaei, M.; Karimi, K.; Horváth, I.S.; Kumar, R. Recent trends in biodiesel production. *Biofuel Res. J.* **2015**, *2*, 258–267. [CrossRef]
6. Menon, V.; Rao, M. Trends in bioconversion of lignocellulose: Biofuels, platform chemicals & biorefinery concept. *Prog. Energy Combust. Sci.* **2012**, *38*, 522–550. [CrossRef]
7. Taherzadeh, M.; Karimi, K. Pretreatment of Lignocellulosic Wastes to Improve Ethanol and Biogas Production: A Review. *Int. J. Mol. Sci.* **2008**, *9*, 1621–1651. [CrossRef]
8. Skoulou, V.; Mariolis, N.; Zanakis, G.; Zabaniotou, A. Sustainable management of energy crops for integrated biofuels and green energy production in Greece. *Renew. Sustain. Energy Rev.* **2011**, *15*, 1928–1936. [CrossRef]
9. Bateni, H.; Karimi, K.; Zamani, A.; Benakashani, F. Castor plant for biodiesel, biogas, and ethanol production with a biorefinery processing perspective. *Appl. Energy* **2014**, *136*, 14–22. [CrossRef]
10. Ogunniyi, D. Castor oil: A vital industrial raw material. *Bioresour. Technol.* **2006**, *97*, 1086–1091. [CrossRef]
11. Kurian, J.K.; Nair, G.R.; Hussain, A.; Raghavan, G.V. Feedstocks, logistics and pre-treatment processes for sustainable lignocellulosic biorefineries: A comprehensive review. *Renew. Sustain. Energy Rev.* **2013**, *25*, 205–219. [CrossRef]
12. Sandin, G.; Røyne, F.; Berlin, J.; Peters, G.M.; Svanström, M. Allocation in LCAs of biorefinery products: Implications for results and decision-making. *J. Clean. Prod.* **2015**, *93*, 213–221. [CrossRef]
13. Shafiei, M.; Karimi, K.; Zilouei, H.; Taherzadeh, M. Economic Impact of NMMO Pretreatment on Ethanol and Biogas Production from Pinewood. *BioMed Res. Int.* **2014**, *2014*, 1–13. [CrossRef] [PubMed]
14. Glisic, S.B.; Orlović, A.M. Review of biodiesel synthesis from waste oil under elevated pressure and temperature: Phase equilibrium, reaction kinetics, process design and techno-economic study. *Renew. Sustain. Energy Rev.* **2014**, *31*, 708–725. [CrossRef]
15. Marchetti, J.M.; Miguel, V.; Errazu, A. Techno-economic study of different alternatives for biodiesel production. *Fuel Process. Technol.* **2008**, *89*, 740–748. [CrossRef]
16. You, Y.-D.; Shie, J.-L.; Chang, C.-Y.; Huang, S.-H.; Pai, C.-Y.; Yu, Y.-H.; Chang, C.H. Economic Cost Analysis of Biodiesel Production: Case in Soybean Oil. *Energy Fuels* **2008**, *22*, 182–189. [CrossRef]
17. Ong, H.; Mahlia, T.M.I.; Masjuki, H.; Honnery, D. Life cycle cost and sensitivity analysis of palm biodiesel production. *Fuel* **2012**, *98*, 131–139. [CrossRef]
18. Quintero, J.A.; Felix, E.R.; Rincón, L.E.; Crisspín, M.; Baca, J.F.; Khwaja, Y.; Cardona, C.A. Social and techno-economical analysis of biodiesel production in Peru. *Energy Policy* **2012**, *43*, 427–435. [CrossRef]
19. Karmee, S.K.; Patria, R.D.; Lin, C.S.K. Techno-Economic Evaluation of Biodiesel Production from Waste Cooking Oil—A Case Study of Hong Kong. *Int. J. Mol. Sci.* **2015**, *16*, 4362–4371. [CrossRef]
20. Khounani, Z.; Nazemi, F.; Shafiei, M.; Aghbashlo, M.; Tabatabaei, M. Techno-economic aspects of a safflower-based biorefinery plant co-producing bioethanol and biodiesel. *Energy Convers. Manag.* **2019**, *201*, 112184. [CrossRef]
21. Aghbashlo, M.; Tabatabaei, M.; Hosseinpour, S. On the exergoeconomic and exergoenvironmental evaluation and optimization of biodiesel synthesis from waste cooking oil (WCO) using a low power, high frequency ultrasonic reactor. *Energy Convers. Manag.* **2018**, *164*, 385–398. [CrossRef]

22. Hoang, A.T.; Tabatabaei, M.; Aghbashlo, M.; Carlucci, A.P.; Ölçer, A.I.; Le, A.T.; Ghassemi, A. Rice bran oil-based biodiesel as a promising renewable fuel alternative to petrodiesel: A review. *Renew. Sustain. Energy Rev.* **2020**, *135*, 110204. [CrossRef]
23. Anna Ekman, O.W.; Elisabeth, J.; Pål, B. Possibilities for sustainable biorefineries based on agricultural residues—A case study of potential straw-based ethanol production in sweden. *Appl. Energy* **2013**, *102*, 299–308. [CrossRef]
24. Quintero, J.A.; Moncada, J.; Cardona, C.A. Techno-economic analysis of bioethanol production from lignocellulosic residues in Colombia: A process simulation approach. *Bioresour. Technol.* **2013**, *139*, 300–307. [CrossRef]
25. Shafiei, M.; Karimi, K.; Taherzadeh, M. Techno-economical study of ethanol and biogas from spruce wood by NMMO-pretreatment and rapid fermentation and digestion. *Bioresour. Technol.* **2011**, *102*, 7879–7886. [CrossRef]
26. Rahimi, V.; Shafiei, M. Techno-economic assessment of a biorefinery based on low-impact energy crops: A step towards commercial production of biodiesel, biogas, and heat. *Energy Convers. Manag.* **2019**, *183*, 698–707. [CrossRef]
27. Bateni, H.; Karimi, K. Biodiesel production from castor plant integrating ethanol production via a biorefinery approach. *Chem. Eng. Res. Des.* **2016**, *107*, 4–12. [CrossRef]
28. Khoshnevisan, B.; Rafiee, S.; Tabatabaei, M.; Ghanavati, H.; Mohtasebi, S.S.; Rahimi, V.; Shafiei, M.; Angelidaki, I.; Karimi, K. Life cycle assessment of castor-based biorefinery: A well to wheel LCA. *Int. J. Life Cycle Assess.* **2017**, *23*, 1788–1805. [CrossRef]
29. Rajaeifar, M.A.; Tabatabaei, M.; Aghbashlo, M.; Hemayati, S.S.; Heijungs, R. Biodiesel Production and Consumption: Life Cycle Assessment (LCA) Approach. In *Biorefining of Biomass to Biofuels*; Springer: Berlin/Heidelberg, Germany, 2018; pp. 161–192.
30. Humbird, D.; Tao, R.D.L.; Kinchin, C.; Hsu, D.; Aden, A.; Schoen, P.; Olthof, J.L.B.; Worley, M.; Sexton, D.; Dudgeon, D. *Process Design and Economics for Biochemical Conversion of Lignocellulosic Biomass to Ethanol, Dilute-Acid Pretreatment and Enzymatic Hydrolysis of Corn Stover*; National Renewable Energy Laboratory: Golden, France, 2011.
31. Zhang, Y.; A Dubé, M.; McLean, D.D.; Kates, M. Biodiesel production from waste cooking oil: 2. Economic assessment and sensitivity analysis. *Bioresour. Technol.* **2003**, *90*, 229–240. [CrossRef]
32. Shafiei, M.; Kabir, M.M.; Zilouei, H.; Horváth, I.S.; Karimi, K. Techno-economical study of biogas production improved by steam explosion pretreatment. *Bioresour. Technol.* **2013**, *148*, 53–60. [CrossRef]
33. Turton, R.; Bailie, R.C.; Whiting, W.B.; Shaeiwitz, J.A. *Analysis, Synthesis and Design of Chemical Processes*; Prentice Hall PTR: Upper Saddle River, NJ, USA, 2003.
34. Wooley, R.; Putsche, V. Development of an ASPEN PLUS Physical Property Database for Biofuels Components. National Renewable Energy Laboratory, TP-425-20685: Golden, CO, USA, 1996.

© 2020 by the authors. Licensee MDPI, Basel, Switzerland. This article is an open access article distributed under the terms and conditions of the Creative Commons Attribution (CC BY) license (http://creativecommons.org/licenses/by/4.0/).

agronomy

MDPI

Article

Effect of Novel *Penicillium verruculosum* Enzyme Preparations on the Saccharification of Acid- and Alkali-Pretreated Agro-Industrial Residues

Susan G. Karp [1], Dmitrii O. Osipov [2], Margarita V. Semenova [2], Alexandra M. Rozhkova [2,*], Ivan N. Zorov [2,3], Olga A. Sinitsyna [3], Carlos R. Soccol [1] and Arkady P. Sinitsyn [2,3]

1 Department of Bioprocess Engineering and Biotechnology, Federal University of Paraná, Curitiba 81531-990, Paraná, Brazil; susan.karp@ufpr.br (S.G.K.); soccol@ufpr.br (C.R.S.)
2 Federal Research Centre Fundamentals of Biotechnology of the Russian Academy of Sciences, 119071 Moscow, Russia; doosipov@gmail.com (D.O.O.); margs@mail.ru (M.V.S.); inzorov@mail.ru (I.N.Z.); apsinitsyn@gmail.com (A.P.S.)
3 Department of Chemistry, Lomonosov Moscow State University, 119991 Moscow, Russia; oasinitsyna@gmail.com
* Correspondence: a.rojkova@fbras.ru

Received: 18 August 2020; Accepted: 4 September 2020; Published: 7 September 2020

check for updates

Abstract: This study aimed at evaluating different enzyme combinations in the saccharification of sugarcane bagasse (SCB), soybean husks (SBH) and oil palm empty fruit bunches (EFB) submitted to mild acid and alkaline pretreatments. Enzyme pools were represented by B1 host (crude cellulase/xylanase complexes of *Penicillium verruculosum*); B1-XylA (*Penicillium canescens* xylanase A expressed in *P. verruculosum* B1 host strain); and F10 (*Aspergillus niger* β-glucosidase expressed in B1 host strain). Enzyme loading was 10 mg protein/g dry substrate and 40 U/g of β-glucosidase (F10) activity. SCB was efficiently hydrolyzed by B1 host after alkaline pretreatment, yielding glucose and reducing sugars at 71 g/L or 65 g/100 g of dry pretreated substrate and 91 g/L or 83 g/100 g, respectively. B1 host performed better also for EFB, regardless of the pretreatment method, but yields were lower (glucose 27–30 g/L, 25–27 g/100 g; reducing sugars 37–42 g/L, 34–38 g/100 g). SBH was efficiently saccharified by the combination of B1 host and B1-XylA, yielding similar concentrations of reducing sugars for both pretreatments (92–96 g/L, 84–87 g/100 g); glucose recovery, however, was higher with alkaline pretreatment (81 g/L, 74 g/100 g). Glucose and reducing sugar yields from initial substrate mass were 42% and 54% for SCB, 36% and 42–47% for SBH and 16–18% and 21–26% for EFB, respectively.

Keywords: sugarcane bagasse; soybean husks; palm empty fruit bunches; recombinant enzymes; pretreatment

1. Introduction

Agro-industrial residues are important sources of lignocellulosic biomass, the main feedstock to be used as a substitute for petrol in a circular bioeconomy. These organic materials are generated in large volumes as a result of agricultural and agro-industrial activities and are in great part subtilized, especially in developing countries that are the main suppliers of commodities. According to Magalhães Jr. et al. [1], around 900 million tons of biomass from agro-industrial residues such as sugarcane bagasse, cereal straws and oil palm solid wastes will be generated in South America in 2025, and their most probable destination, if not left in the field, is burning for energy generation and composting, the most traditional processing methods. However, considering that more than 60% of their mass can be converted to fermentable sugars, these residues could be valorized through microbial processes to produce second generation (2G) ethanol and other biofuels, organic acids, biomaterials and many other commercial bioproducts.

Nowadays, 2G ethanol is the product with the most mature stage of technological development obtained in lignocellulose biorefineries. However, the cost of 2G ethanol is still not competitive with that of petroleum-based liquid fuels [2], in part because of the cost of enzymes that are necessary to convert fibrous carbohydrates into fermentable sugars. Another step that impacts the overall cost and lacks technological maturity is the pretreatment, necessary to prepare the recalcitrant structure of lignocellulosic biomass to enzymatic hydrolysis or saccharification. In general, the aim of the pretreatment is to enhance the recovery of glucose and other reducing sugars from lignocellulosic biomass in the saccharification step, and this is usually accomplished through the decrease of the crystallinity of cellulose, degree of polymerization, lignin content and moisture content, associated to an increase of available surface area [3].

Several methods have been reported to pretreat lignocellulosic biomass, most of them thermochemical processes. These include steam explosion with or without alkali washing, dilute acid hydrolysis, alkaline pretreatment, organosolv, ammonia fiber expansion, liquid hot water, wet oxidation and others [3]. Alkaline pretreatment is especially efficient in solubilizing lignin, while mild acid pretreatment is more directed to the hemicellulose fraction.

In 2018, the global sugarcane production was of 1.9 billion tons, and Brazil was the main producer accounting for 39% of the total production, followed by India with 20% [4]. In the harvest of 2018/2019, India has surpassed Brazil producing more than 700 million tons, but for 2019/2020 the projections are very similar for both countries [5]. The Brazilian production estimative for the harvest of 2019/2020 is of around 643 million tons [6]. Sugarcane bagasse is a porous residue of cane stalks generated after the crushing and extraction of sugarcane juice. It is composed mainly by cellulose (32–44%), hemicellulose (27–32%), lignin (19–24%) and ashes (4.5–9%). Sugar mills generate approximately 270–280 kg of bagasse (with 50% moisture) per metric ton of sugarcane [7,8].

Soybean is one of the most important sources of vegetable protein for human and animal nutrition and one of the most economical sources of oil for food and biofuel production. According to the United States Department of Agriculture [9], the world production of soybean for 2019/2020 is estimated at 338 million metric tons. The world's leading producer is Brazil and the second is the United States, responsible for 37% and 27% of the global production, respectively. From one ton of soybean with around 13% moisture, 50 kg of soybean husks are obtained [10]. The composition of soybean husks depends on the dehulling process, so they may contain varying amounts of cellulose (29–51%), hemicelluloses (10–25%), lignin (1–4%), pectins (4–8%), proteins (11–15%) and minor extractives [11].

The global production of oil palm is estimated at 73 million metric tons for the period of 2019/2020, Indonesia and Malaysia being the leading producers with 58% and 26% of the total, respectively [9]. The milling of oil palm to extract the oil generates a solid fibrous residue, named oil palm empty fruit bunches (EFB). It is estimated that for each ton of fresh fruit bunches processed, 220 kg of EFB are generated. This residue is composed of around 30, 25 and 25% (mass percentages) of cellulose, hemicellulose and lignin, respectively [12].

The present work aimed at evaluating the effect of different enzyme preparations and combinations in the saccharification of these three important agro-industrial residues, the sugarcane bagasse, soybean husks and oil palm empty fruit bunches, pretreated by mild acid and alkali. The enzyme pools were represented by cellulase and xylanase complexes of *Penicillium verruculosum* (B1 host preparation), by the B1-XylA preparation obtained through the recombinant expression of *Penicillium canescens* xylanase A in the *P. verruculosum* B1 host strain, and by the F10 preparation obtained by recombinant expression of *Aspergillus niger* β-glucosidase in the B1 host strain. All preparations were crude, which could significantly reduce the cost associated to the enzymatic saccharification step.

2. Materials and Methods

2.1. Substrates

Sugarcane bagasse (SCB) was donated by the Usina Santa Terezinha (Maringá, Paraná, Brazil), soybean husks (SBH) were purchased from the company Imcopa (Araucária, Paraná, Brazil) and oil palm empty fruit bunches (EFB) were provided by the company Biopalma Vale S.A. (Mojú, Pará, Brazil). Substrates were dried at 70 °C overnight, grinded in a knife mill and particle sizes <3 mm were selected by sieving. SCB and EFB presented a small fibers aspect and SBH presented a powder aspect. Table 1 presents the composition of each substrate.

Table 1. Substrates' composition (% in dry basis).

Substrate	SCB	SBH	EFB
Cellulose	36	37	28
Hemicellulose	31	27	24
Lignin	23	8.7	20
Lipids	2.3	1.9	6.3
Proteins	2.2	14	3.4
Ash	2.7	4.2	3.2
Extractives/others	2.2	6.8	15

SCB—sugarcane bagasse; SBH—soybean husks; EFB—empty fruit bunches of oil palm. Analytical methods were: NREL TP-510-42618 (cellulose, hemicellulose, lignin, others—pectin and acetyl); gravimetric analysis after solvent extraction in Soxhlet using hexane (lipids); Kjeldahl (proteins); gravimetric analysis after calcination at 555 °C for 6 h (ash); NREL TP-510-42619 (extractives); "others" are mostly represented by pectin in SBH and acetyl in EFB.

2.2. Enzyme Preparations

Three different enzyme preparations (Table 2) were used in this study in different combinations. Dry B1 host preparation represents a complex of cellulases and xylanases obtained by highly productive recombinant *P. verruculosum* strain after UV-mutagenesis [13,14]; dry B1-XylA preparation was obtained by recombinant *P. verruculosum* strain after heterologous expression of *P. canescens* xylanase A [15]; and dry F10 preparation was obtained by recombinant *P. verruculosum* strain after heterologous expression of *Aspergillus niger* β-glucosidase (cellobiase) [16].

Table 2. Enzyme preparations.

Name	Protein, mg/g	CMC, U/g	β-Glucan, U/g	Xylan, U/g	pNPG, U/g
B1 host	970	16,542	15,062	17,532	1074
B1-XylA	441	2240	2240	60,000	610
F10	655	7007	6797	3800	39,852

Activities of enzyme preparations (Table 2) toward carboxymethylcellulose (CMC), barley β-glucan and birch wood xylan were determined by detection of reducing sugars release using Somogyi-Nelson assay [17–19]. Enzyme activities were assayed for 10 min at pH 5 (0.05 M Na-acetate buffer) and 50 °C using a substrate concentration of 0.5% in the reaction mixture [20]. CMCase, β-glucanase and xylanase activities were expressed in international units. One unit of activity corresponds to the quantity of enzyme releasing 1 μmol of reducing sugars (in glucose equivalents) per minute.

Enzyme activity toward *p*-nitrophenyl glucopyranoside (pNPG) was determined by detection of *p*-nitrophenyl release by a photometric assay. Enzyme activity was assayed for 10 min at pH 5 (0.05 M Na-acetate buffer) and 40 °C using a substrate concentration of 10 mM in the reaction mixture. One β-glucosidase unit of activity is the amount of enzyme which liberates 1 μmol of *p*-nitrophenol per minute [21].

2.3. Pretreatment Conditions

Alkaline pretreatment was conducted with a substrate concentration of 10% (m/v) of solids, NaOH 2% (m/v), at 121 °C for 1 h. After, the pH of the reaction medium was adjusted to 4.5 with HCl, the solids were separated from the liquid phase in a glass filter and washed abundantly with distilled water. The excess water was removed, and a sample was taken to determine the moisture content of the pretreated substrates by gravimetric analysis.

Acid pretreatment was conducted with a substrate concentration of 10% (m/v) of solids, H_2SO_4 1% (m/v), at 121 °C for 1 h. Afterward, the pH of the reaction medium was adjusted to 4.5 with NaOH, and the solids were separated from the liquid phase in a glass filter and washed abundantly with distilled water. The excess water was removed, and a sample was taken to determine the moisture content of the pretreated substrates by gravimetric analysis.

2.4. Saccharification Conditions

Three different combinations of enzyme preparations were tested, as presented in Table 3. Substrate concentration (in dry basis) was set at 100 g/L, in sodium acetate buffer (1 M, pH 4.5), and the reaction volume was 20 mL. The enzyme dosage for B1 host, B1-XylA and B1 host + B1-XylA was 10 mg protein/g dry substrate. In all cases, F10 β-glucosidase preparation was added to the reaction mixture in the dosage 40 U of pNPG /g of dry substrate. Hydrolysis flasks were incubated at 45 °C, 250 rpm and samples were taken after 3 h, 24 h and 48 h. Experiments were conducted in duplicates, with repetition.

Table 3. Enzyme combinations.

Condition	Enzyme Combinations
B1 host	B1 host 10 mg/g substrate + F10 40 U/g substrate
B1-XylA	B1-XylA 10 mg/g substrate + F10 40 U/g substrate
B1 host + B1-XylA	B1 host 8 mg/g substrate + B1-XylA 2 mg/g substrate + F10 40 U/g substrate

2.5. Analytical Procedures

2.5.1. Glucose Concentration

Glucose concentration was determined according to the glucose oxidase/peroxidase method [22]. One hundred microliters of solids-free, properly diluted samples were mixed with 1 mL of the glucose oxidase/peroxidase reagent (R1 + R2 40:1 *v/v*, from Impact Ltd., Moscow, Russia) and incubated for 15 min at 40 °C. Absorbance was read at 490 nm in a UV-vis spectrophotometer (Agilent Technologies Inc., Santa Clara, CA, USA).

2.5.2. Reducing Sugars Concentration

The concentration of total reducing sugars was determined according to the Somogyi Nelson method [17–19], using glucose as a standard for calibration. Two hundred microliters of solids-free, properly diluted samples were mixed with 200 μL of the Somogyi reagent and incubated for 40 min at 100 °C. After cooling to room temperature, 200 μL of the Nelson reagent were added and incubated for 15 min. Then, 400 μL of acetone and 1 mL of water were added, and the absorbance was measured at 610 nm in a UV-vis spectrophotometer (Agilent Technologies Inc., Santa Clara, CA, USA).

2.5.3. High Performance Liquid Chromatography (HPLC) Analysis of Monosaccharides Composition

The composition of low molecular weight sugars (xylose and fructose) was determined via ion-exchange chromatography on an Agilent 1100 Series HPLC system (Agilent, Santa Clara, CA, USA) with a Diaspher-110-Amin column 5 μm 4 × 250 mm; the eluent was acetonitrile–water at 75:25 (*v/v*), the flow rate was 1 mL/min and the sample's volume was 10–100 μL. To prepare samples acetonitrile (0.8 mL) was added to 0.2 mL of hydrolysate, followed by centrifugation for 5 min at

9000× *g*. Solutions of xylose and fructose (Megazyme, Victoria, Australia) at the concentration of 1 g/L were used as standards.

3. Results

The mass of recovered solids after acid and alkaline pretreatments was determined for each substrate. The concentration of monosaccharides and reducing sugars was also quantified in the liquid fraction of the pretreatment processes, in order to determine the amount of sugars lost in this step. Results are presented in Table 4.

Table 4. Recovery of solids and reducing sugars' loss after acid and alkaline pretreatments.

Substrate-Pretreatment	Solids Recovery in Dry Mass Basis (%)	Sugar Loss in Dry Mass Basis (%)
SCB—Acid	63.6	24.1
SCB—Alkaline	65.2	N. d.
SBH—Acid	56.1	11.36
SBH—Alkaline	48.7	N. d.
EFB—Acid	63.4	18.14
EFB—Alkaline	67.4	N. d.

Note: N. d.—Not detectable; SCB—sugarcane bagasse; SBH—soybean husks; EFB—empty fruit bunches of oil palm.

No detectable amount of sugars was found in the liquid fractions of the alkaline pretreatments in a 1:10 (*v/v*) dilution, and since the dark color of the liquid interfered in the absorbance reading, it was not possible to test lower dilutions. The acid pretreatment resulted in glucose, xylose and reducing sugars mass losses of 2.71%, 20.0% and 24.1% in SCB, 0.61%, 1.25% and 11.36% in SBH and 0.33%, 14.5% and 18.14% in EFB, respectively.

The concentrations of total reducing sugars and glucose released along enzymatic saccharification of untreated, acid pretreated and alkaline pretreated substrates are presented in Figures 1–3, respectively.

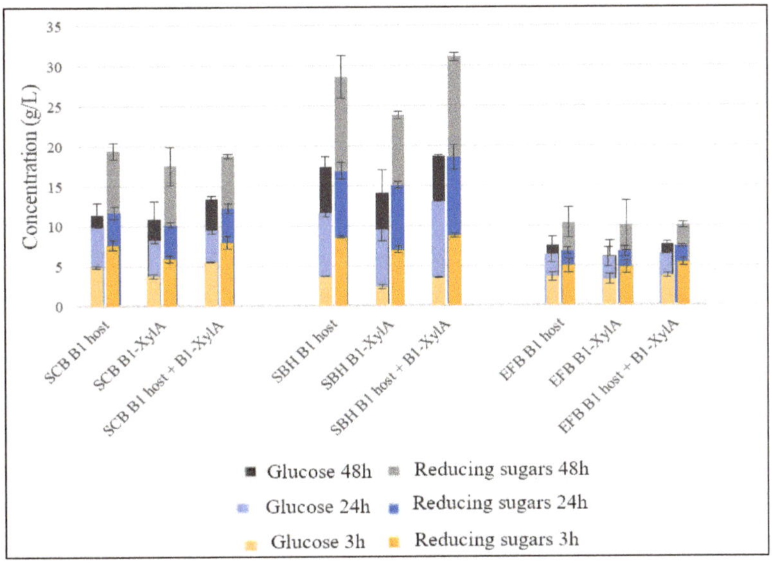

Figure 1. Glucose and reducing sugars released after 3 h, 24 h and 48 h of enzymatic saccharification of unpretreated sugarcane bagasse (SCB), soybean hulls (SBH) and palm empty fruit bunches (EFB) using different enzyme combinations (B1 host, B1-XylA and B1 host + B1-XylA).

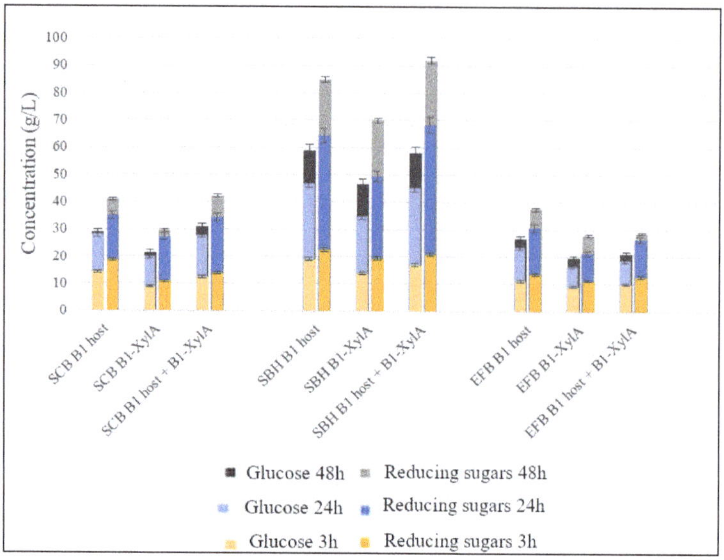

Figure 2. Glucose and reducing sugars released after 3 h, 24 h and 48 h of enzymatic saccharification of sugarcane bagasse (SCB), soybean hulls (SBH) and palm empty fruit bunches (EFB) after acid pretreatment using different enzyme combinations (B1 host, B1-XylA and B1 host + B1-XylA).

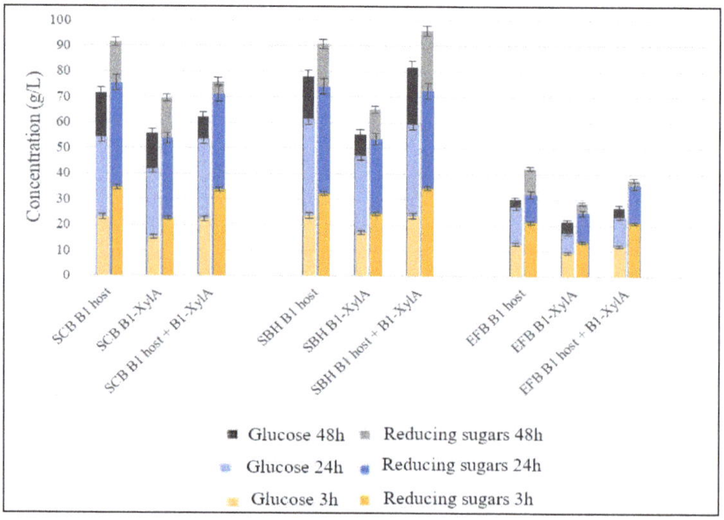

Figure 3. Glucose and reducing sugars released after 3 h, 24 h and 48 h of enzymatic saccharification of sugarcane bagasse (SCB), soybean hulls (SBH) and palm empty fruit bunches (EFB) after alkaline pretreatment using different enzyme combinations (B1 host, B1-XylA and B1 host + B1-XylA).

Untreated samples of SCB released between 17 and 19 g/L of reducing sugars, between 11 and 13 g/L of glucose and nearly 2 g/L of xylose after 48 h; fructose was not detected. The ratio between glucose and xylose was around 6, and there was not a significant difference between enzyme preparations. Untreated SBH released, after 48 h, an average of 31 g/L of reducing sugars using B1 host + B1-XylA and around 29 g/L using B1 host, without significant difference between these enzyme combinations. Glucose concentration, however, was higher with B1 host + B1-XylA (almost 19 g/L).

Small amounts of xylose and fructose (3–4 g/L of xylose, 2–3 g/L of fructose) were determined with little advantage of B1 host + B1-XylA. Untreated EFB released 10 g/L of reducing sugars and around 7.5 g/L of glucose after 48 h, using B1 host and B1 host + B1-XylA; when B1-XylA alone was used, the glucose concentration after 48 h was 6 g/L. Xylose concentration was of about 3 g/L for all enzyme combinations and no fructose was detected.

Mild acid pretreatment was especially efficient in SBH, allowing to obtain 92 g/L of reducing sugars after 48 h when the enzyme combination B1 host + B1-XylA was used. No residual fibers were observed. In 24 h, 67 g/L of reducing sugars were obtained. Glucose concentration after 48 h was equivalent between B1 host and B1 host + B1-XylA (58 g/L). These results represent an improvement of 3× as compared to the untreated samples. Acid pretreated SCB released, after 48 h, around 40 g/L of reducing sugars and 30 g/L of glucose when using B1 host or B1 host + B1-XylA. In 24 h, concentrations of reducing sugars were in the order of 34 g/L; thus, no pronounced time-dependent increase was observed. These results represent an improvement of around 2× and 3× for reducing sugars and glucose, respectively, as compared to the untreated samples; however, all samples still contained residual fibers after 48 h, indicating that the carbohydrates were not completely accessed. Acid pretreated EFB showed better results for B1 host as compared to other enzyme combinations after 48 h (37 g/L of reducing sugars and 27 g/L of glucose), an improvement of 3.7× as compared to untreated samples. Similar to SCB, residual fibers were observed after 48 h of hydrolysis.

Alkaline pretreatment was the most effective for SCB, and the enzyme preparation B1 host was the most efficient among all. Concentrations of 91 g/L reducing sugars and 71 g/L glucose were obtained after 48 h. After 24 h, concentrations were 75 g/L and 54 g/L, respectively. This represents an improvement of 5× and 6× in relation to untreated samples and of around 2× in relation to acid pretreated samples. Alkaline pretreated SBH released 96 g/L of reducing sugars and 81 g/L of glucose, after 48 h, when using the enzyme preparation B1 host + B1-XylA. This preparation also provided the best results in acid-pretreated samples of SBH. In 24 h, enzyme preparations B1 host and B1 host + B1-XylA released around 73 g/L of reducing sugars and around 60 g/L of glucose. When comparing mild acid and alkaline pretreatments for SBH, a slight increase was observed for reducing sugars, from 92 to 96 g/L, and an important increase was observed for glucose, from 58 to 81 g/L. Alkaline pretreated EFB samples released 42 g/L of reducing sugars and 30 g/L of glucose after 48 h, when using B1 host. These results are consistent with those obtained for acid pretreated EFB, indicating a similar effect of both pretreatments and a better effect of the enzyme preparation B1 host in this substrate. However, it is clear that the potential of the substrate was not fully accessed yet. It is important to remark that, between 24 and 48 h of enzymatic hydrolysis, alkaline pretreated sugarcane bagasse and soybean husks were completely liquefied, while empty fruit bunches remained in great part as a solid fraction after 48 h of hydrolysis.

Table 5 presents the summary of the best saccharification results obtained for each substrate and pretreatment, and the corresponding enzyme combinations.

Table 5. Summary of best results of enzymatic saccharification after 48 h.

Substrate-Pretreatment	Enzyme Combination	Glucose Concentration (g/L)	Reducing Sugars Concentration (g/L)	Glucose Yield (g/100 g Substrate)	Reducing Sugars Yield (g/100 g Substrate)
SCB—Untreated	B1 host, B1-XylA or B1 host + B1-XylA	12	18	11	16
SCB—Acid	B1 host or B1 host + B1-XylA	30	40	17 (27)	23 (36)
SCB—Alkaline	B1 host	71	91	42 (65)	54 (83)
SBH—Untreated	B1 host + B1-XylA	19	31	17	28
SBH—Acid	B1 host + B1-XylA	58	92	30 (53)	47 (84)
SBH—Alkaline	B1 host + B1-XylA	81	96	36 (74)	42 (87)
EFB—Untreated	B1 host or B1 host + B1-XylA	7.5	10	6.8	9
EFB—Acid	B1 host	27	37	16 (25)	21 (34)
EFB—Alkaline	B1 host	30	42	18 (27)	26 (38)

Note: The yields of glucose and reducing sugars were calculated based on the total initial mass of dry substrates (before pretreatment); values in parentheses represent the yields obtained from the mass recovered after pretreatment, in dry basis; it was considered that, after enzymatic hydrolysis, one molecule of water is incorporated in each monomer. SCB—sugarcane bagasse; SBH—soybean husks; EFB—empty fruit bunches of oil palm.

4. Discussion

The association of efficient enzyme cocktails and effective pretreatment strategies has a very important function in making the saccharification of lignocellulosic biomass feasible, especially of the most recalcitrant biomass types. Sugarcane bagasse and palm empty fruit bunches are very fibrous materials, while soybean husks are softer and easier to grind. SCB is porous and absorbs water in approximately 10 times its mass, while palm EFB is harder and less porous. These characteristics were reflected in the efficiency of pretreatments and saccharification.

SBH was easily hydrolyzed after both acid and alkaline pretreatments, and even the untreated SBH samples released a considerable amount of reducing sugars (around 30 g/L). On the other hand, EFB was the most recalcitrant substrate, with no difference between acid and alkaline pretreatments in terms of reducing sugars yield. Sugarcane bagasse was the substrate that showed the most pronounced effect of pretreatment type, being the alkaline pretreatment significantly more effective than the acid one. This is probably related to the ability of alkali to solubilize lignin, which was successfully achieved considering the porous nature of SCB.

Acid pretreatment is highly effective in disrupting the lignocellulosic matrix by the cleavage of glycosidic bonds. This process mainly solubilizes the hemicellulosic portion of the biomass, and part of the lignin [23]. Alkaline pretreatment removes lignin with high efficiency and cleaves glycosidic and ester side chains, contributing to decrystallization, increased porosity and swelling of cellulose [2]. These modifications facilitate in great extent the access of enzymes to the carbohydrate molecules. An approach to take advantage of both mechanisms is the sequential acid-alkaline pretreatment. This strategy has been applied to corn stover and corn cobs resulting in reducing sugar yields higher than 90% [23]. However, the need to perform two separate steps of thermochemical treatments increases process costs and time, discouraging the application in an industrial scale.

In our experiments, sugar losses between 11 and 24% were detected after acid pretreatment, as presented in Table 4. Alkaline pretreatment was more selective to lignin, as observed from the brown color of the resulting liquid phase. Regardless of the pretreatment strategy, there were significant mass losses for all substrates, which were in the order of 32–37% for SCB and EFB. SBH was the substrate that presented the most pronounced mass loss as a result of pretreatment, either acid (43.9%) or alkaline (51.3%). Considering the low lignin content of this substrate, we assumed that one of the factors contributing to this mass loss was the small particle size of the grinded material. Although all substrates were grinded and particle sizes <3 mm were selected, SCB and EFB presented a small-fibers aspect after mechanical processing and selection, while SBH presented a powder aspect. Another point to be considered is the relatively high concentration of proteins (14%, Table 1) that can be extracted by the thermochemical pretreatments. Rojas et al. [24] reported a mass yield of 40% after acid pretreatment of SBH with H_2SO_4 3% (*v/v*), 25% of solids, at 120 °C for 40 min, and this mass loss of 60% was mainly attributed to the solubilization of protein, pectin and hemicellulose. The high solubilization of biomass components and the possible loss of small particles during pretreatment in the present work contributed to the relatively low yields of reducing sugars recovery from the initial mass of SBH (42–47%, Table 5), despite the high efficiency of enzymatic hydrolysis.

The enzymatic saccharification of sugarcane bagasse has been widely studied in the last decade. Today, sugarcane bagasse is the second most important agro-industrial residue used as a feedstock for 2G ethanol, after corn residues. Even so, the scientific literature still reports challenges to be overcome in the conversion of substrates and sugar release, especially when considering industrial applications. Prajapati et al. [2] applied a cocktail of cellulases and hemicellulases from *A. tubingensis* to hydrolyze alkali treated sugarcane bagasse, with 6–8% (*w/v*) solids, for 96 h at 45 °C, applying an enzyme preparation containing FPase activity (1.03 U/mL), β-glucosidase (0.6 U/mL), endo-β-glucanase (6.8 U/mL), α-galactosidase (1.6 U/mL), β-xylosidase (0.17 U/mL), β-mannosidase (0.05 U/mL), endo-β-mannanase (13.7 U/mL) and endo-β-xylanase (7.26 U/mL). The maximum concentration of released sugars was around 20 g/L of a mixture of glucose, xylose and arabinose. Scarpa et al. [25] reached a glucose concentration of 7.32 g/L from sugarcane bagasse submitted to hydrothermal alkaline pretreatment, at a solids load of 13.5% (*w*/*v*, dry basis) and an endoglucanase load of 288 U/g cellulose, for 130 h and 57 °C. These are considered low concentrations for industrial fermentations, especially in the segment of bioethanol production.

Martin et al. [26] evaluated the saccharification of sugarcane bagasse pretreated with a mixture of glycerol and sulfuric acid (79.6% glycerol, 0.6 or 1.1% H_2SO_4) or with sulfuric acid alone (1.1% H_2SO_4), at 188–194 °C, for 100–140 min. The authors also evaluated six enzyme preparations, three commercial *Trichoderma*-based cocktails and three preparations developed at the Bach Institute of Biochemistry (Russian Academy of Sciences), namely PV (host strain cellulase/xylanase cocktail—the same B1 host preparation used in our experiments), PV-Xyl PCA (produced by a recombinant *P. verruculosum* strain after heterologous expression of *P. canescens* xylanase A), and PV-Hist BGL (produced by a recombinant *P. verruculosum* strain after heterologous expression of *A. niger* β-glucosidase). Both PV-Xyl and PV-Hist BGL preparations also contained the cellulase complex of *P. verruculosum*. Better results of enzymatic convertibility were obtained with glycerol-treated bagasse. After 48 h of hydrolysis, around 30 g/L of reducing sugars, mostly represented by glucose, were obtained (using the PV enzyme loading of 10 mg protein/g substrate and 50 g/L of substrate), corresponding to a cellulose conversion value of almost 80%. The preparation PV-Xyl is the same as the B1-XylA preparation used in our experiments. It has already been tested for bagasse, aspen and pine wood and results were reported by Osipov et al. [15]. These results corroborate the suitability of the enzyme cocktails obtained from *P. verruculosum* to hydrolyze sugarcane bagasse. Since these cocktails are crude preparations, the economic feasibility of the process can be significantly enhanced as compared to the use of commercial purified preparations.

Hickert et al. [27] performed the saccharification of soybean husks using an enzyme preparation obtained from *Penicillium echinulatum*. Pretreatment was performed by dilute acid hydrolysis (121 °C, 40 min, solid-liquid ratio of 1:10, 1% *v/v* sulphuric acid) and it was followed by enzymatic hydrolysis using a solid-liquid proportion of 1:20 (dry matter in citrate phosphate buffer pH 4.8), *P. echinulatum* S1M29 enzyme preparation (enzyme loading of 10, 15, or 20 FPU/g), at 120 rpm, 50 °C for 96 h. The liquid fractions of acid pretreatment and enzymatic hydrolysis were mixed, yielding a solution containing (g/L): glucose 38, xylose 21, arabinose 4, mannose 6 and cellobiose 7. The efficiency of saccharification was 72%. Qing et al. [28] studied the saccharification of SBH submitted to acid and alkaline pretreatments (1% *v/v* H_2SO_4 or NaOH, 120 °C, 1 h) using the commercial preparation Accelerase 1500 (60 FPU/g), and results were different from those of the present work, since the alkaline pretreatment promoted considerably higher enzymatic conversion than the acid one (80% versus 65%). There are no literature reports on the application of the enzyme preparations obtained from *P. verruculosum*, used in the present work, to hydrolyze SBH or EFB.

The saccharification of oil palm EFB was evaluated by Medina et al. [12], comparing acid-alkaline pretreatment, steam explosion and steam explosion with alkaline delignification. The enzymatic digestibility of the pretreated substrate was evaluated with Celluclast® 1.5 L and Novozym 188 (mass ratio of 1:0.3), loaded at 60 FPU per g EFB, with pH 4.8 (0.1 M sodium citrate buffer) and maintained at 55 °C, 130 rpm for 5 days. The mass-volume ratio of EFB was 2.5% (*w*/*v*). The best result of enzymatic digestibility (72% after 5 days) was obtained with the sequential acid-alkaline pretreatment (1% H_2SO_4, 2.5% NaOH, 121 °C), while digestibility results obtained with steam explosion coupled with alkaline delignification were lower than 50%, and with steam explosion alone, lower than 16%. These results, together with the findings of the present research, indicate that the recalcitrant nature of EFB cannot be overcome by traditional industrial pretreatment methods such as steam explosion and single-step thermochemical pretreatments, and that both pretreatment and saccharification are bottlenecks to be optimized for this substrate.

5. Conclusions

The effects of different enzyme combinations were demonstrated in the saccharification of sugarcane bagasse, soybean husks and oil palm empty fruit bunches submitted to mild acid and alkaline pretreatments. Sugarcane bagasse was efficiently saccharified after alkaline pretreatment, while soybean husks were efficiently saccharified regardless of the pretreatment method. Palm empty fruit bunches also showed similar responses of enzymatic saccharification independently of the pretreatment strategy; however, with much lower sugar yields. The best enzyme choice for sugarcane bagasse was B1 host, representing a cellulase/xylanase complex of *P. verruculosum*; for soybean husks, the combination of B1 host (80%) and B1-XylA (20%), this last one obtained through the recombinant expression of *P. canescens* xylanase A in the *P. verruculosum* B1 host strain, gave the best results; for palm empty fruit bunches, the B1 host preparation was the most efficient.

Author Contributions: Conceptualization, C.R.S. and A.P.S.; methodology, D.O.O., I.N.Z. and O.A.S.; validation, S.G.K.; formal analysis, D.O.O. and I.N.Z.; investigation, S.G.K. and M.V.S.; data curation, D.O.O., M.V.S. and O.A.S.; writing—original draft preparation, S.G.K.; writing—review and editing, A.M.R. and C.R.S.; supervision, A.P.S.; project administration, A.M.R. and C.R.S. All authors have read and agreed to the published version of the manuscript.

Funding: This research was funded in part by the Coordenação de Aperfeiçoamento de Pessoal de Nível Superior—Brasil (CAPES)—Finance code 88887.472495/2019-00, in the frame of the project Biotechnology Processes for the Production of Advanced Biofuels (BioADD), CAPES-PRINT UFPR, and in part by the Russian Foundation for Basic Research (RFBR No. 18-54-80027). The APC was funded by the Russian Foundation for Basic Research (RFBR No. 18-54-80027).

Acknowledgments: Authors are thankful to Julio César de Carvalho, Luciana Porto de Souza Vandenberghe, Adenise Lorenci Woiciechowski and Luiz Alberto Júnior Letti for administrative and technical support, and to the companies Biopalma Vale S.A., Usina Santa Terezinha and Imcopa for providing the substrates.

Conflicts of Interest: The authors declare no conflict of interest. The funders had no role in the design of the study; in the collection, analyses, or interpretation of data; in the writing of the manuscript, or in the decision to publish the results.

References

1. Magalhães, A.I. Jr.; Carvalho, J.C.; Pereira, G.V.M.; Karp, S.G.; Câmara, M.C.; Medina, J.D.C.; Soccol, C.R. Lignocellulosic biomass from agro-industrial residues in South America: Current developments and perspectives. *Biofuels Bioprod. Bioref.* **2019**, *13*, 1505–1519. [CrossRef]
2. Prajapati, B.P.; Jana, U.K.; Suryawanshi, R.K.; Kango, N. Sugarcane bagasse saccharification using *Aspergillus tubingensis* enzymatic cocktail for 2G bio-ethanol production. *Renew. Energ.* **2020**, *152*, 653–663. [CrossRef]
3. Karp, S.G.; Woiciechowski, A.L.; Soccol, V.T.; Soccol, C.R. Pretreatment strategies for delignification of sugarcane bagasse: A review. *Braz. Arch. Biol. Technol.* **2013**, *56*, 679–689. [CrossRef]
4. FAO–Food and Agriculture Organization of the United Nations. FAOSTAT Crops. Available online: http://www.fao.org/faostat/en/#data/QC (accessed on 10 May 2020).
5. USDA–United States Department of Agriculture. Sugar: World Markets and Trade. Available online: https://apps.fas.usda.gov/psdonline/circulars/sugar.pdf (accessed on 10 May 2020).
6. CONAB–Companhia Nacional de Abastecimento. Safra Brasileira de Cana de Açúcar. Available online: https://www.conab.gov.br/info-agro/safras/cana (accessed on 10 May 2020).
7. Rodrigues, R.C.L.B.; Felipe, M.G.A.; Sil, J.B.A.; Vitolo, M. Response surface methodology for xylitol production from sugarcane bagasse hemicellulosic hydrolyzate using controlled vacuum evaporation process variables. *Proc. Biochem.* **2003**, *38*, 1231–1237. [CrossRef]
8. Soccol, C.R.; Vandenberghe, L.P.S.; Medeiros, A.B.P.; Karp, S.G.; Buckeridge, M.; Ramos, L.P.; Pitarelo, A.P.; Ferreira-Leitão, V.; Gottschalk, L.M.F.; Ferrara, M.A.; et al. Bioethanol from lignocelluloses: Status and perspectives in Brazil. *Bioresour. Technol.* **2010**, *101*, 4820–4825. [CrossRef]
9. USDA–United States Department of Agriculture. World Agricultural Production. Available online: https://apps.fas.usda.gov/psdonline/circulars/production.pdf (accessed on 10 May 2020).
10. Karp, S.G.; Igashiyama, A.H.; Siqueira, P.F.; Carvalho, J.C.; Vandenberghe, L.P.S.; Thomaz-Soccol, V.; Coral, J.; Tholozan, J.L.; Pandey, A.; Soccol, C.R.; et al. Application of the biorefinery concept to produce L-lactic acid from the soybean vinasse at laboratory and pilot scale. *Bioresour. Technol.* **2011**, *102*, 1765–1772. [CrossRef]
11. Liu, H.-M.; Li, H.-Y. Application and conversion of soybean hulls. In *Soybean-The Basis of Yield, Biomass and Productivity*, 1st ed.; Kasai, M., Ed.; IntechOpen: London, UK, 2017. [CrossRef]

12. Medina, J.D.C.; Woiciechowski, A.L.; Zandona Filho, A.; Brar, S.K.; Magalhães Júnior, A.I.; Soccol, C.R. Energetic and economic analysis of ethanol, xylitol and lignin production using oil palm empty fruit bunches from a Brazilian factory. *J. Clean. Prod.* **2018**, *195*, 44–55. [CrossRef]
13. Morozova, V.V.; Gusakov, A.V.; Andrianov, R.M.; Pravilnikov, A.G.; Osipov, D.O.; Sinitsyn, A.P. Cellulase complex of the fungus *Penicillium verruculosum*: Properties of major endoglucanases and cellobiohydrolases. *Biotechnol. J.* **2010**, *5*, 871–880. [CrossRef]
14. Gusakov, A.V.; Sinitsyn, A.P. Cellulases from *Penicillium* species for producing fuel from biomass. *Biofuels* **2012**, *3*, 463–477. [CrossRef]
15. Osipov, D.O.; Rozhkova, A.M.; Matys, V.Y.; Koshelev, A.V.; Okunev, O.N.; Rubtsova, E.A.; Pravilnikov, A.G.; Zorov, I.N.; Oveshnikov, I.N.; Davidov, E.R.; et al. Production of biocatalysts on the basis of recombinant heterologous xylanase producer strains in the *Penicillium verruculosum* fungus: Their application in the hydrolysis of timber and wood processing industry wastes. *Catal. Ind.* **2011**, *3*, 34–40. [CrossRef]
16. Dotsenko, G.S.; Gusakov, A.V.; Rozhkova, A.M.; Korotkova, O.G.; Sinitsyn, A.P. Heterologous beta-glucosidase in a fungal cellulase system: Comparison of different methods for development of multienzyme cocktails. *Proc. Biochem.* **2015**, *50*, 1258–1263. [CrossRef]
17. Nelson, N. A photometric adaptation of the Somogyi method for the determination of sugars. *J. Biol. Chem.* **1944**, *153*, 375–379.
18. Somogyi, M. A new reagent for the determination of sugars. *J. Biol. Chem.* **1945**, *160*, 61–68.
19. Somogyi, M. Notes on sugar determination. *J. Biol. Chem.* **1952**, *195*, 19–23.
20. Sinitsyna, O.A.; Bukhtoyarov, E.F.; Gusakov, A.V.; Okunev, O.N.; Bekkarevitch, A.O.; Vinetsky, Y.P.; Sinitsyn, A.P. Isolation and properties of major components of *Penicillium canescens* extracellular enzyme complex. *Biochemistry* **2003**, *68*, 1200–1209. [CrossRef] [PubMed]
21. Bulakhov, A.G.; Volkov, P.V.; Rozhkova, A.M.; Gusakov, A.V.; Nemashkalov, V.A.; Sinitsyn, A.P. Using an inducible promoter of a gene encoding *Penicillium verruculosum* glucoamylase for production of enzyme preparations with enhanced cellulase performance. *PLoS ONE* **2017**, *12*, e0170404. [CrossRef]
22. Kaputska, L.A.; Annala, A.E.; Swanson, W.C. The peroxidase-glucose oxidase system: A new method to determine glucose liberated by carbohydrate degrading soil enzymes. *Plant Soil* **1981**, *63*, 487–490. [CrossRef]
23. Woiciechowski, A.L.; Dalmas Neto, C.J.; Vandenberghe, L.P.S.; Carvalho Neto, D.P.; Sydney, A.C.N.; Letti, L.A.J.; Karp, S.G.; Torres, L.A.Z.; Soccol, C.R. Lignocellulosic biomass: Acid and alkaline pretreatments and their effects on biomass recalcitrance–Conventional processing and recent advances. *Bioresour. Technol.* **2020**, *304*, 122848. [CrossRef]
24. Rojas, M.J.; Siqueira, P.F.; Miranda, L.C.; Tardioli, P.W.; Giordano, R.L.C. Sequential proteolysis and cellulolytic hydrolysis of soybean hulls for oligopeptides and ethanol production. *Ind. Crop. Prod.* **2014**, *61*, 202–210. [CrossRef]
25. Scarpa, J.C.P.; Marques, N.P.; Monteiro, D.A.; Martins, G.M.; de Paula, A.V.; Boscolo, M.; Silva, R.; Gomes, E.; Bocchini, D.A. Saccharification of pretreated sugarcane bagasse using enzymes solution from *Pycnoporus sanguineus* MCA 16 and cellulosic ethanol production. *Ind. Crop. Prod.* **2019**, 111795. [CrossRef]
26. Martin, C.; Volkov, P.V.; Rozhkova, A.M.; Puls, J.; Sinitsyn, A.P. Comparative study of the enzymatic convertibility of glycerol- and dilute acid-pretreated sugarcane bagasse using *Penicillium*-and *Trichoderma*-based cellulase preparations. *Ind. Crop. Prod.* **2015**, *77*, 382–390. [CrossRef]
27. Hickert, L.R.; Cruz, M.M.; Dillon, A.J.P.; Fontana, R.C.; Rosa, C.A.; Ayub, M.A.Z. Fermentation kinetics of acid–enzymatic soybean hull hydrolysate in immobilized-cell bioreactors of *Saccharomyces cerevisiae*, *Candida shehatae*, *Spathaspora arborariae*, and their co-cultivations. *Biochem. Eng. J.* **2014**, *88*, 61–67. [CrossRef]
28. Qing, Q.; Guo, Q.; Zhou, L.; Gao, X.; Lu, X.; Zhang, Y. Comparison of alkaline and acid pretreatments for enzymatic hydrolysis of soybean hull and soybean straw to produce fermentable sugars. *Ind. Crop. Prod.* **2017**, *109*, 391–397. [CrossRef]

© 2020 by the authors. Licensee MDPI, Basel, Switzerland. This article is an open access article distributed under the terms and conditions of the Creative Commons Attribution (CC BY) license (http://creativecommons.org/licenses/by/4.0/).

agronomy

MDPI

Article

Comparative Study of the Convertibility of Agricultural Residues and Other Cellulose-Containing Materials in Hydrolysis with *Penicillium verruculosum* Cellulase Complex

Dmitrii O. Osipov [1,*], Gleb S. Dotsenko [1], Olga A. Sinitsyna [2], Elena G. Kondratieva [2], Ivan N. Zorov [1,2], Igor A. Shashkov [1], Aidar D. Satrutdinov [1] and Arkady P. Sinitsyn [1,2]

1 Federal Research Centre "Fundamentals of Biotechnology", Russian Academy of Sciences, 119071 Moscow, Russia; gsdotsenko@gmail.com (G.S.D.); inzorov@mail.ru (I.N.Z.); igorshashkov@bk.ru (I.A.S.); sat-aidar@mail.ru (A.D.S.); apsinitsyn@gmail.com (A.P.S.)

2 Department of Chemistry, M. V. Lomonosov Moscow State University, 119991 Moscow, Russia; oasinitsyna@gmail.com (O.A.S.); elgenkon@inbox.ru (E.G.K.)

* Correspondence: d.osipov@fbras.ru

Received: 29 September 2020; Accepted: 2 November 2020; Published: 4 November 2020

check for updates

Abstract: Non-edible cellulose-containing biomass is a promising and abundant feedstock for simple sugar production. This study presents the results of different cellulose-containing materials (CCM) hydrolysis experiments with *P. verruculosum* enzyme complexes in laboratory conditions. Among the non-pretreated substrates, only a few had a relatively high convertibility—soy bean husks (31%) and sugar beat pulp (20%)—while wheat straw, oat husks, sunflower peals, and corn stalks had a low convertibility of 3% to 12%. This indicates that a major part of CCM needs pretreatment. Steam-exploded (with $Ca(OH)_2$) soy bean and oat husks (76% and 58%), fine ball-milled aspen wood and nitric acid-pretreated aspen wood (62% and 78%), and steam-exploded (with sulfuric acid) corn stalks (55%) had a high convertibility. Woody biomass pretreated with pulp and paper mills also had a high convertibility (56–78%)—e.g., never dried kraft hardwood and softwood pulp (both bleached and unbleached). These results demonstrate that effective cellulose-containing material processing into simple sugars is possible. Simple sugars derived from CCM using *P. verruculosum* preparation are a promising feedstock for the microbiological production of biofuels (bioethanol and biobutanol), aminoacids, and organic acids (e.g., lactic acid for polylactic acid production).

Keywords: cellulose-containing materials; *Penicillium verruculosum*; recombinant enzymes; pretreatment

1. Introduction

All natural cellulose fibers have their origin in the plant cell wall. Different plants' cell walls are radically different, but the cellulose content remains constant in the range of 35% to 50% of plant dry weight [1]. With rare exceptions, cellulose fibers are encased in a matrix composed of hemicelluloses and lignin, which constitute approximately 20–35% and 5–30% of plant dry weight [1–3]. Plant cell wall composition and architecture are fundamental characteristics of cellulose-containing biomass. While some other characteristics, such as size and moisture content, could be easily adjusted, carbohydrate and lignin content, pore density, and cellulose crystallinity are much harder to change and make the utilization of cellulose-containing materials more complex than that of pure cellulose.

The main requirements for biotechnology feedstock are low cost, availability, and high sugar yield [4,5]. The cost and availability depend on the logistics. In turn, the sugar yield in enzymatic hydrolysis or the convertibility of cellulose-containing materials are determined by fundamental characteristics. It is necessary to identify cellulose-containing materials that match the requirements for biotechnology processes. It is reasonable to obtain simple sugars from low-cost non-food feedstock firstly due to the lack of competition with edible biomass and secondly to resolve some ecological problems—i.e., environmental pollution, greenhouse gas emission, or the alienation of productive land for landfills.

The first group of widely available potential feedstock is agricultural waste which consists mainly of herbaceous biomass (straw, husks, cobs, peels, leaves) with a low lignin content. The worldwide production of crop plant residues is about 3.4 billion tons yearly [6], including 100 million dry tons of corn stover [7], 280 million dry tons of bagasse [8], and 354 million dry tons of wheat straw [9]. As sugarcane bagasse in tropical countries, sugar beet pulp is a major by-product of the sugar industry in countries with a continental climate. The annual production of sugar beet pulp is 21–22 million wet tons, and its main use is as animal feed [10]. More than 100 million tons of agricultural plant residues are generated every year in the Russian Federation [11]. In general, the dry weight residue/grain ratio for cereal varies from 1 to 2 [12]. This agricultural biomass requires reutilization, otherwise it is left on field, dumped, and incinerated.

The annual harvest of wood biomass for saw logs and paper and pulp is 960 million tons. The estimated forestry processing wastes in the EU are about 88.2 million tons per year [13]; in the Russian Federation, they are about 35.5 million m^3/yr [14]. Woodworking industry residues are widespread and often have no need to be additionally milled, but felling residues have to. Operating felling residues are important for sustainable forest use and preventing forest soil contamination. Aspen (*Populus tremula*) has a great potential in future bioeconomy as an energy crop due to its fast growth or as a felling residue because of its poor wood.

Food production generates a huge amount of residues and by-products that are usually used as animal feed due to residual protein, starch, and sugar [15,16] but could find application as a biotechnology feedstock. A total of 90 to 150 million tons of cultivated wheat is converted into bran every year [17]. Starch-containing bran was studied for bioethanol production [18] and protein extraction [19]. The main coproducts of grain-derived fuel and food-grade ethanol production are distillers' dry grain and solubles, which are limited in monogastric livestock. The growing production is expected to drive its supple up and prices down, making it a promising cellulose-containing resource [20].

It is common opinion that raw wet and dry cellulose-containing materials need to be pretreated prior enzymatic saccharification due to their rigidity. The most studied types of pretreatment are dry or wet mechanical milling, steam explosion and its varieties, dilute acid or alkaline pretreatment, and pretreatment with organic solvents. Irrespective of the pretreatment process, the cellulose-containing material's fundamental characteristics change. While relatively novel pretreatment processes require additional optimization and upscaling, pulp and paper mills are already fine-tuned for the production of delignified cellulose-rich materials located close to forest resources and could be quickly integrated into forest biorefinery. The estimated global production of mechanical and semi-chemical pulp is 36.3, with chemical pulp at 142.4, Kraft pulp (bleached and unbleached) at 140, and sulfite pulp at 2.4 million t/yr in 2017 [21]. The only technology missing in pulp mill production chain is enzymatic hydrolysis. The production of cellulosic sugars and further biofuels and biochemicals on a reorganized mill is economically feasible because of the compatible supply, facilities, and workforce [22]

In some cases, it is reasonable to process unconventional local cellulose-containing materials, such as municipal solid waste or algal biomass [23].

The aim of this research is to evaluate the enzymatic convertibility of cellulose-containing materials of different origin—including agricultural and industrial byproducts and waste in enzymatic hydrolysis with *P. verruculosum* cellulolytic complexes—to simple sugars which could be used as raw materials in different sectors of bioindustry.

2. Materials and Methods

2.1. Samples of CCM

The sugar beet pulp was from Kanevsky sugar refinery (Krasnodar, Russia) and wheat straw from Khimavtomatica SPA (Moscow, Russia). Dried and steam explosion-pretreated bagasse was provided by DNL (Rotterdam, The Netherlands), Florida Crystals (West Palm Beach, FL, USA), and BC International. Sunflower peels were from East Siberian Biotechnology Center PLC (Tulun, Russia). Raw and pretreated soy husks and oat bran were provided by ADM (Decatur, IL, USA). Dried and steam explosion-pretreated corn stalks were obtained from NREL (Golden, CO, USA), Dyadic International (Jupiter, FL, USA), and ADM (Decatur, IL, USA). Corn bran was obtained from Gruma S.A.B. de S.V. (San Pedro, Mexico). Wheat bran and brewing waste were from Bigor LLC (Moscow, Russia). Wet and dried distillers grains were provided by ADM (Decatur, Illinois). Original and ball-milled wood samples (pine, aspen, larch, hevea) were from the East Siberian Biotechnology Center (Tulun, Russia). Samples of bleached and unbleached softwood and hardwood pulp were produced from spruce and birch-aspen mixtures, respectively, from Arkhangelsk Pulp and Paper Mill (Arkhangelsk, Russia).

2.2. Substrates

Carboxymethylcellulose (CMC, sodium salt, medium viscosity), birch glucuronoxylan, *p*-nitrophenyl-β-D-glucopyranoside, and cellobiose were purchased from Sigma (St. Louis, MO, USA); Avicel PH105 cellulose was from Serva (Heidelberg, Germany); filter paper No.1 was from Whatman (Little Chalfont, UK).

2.3. Enzyme Preparations

B151 and F10 preparations represent freeze-dried ultra-concentrated cultural filtrates of *Penicillium verruculosum* strains B151 (represents cellulase and xylanase complex) [24] and F10 (recombinant strain after the heterologous expression of *Aspergillus niger* β-glucosidase (cellobiase)) cultivated in 1l fermenters at the Institute of Biochemistry of RAS (Moscow, Russia) [25].

2.4. Analytical Procedures

2.4.1. Enzyme Activity Assays

The filter paper activity was determined using a standard method with 1 cm × 5 cm (50 mg) Whatman No. 1 filter paper strips in 0.05 sodium citrate buffer pH 4.8 at 50 °C for 1 h and DNS reagent for measuring reducing sugars [26]. The CMCase and xylanase activities were determined by a reducing sugar release at pH 5.0 and 50 °C after 10 min using a substrate concentration of 5 mg/mL in the reaction mixture [27]. Avicelase activity was determined by the reducing sugar release at pH 5.0 and 40 °C after 60 min of enzyme reaction with Avicel PH105 (5 mg/mL) [28]. Reducing sugars were assayed by the Nelson-Somogyi spectrophotometric method based on molybdenum blue formation in the reaction of molybdic acid reduction by cuprous oxide. A total of 0.2 mL of reducing sugars containing sample solution with 0.2 mL of Somogyi reagent were mixed and incubated at 100 °C for 1 h, then sequentially 0.2 mL of Nelson, 0.4 mL of acetone and 1 mL of distilled water were added. Adsorption was measured at 610 nm [29]. One unit of activity corresponded to the quantity of enzyme releasing 1 µmol of reducing sugars (in glucose equivalents) for one minute. The activity against p-NP-β-glucospiranoside was determined at pH 5.0 and 40 °C by measuring the p-nitrophenol released, as described elsewhere [28]. One β-glucosidase unit of activity is the quantity of enzyme which liberates 1 micromole of *p*-nitrophenol in one minute. Cellobiase was determined by the incubation of 2.5 mM of cellobiose solution with the enzymes at 40 °C and pH 5.0 and measuring the released glucose with the glucose oxidase/peroxidase method [30]. One unit of cellobiase activity is the quantity of enzyme which liberates 1 micromole of glucose in one minute. The activities of the enzyme preparations are given in the Table 1. Protein concentration was determined according to Lowry protein assay [31] using bovine serum albumin as a standard.

Table 1. Properties of *P. verruculosum* enzyme preparations (EP).

EP	Protein, mg/g	FP, U/g	Avicel, U/g	CMC, U/g	pNPG, U/g	Cellobiose, U/g	Xylan, U/g
B151	970 ± 26	760 ± 25	578 ± 17	16,542 ± 340	1074 ± 52	603 ± 47	17,532 ± 505
F10	655 ± 14	147 ± 5	853 ± 21	7007 ± 73	39,852 ± 926	46,663 ± 1134	3800 ± 39

2.4.2. Enzymatic Hydrolysis

CCM hydrolysis tests were conducted in 50 mL vessels incubated at 50 °C and 250 rpm on an Innova 40 shaker (Edison, NJ, USA) for 48 h. A reaction mixture of 20 mL total contained 100 g/L of CCM dry matter in 0.1 M of Na-acetate buffer at pH 5.0 with 10 mg/g of dry matter protein loading of B151 enzyme preparation. To overcome the cellobiose inhibition effect, an amount (40 U/g dry substrate) of cellobiase activity of F10 enzyme preparation was added to the mixture. Then, 1 mM of NaN_3 and 100 µg/mL of ampicillin were used to prevent contamination. Nelson-Somogyi assay [29] was used to determine the concentration of reducing sugars in the reaction mixture.

CCM enzymatic convertibility was defined as a degree of conversion (48 h) to reducing sugars (in glucose equivalent) as a percentage per initial concentration of substrate (*w*/*w*).

2.5. Pretreatment Conditions

Wheat straw; sunflower peels; bagasse; and aspen, larch, and pine sawdust were fine milled using an AGO-2 laboratory ball mill (<20 µm). Sugar beet pulp were pretreated in a double-screw extruder at 182 °C, 30 atm, for 50 s.

Aspen wood was milled (300 µm sieve) using an IM450 industrial impeller mill (SE TechPribor, Shchyokino, Russia). Milled aspen wood was pretreated in different ways: with diluted (0.9–12.7%) sulfuric acid at 120–180 °C for 15 to 180 min, with 0.18–0.54% sulfuric acid solution in ethanol and butanol, and with diluted (0.15–4.8%) nitric acid at 100–160 °C and elevated pressure for 60 min. The diluted acid pretreatment was conducted in a 100 mL pressurized steel cylinder. A cylinder with milled aspen wood and an acid solution with a solid-to-liquid ratio of 1:5 was placed in a temperature-controlled oil bath. Elevated pressure was created by nitrogen injection. By the end of the pretreatment process, the reactors were cooled in cold water, then the slurry was filtered and the remaining solids were washed with water or alcohol–water mixture.

3. Results and Discussion

Unpretreated agricultural residues are resistant to cellulolytic enzymes and characterized by a low reducing sugar yield in hydrolysis with a mixture of *P. verruculosum* B151 cellulase complex and F10 β-glucosidase. The wheat straw convertibility was 12%. For the sugar beet pulp, it was 20%, for oat husks it was 5%, for sunflower peels it was 3%, for corn stalks it was 10%, and for bagasse it was 17%. Only the soy husk convertibility was relatively high, at 31% (Table 2).

Table 2. Convertibility of different cellulose-containing materials in hydrolysis by a mixture of *P. verruculosum* B151 cellulase complex and F10 β-glucosidase.

Substrate	Convertibility, %
Agricultural residues	
Wheat straw	12
Wheat straw pretreated by dry fine ball-milling (<20 µm)	45
Wheat straw pretreated by 1% NaOH, 85 °C	55
Wheat straw pretreated by steam explosion	75
Wheat straw steam pretreated by steam explosion with $Ca(OH)_2$	69

Table 2. *Cont.*

Substrate	Convertibility, %
Sugar beet pulp	20
Sugar beet pulp extruded	27
Oat husks	5
Oat husks pretreated by steam explosion with $Ca(OH)_2$	76
Soy bean husks	38
Soy bean husks pretreated by steam explosion with $Ca(OH)_2$	58
Sunflower peels	3
Sunflower peels pretreated by dry fine ball-milling (<20 μm)	7
Corn stalks	10
Corn stalks pretreated by steam explosion with H_2SO_4	55
Corn stalks pretreated by steam explosion with $Ca(OH)_2$	36
Sugar cane bagasse	18
Sugar cane bagasse pretreated by dry fine ball-milling (<20 μm)	42
Sugar cane bagasse pretreated by steam explosion with $Ca(OH)_2$	41
Sugar cane bagasse pretreated by steam explosion with H_2SO_4	34
Food-industry waste	
Brewing waste (rye-wheat)	10
Wheat bran (destarched)	14
Corn bran (destarched)	12
Distillers grains wet (WDG)	18
Distillers grains dried (DDG)	16
Pulp and paper industry products	
Never dried bleached softwood kraft pulp	78
Never dried unbleached softwood kraft pulp	68
Dried bleached softwood kraft pulp	58
Dried unbleached softwood kraft pulp	48
Never dried bleached hardwood kraft pulp	66
Never dried unbleached hardwood kraft pulp	56
Dried bleached hardwood kraft pulp	50
Dried unbleached hardwood kraft pulp	42
Wood industry waste and forestry residues	
Pine sawdust	8
Pine sawdust (deresinated) pretreated by dry fine ball-milling (<20 μm)	45
Larch sawdust	6
Larch sawdust pretreated by dry fine ball-milling (<20 μm)	22
Aspen sawdust	8
Aspen sawdust pretreated by dry fine ball-milling (<20 μm)	50
Hevea sawdust	4
Hevea sawdust pretreated by dry fine ball-milling (<20 μm)	14

Mechanical pretreatments such as fine ball milling or extruding have resulted in increasing the convertibility of these substrates 1.35–3.75 fold. The data of enzymatic hydrolysis obtained for mechanically pretreated materials have shown that the ball milling (which gives an average particle size of less than 20 μm) of wheat straw and bagasse results in increased convertibility to 45% and 42%, respectively. The ball milling of sunflower peels leads to a very limited improvement in convertibility at 7%, which indicates that they are practically not digestible by cellulolytic enzymes. Sugar beet pulp pretreatment by extrusion has also shown a limited improvement in convertibility (to 27% only).

The wheat straw convertibility after delignification with hot alkaline solution increased 4.6-fold (up to 55%).

Steam explosion pretreatment with different additives (H_2SO_4, $Ca(OH)_2$) demonstrated that wheat straw, out husks, soy bean husks, corn stalks, and bagasse were easy to hydrolyze with enzymes, and this pretreatment enhanced the convertibility up to 69–75%, 76%, 58%, 36–55%, and 34–41%, respectively. Supplementation with H_2SO_4 and $Ca(OH)_2$ has shown an opposite result for different materials: calcium hydroxide is preferable for bagasse pretreatment (7% higher sugar yield), while a corn stalk pretreatment required sulfuric acid (19% higher sugar yield). Wheat straw steam pretreatment required no additives.

The enzymatic hydrolysis of unpretreated food-industry waste has shown that this kind of cellulose-containing material is far from being a potential source of simple sugars for biotechnology. Thus, the convertibility of brewing waste, destarched corn, and wheat bran was very low, at 10%, 12%, and 14%, respectively. The convertibility of wet and dry distillers grains hydrolysis was slightly higher, at 18% wet and 16% dry, respectively.

Pulp and paper production is a large-tonnage and streamlined industry. The range of products in this area is very wide. They differ in the raw materials (hardwood or softwood, others) used and the way they are produced (wood cooking, bleaching). Creating an integrated biorefinery plant around existing pulp and paper mills would enhance their marketability by efficient converting all biomass components into value-added products [32]. The Kraft pulping process is a promising pretreatment technology for biocatalytic conversion of cellulose and hemicelluloses to glucose and other monosaccharides [33,34]. The convertibility of newer dry as well as dried kraft fibers representing by bleached and unbleached soft wood and hardwood pulp was evaluated (Table 1). The highest convertibility was demonstrated by wet bleached softwood pulp, at 78%; the reason for the high convertibility is the almost complete removal of lignin by the Kraft process (a remaining lignin content was 2–3% [33]). The convertibility of wet bleached hardwood pulp was 58%; this is lower compared with bleached softwood pulp because of the xylan influence [33]. Unbleached wet softwood and hardwood pulp had approximately a 1.1 times lower convertibility compared with similar types of bleached pulp; the decrease in convertibility of unbleached pulp is explained by the higher lignin content.

The drying and subsequent hornification of all kraft pulps types had a significant effect on convertibility as it reduces swelling and the cellulose fiber accessibility [35] and causes a collapse in the pore structure [36]. The convertibility was reduced by 1.3–1.4 times compared to wet pulp.

The recycling and utilization of wood industry wastes and forestry residues is crucial for wood processing and environmental security. In total, five types of wood species were included in this study. The convertibility of pine, larch, aspen, and hevea sawdust was low and found to be in the range of 4–8% (Table 1). Mechanical pretreatment (dry fine ball-milling that results in an average particle size less than 20 μm) has resulted in a significant increase in the convertibility of pine wood and aspen wood to up to 45–50%; in the case of larch, it is up to 22%. The increase in convertibility was due to defibrillation and reduction in the crystallinity of fibers and increasing surface area related to reducing particle size [35]. Despite its high efficiency, fine ball-milling has serious disadvantages, such as being energy consuming and difficulties in scaling up [37]. In view of the rising energy prices and power intensity, fine milling is not economically reasonable [38].

To counteract these disadvantages, less intensive milling processes can be combined with chemical and physicochemical pretreatments such as dilute acid and organosolv pretreatments. We have studied the convertibility of aspen wood subjected to pretreatment by different water and organic solutions of mineral acids. Relatively low temperatures of pretreatment process were selected to prevent the unfavorable degradation of carbohydrates and inhibitors formation.

Pretreatment of aspen wood by dilute acids: Five samples were obtained using dilute sulfuric acid pretreatment of aspen wood particles (200–300 μm). The results demonstrate a linear correlation between acid concentration and substrate convertibility (Table 3). There was no difference found (58.4% convertibility) for samples processed with 12.7% and 8.7% sulfuric acid, which could mean that the maximum available polysaccharides for enzymatic hydrolysis due to the solubilization of hemicelluloses are limited. Further reduction in the acid concentration to 4.4% and 1.8% results in the convertibility decreasing to 44.2% and 42.9%, respectively. Further reduction in the acid concentration to 0.9% provides a convertibility of 41.6%, which is just 1.4 times lower than the result obtained by 12.7% acid. Such a reduction in chemical consumption can be economically feasible even with a lower biomass convertibility.

Table 3. Convertibility of aspen wood pretreated using different types of acid-containing solutions at elevated temperatures.

Conditions				Convertibility, %
Aspen wood (200–300 μm)				8
Dilute sulfuric acid pretreatment				
Temperature, °C	Time, h	Acid concentration, %		
140	1	12.7		58.4
		8.7		58.4
		4.4		44.2
		1.8		42.9
		0.9		41.6
Dilute nitric acid pretreatment				
Temperature, °C	Time, h	Pressure, at	Acid concentration, %	
100	1	6	4.8	61.2
130		5	1.1	62.8
125		9	4.8	60.6
125		14	4.8	65.1
125		18	4.8	78.7
125		22	4.8	78.6
150		5	0.2	49.3
160		5	0.5	45.6
160		5	0.7	48.2
160		5	0.3	46.9
160		5	0.2	43.7
Sulfuric acid organosolv				
Temperature, °C	Time, h	Organic phase, %	Acid concentration, %	
140	1	50% EtOH	0.36	38.7
		65% EtOH	0.54	48.3
		80% EtOH	0.54	43.9
		50% BtOH	0.36	36.4
		65% BtOH	0.54	58.7
		80% BtOH	0.54	51.1
		25% EtOH, 25% BtOH	0.54	54.3
		40% EtOH, 10% BtOH	0.54	51.1
		20% EtOH, 40% BtOH	0.36	51.5
		20% EtOH, 40% BtOH	0.18	45.9
		25% EtOH, 25% BtOH	0.18	37.6
		10% EtOH, 40% BtOH	0.18	36.0

Enhancing the hydrolysis of lignocellulose biomass for the efficient conversion of cellulose and hemicellulose by pretreatment using nitric acid has not been highly studied compared to sulfuric acid. To estimate the influence of nitric acid on the convertibility of 200–300 μm aspen wood in severe and mild conditions, another series of experiments was carried out.

The experimental results demonstrate that the reducing sugars yields after a relatively mild pretreatment at 100–130 °C, 5–6 at, and 1–4.8% nitric acid were purely comparable, at about 60% (Table 3). Maximum convertibility was achieved with a subsequent elevation of gas pressure. The aspen wood convertibility enhanced significantly from 60.6% to 78.7% as the pressure increased from 9 to 18–22, while the temperature and acid concentration remained constant (125 °C, 4.8%). These nitric acid concentration and pressure values found in this study are optimal for pretreatment, since increasing the temperature up to 160 °C results in a convertibility below 50% for acid concentrations of 0.3–0.7%. There was no additional effect on the reducing sugars yield when the pressure was raised from 18 to 22 at.

It could be concluded that aspen wood pretreatment with dilute nitric acid (convertibility 78.7% at an acid concentration of 4.8%) is more efficient than with dilute sulfuric acid (convertibility 58.7% at an acid concentration of 12.7%) but requires more complex equipment.

Organosolve pretreatment of aspen wood: The results obtained in this study demonstrate that dilute acid pretreatment is very effective. However, after such pretreatment lignin solubilized poorly [39], even though the hemicellulose matrix is being dissolved. Thus, the next panel of experiments was aimed to discover best conditions for fractionation and recovery of lignin, cellulose, and hemicelluloses. All the experiments were conducted at constant temperature 140 °C for the same time 1 h, but at different concentrations of organic solvents (ethanol and *n*-buthanol) and sulfuric acid (0.18–0.54%) as a catalyst.

The data display that 50% (*v*/*v*) alcohol–water mixtures have the same efficiency of 36–37% with 0.36% acid. Using a 0.54% acid concentration, 65% *n*-butanol is more preferable than 65% ethanol, ensuring a convertibility of 58.7% and 48.3%, respectively (Table 3). Organosolve pretreatment with 80% alcohol results in a lower convertibility (43.9% and 54.1% for ethanol and *n*-butanol respectively), but *n*-butanol still provided a better enzymatic digestibility. This likely could be explained as the swelling of cellulose fibers decreased as the ethanol concentration increased [40].

During organosolve pretreatment with *n*-butanol, three fractions were obtained: black liquor containing dissolved lignin, hemicellulose-enriched liquid fraction, and cellulose-containing solid fraction [41]. This spatial separation also could explain better biomass hydrolysability after *n*-butanol pretreatment. Previously [42], it has been found that the swelling of cellulose in organic solvent strongly depends on the species of organic solvents—the solvent basicity, the molar volume, and the hydrogen bonding capability—thus, *n*-butanol is more significant in two-component mixtures. This was shown in experiments with alcohol concentrations of 25% EtOH + 25% BtOH (convertibility 54%) and 40% EtOH + 10% BtOH (convertibility 51%). However, in general, more concentrated *n*-butanol single-component mixtures are preferable to two-component mixtures.

Decreasing acid concentrations from 0.36% to 0.18% resulted in a lower convertibility using the same mixture composition (20% EtOH + 40% BtOH)—52% and 46%, respectively. At an acid concentration of 0.18%, this mixture composition has no effect on the convertibility.

These results indicate that alcohol-water mixtures allow using less concentrated acids, while biomass components could be fractionated and alcohols could be recirculated. In this study, the optimal conditions for pretreatment were found: 140 °C, 1 h, 0.54% sulfuric acid, 65% *n*-butanol. These conditions lead to a 7.4-fold increase in the convertibility or aspen wood (59% compared to 8% of untreated substrate).

4. Conclusions

In total, the convertibility in enzymatic hydrolysis by a mixture of *P. verruculosum* B151 cellulase complex and F10 β-glucosidase of 69 samples of original and pretreated CCM was tested in this study. A major part of the non-pretreated substrates had a low convertibility of 3% to 18%. There were only a few substrates with a higher convertibility among the original unpretreated samples—e.g., soy bean husks had a convertibility of 31%, with sugar beat pulp at 20%. The low level of convertibility of CCM indicates the necessity of pretreatment [43,44].

Among the pretreated feedstocks, steam-exploded (with $Ca(OH)_2$) soy bean and oat husks (76% and 58%), fine ball-milled aspen wood and nitric acid-pretreated aspen wood (62% and 78%), and steam-exploded (with sulfuric acid) corn stalks (55%) had a sufficiently high convertibility. It should be noted that types of cellulosic feed stocks that are the source of pulp and paper industry, such as newer dried kraft hardwood and softwood pulp (both bleached and unbleached), had a high convertibility

(56–78%). This kind of CCM from our point of view had matching characteristics with the requirements for biomass enzymatic conversion to simple sugars and downstream processes, since they have a high convertibility, bulk availability, and waste disposal problems. The softwood and hardwood kraft pulp had a higher convertibility and are the most promising types of cellulose-containing materials from our point of view.

Author Contributions: Conceptualization, A.P.S.; Data curation, D.O.O. and G.S.D.; Formal analysis, D.O.O., G.S.D., and I.N.Z.; Investigation, D.O.O., G.S.D., E.G.K., I.A.S., and A.D.S.; Methodology, D.O.O., O.A.S., and I.N.Z.; Supervision, A.P.S.; Writing—original draft, D.O.O.; Writing—review and editing, A.P.S. All authors have read and agreed to the published version of the manuscript.

Funding: The work was partially supported by the Russian Fund of Basic Research (#18-54-80027).

Conflicts of Interest: The authors declare no conflict of interest. The funders had no role in the design of the study; in the collection, analysis, or interpretation of data; in the writing of the manuscript; or in the decision to publish the results.

References

1. Lynd, L.R.; Wyman, C.E.; Gerngross, T.U. Biocommodity Engineering. *Biotechnol. Prog.* **1999**, *15*, 777–793. [CrossRef] [PubMed]
2. Van Soest, P.J. *Nutritional Ecology of the Ruminant*; Cornell University Press: Ithaca, NY, USA, 1994; ISBN 1501732358.
3. Sarkanen, K.V.; Ludwig, C.H. *Liguins. Occurrence, Formation, Structure, and Reactions*; John Wiley & Sons, Inc.: New York, NY, USA, 1971; ISBN 0471754226.
4. McMillan, J.D. Bioethanol production: Status and prospects. *Renew. Energy* **1997**, *10*, 295–302. [CrossRef]
5. Sánchez, Ó.J.; Cardona, C.A. Trends in biotechnological production of fuel ethanol from different feedstocks. *Bioresour. Technol.* **2008**, *99*, 5270–5295. [CrossRef] [PubMed]
6. Van den Born, G.J.; van Minnen, J.G.; Olivier, J.G.J.; Ros, J.P.M. Integrated analysis of global biomass flows in search of the sustainable potential for bioenergy production. *PBL* **2014**, *1509*, 24.
7. Kadam, K.; McMillan, J. Availability of corn stover as a sustainable feedstock for bioethanol production. *Bioresour. Technol.* **2003**, *88*, 17–25. [CrossRef]
8. Pippo, W.A.; Luengo, C.A. Sugarcane energy use: Accounting of feedstock energy considering current agro-industrial trends and their feasibility. *Int. J. Energy Environ. Eng.* **2013**, *4*, 10. [CrossRef]
9. Kim, S.; Dale, B.E. Global potential bioethanol production from wasted crops and crop residues. *Biomass Bioenergy* **2004**, *26*, 361–375. [CrossRef]
10. Kuramshin, A.I.; Kuramshina, E.A. Preparation of microcrystalline cellulose from sugar beet pulp and its properties. *Uchenye Zap. Kazan. Univ.* **2015**, *157*, 18–26.
11. Porous ecology. Agriculture Produces 250 Million Tons of Waste per Year. Available online: https://www.agroinvestor.ru/technologies/article/29525-dyryavaya-ekologiya (accessed on 22 March 2018).
12. Graham, R.; Nelson, R.; Sheehan, J.; Perlack, R.D.; Wright, L. Current and Potential U.S. Corn Stover Supplies. *Agron. J.* **2007**, *99*, 1–11. [CrossRef]
13. Searle, S.; Malins, C. *Availability of Cellulosic Residues and Wastes in the EU*; White Paper; International Council on Clean Transportation: Washington, DC, USA, 2013.
14. Mokhirev, A.P.; Zyryanov, M.A. Logging operation technology with felling residue sorting. *Syst. Methods. Technol.* **2015**, *3*, 118–122.
15. Mustafa, A.F.; McKinnon, J.J.; Christensen, D.A. Chemical characterization and in vitro crude protein degradability of thin stillage derived from barley- and wheat-based ethanol production. *Anim. Feed Sci. Technol.* **1999**, *80*, 247–256. [CrossRef]
16. Davis, L.; Jeon, Y.-J.; Svenson, C.; Rogers, P.; Pearce, J.; Peiris, P. Evaluation of wheat stillage for ethanol production by recombinant Zymomonas mobilis. *Biomass Bioenergy* **2005**, *29*, 49–59. [CrossRef]
17. Prückler, M.; Siebenhandl-Ehn, S.; Apprich, S.; Höltinger, S.; Haas, C.; Schmid, E.; Kneifel, W. Wheat bran-based biorefinery 1: Composition of wheat bran and strategies of functionalization. *LWT-Food Sci. Technol.* **2014**, *56*, 211–221. [CrossRef]
18. Singh, D.P.; Trivedi, R.K. Biofuel from Wastes an Economic and Environmentally Feasible Resource. *Energy Procedia* **2014**, *54*, 634–641. [CrossRef]

19. Liu, J.; Guan, X.; Zhu, D.; Sun, J. Optimization of the enzymatic pretreatment in oat bran protein extraction by particle swarm optimization algorithms for response surface modeling. *LWT-Food Sci. Technol.* **2008**, *41*, 1913–1918. [CrossRef]
20. Bals, B.; Dale, B.; Balan, V. Enzymatic Hydrolysis of Distiller's Dry Grain and Solubles (DDGS) Using Ammonia Fiber Expansion Pretreatment. *Energy Fuels* **2006**, *20*, 2732–2736. [CrossRef]
21. FAO. *Forest Products 2017*; Rowman & Littlefield: Lanham, MD, USA, 2017; ISBN 9789251317174.
22. Jin, Y.; Jameel, H.; Chang, H.; Phillips, R. Green Liquor Pretreatment of Mixed Hardwood for Ethanol Production in a Repurposed Kraft Pulp Mill. *J. Wood Chem. Technol.* **2010**, *30*, 86–104. [CrossRef]
23. Mandels, M.; Hontz, L.; Nystrom, J. Enzymatic hydrolysis of waste cellulose. *Biotechnol. Bioeng.* **1974**, *16*, 1471–1493. [CrossRef]
24. Solov'eva, I.V.; Okunev, O.N.; Vel'kov, V.V.; Koshelev, A.V.; Bubnova, T.V.; Kondrat'eva, E.G.; Skomarovskii, A.A.; Sinitsyn, A.P. The selection and properties of Penicillium verruculosum mutants with enhanced production of cellulases and xylanases. *Mikrobiologiia* **2005**, *74*, 172–178. [CrossRef]
25. Dotsenko, A.S.; Rozhkova, A.M.; Gusakov, A.V. Properties and N-glycosylation of recombinant endoglucanase II from Penicillium verruculosum. *Moscow Univ. Chem. Bull.* **2015**, *70*, 283–286. [CrossRef]
26. Ghose, T.K. Measurement of cellulase activities. *Pure Appl. Chem.* **1987**, *59*, 257–268. [CrossRef]
27. Sinitsyna, O.A.; Bukhtoyarov, F.E.; Gusakov, A.V.; Okunev, O.N.; Bekkarevitch, A.O.; Vinetsky, Y.P.; Sinitsyn, A.P. Isolation and Properties of Major Components of Penicillium canescens Extracellular Enzyme Complex. *Biochemical* **2003**, *68*, 1200–1209. [CrossRef] [PubMed]
28. Gusakov, A.; Sinitsyn, A.; Salanovich, T.; Bukhtojarov, F.; Markov, A.; Ustinov, B.; Zeijl, C.; Punt, P.; Burlingame, R. Purification, cloning and characterisation of two forms of thermostable and highly active cellobiohydrolase I (Cel7A) produced by the industrial strain of Chrysosporium lucknowense. *Enzyme Microb. Technol.* **2005**, *36*, 57–69. [CrossRef]
29. Somogyi, M. Notes on sugar determination. *J. Biol. Chem.* **1952**, *195*, 19–23.
30. Sinitsyn, A.P.; Chernoglazov, V.M.; Gusakov, A.V. *Methods for Study and Properties of Cellulolytic Enzymes*; VINITI: Moscow, Russia, 1993.
31. Lowry, O.H.; Rosebrough, N.J.; Farr, A.L.; Randall, R.J. Protein measurement with the Folin phenol reagent. *J. Biol. Chem.* **1951**, *193*, 265–275.
32. Van Heiningen, A. Converting a Kraft pulp mill into an integrated forest biorefinery. *Pulp Pap. Canada* **2006**, *107*, 38–43.
33. Aksenov, A.S.; Tyshkunova, I.V.; Poshina, D.N.; Guryanova, A.A.; Chukhchin, D.G.; Sinelnikov, I.G.; Terentyev, K.Y.; Skorik, Y.A.; Novozhilov, E.V.; Synitsyn, A.P. Biocatalysis of industrial kraft pulps: Similarities and differences between hardwood and softwood pulps in hydrolysis by enzyme complex of penicillium verruculosum. *Catalysts* **2020**, *10*, 536. [CrossRef]
34. Novozhilov, E.V.; Aksenov, A.S.; Demidov, M.L.; Chukhchin, D.G.; Dotsenko, G.S.; Osipov, D.O.; Sinitsyn, A.P. Application of complex biocatalysts based on recombinant Penicillium verruculosum enzyme preparations in the hydrolysis of semichemical hardwood pulp. *Catal. Ind.* **2014**, *6*, 348–354. [CrossRef]
35. Aldaeus, F.; Larsson, K.; Srndovic, J.S.; Kubat, M.; Karlström, K.; Peciulyte, A.; Olsson, L.; Larsson, P.T. The supramolecular structure of cellulose-rich wood pulps can be a determinative factor for enzymatic hydrolysability. *Cellulose* **2015**, *22*, 3991–4002. [CrossRef]
36. Grous, W.R.; Converse, A.O.; Grethlein, H.E. Effect of steam explosion pretreatment on pore size and enzymatic hydrolysis of poplar. *Enzyme Microb. Technol.* **1986**, *8*, 274–280. [CrossRef]
37. Laser, M.; Schulman, D.; Allen, S.G.; Lichwa, J.; Antal, M.J.; Lynd, L.R. A comparison of liquid hot water and steam pretreatments of sugar cane bagasse for bioconversion to ethanol. *Bioresour. Technol.* **2002**, *81*, 33–44. [CrossRef]
38. Hendriks, A.T.W.M.; Zeeman, G. Pretreatments to enhance the digestibility of lignocellulosic biomass. *Bioresour. Technol.* **2009**, *100*, 10–18. [CrossRef]
39. Nitsos, C.; Rova, U.; Christakopoulos, P. Organosolv fractionation of softwood biomass for biofuel and biorefinery applications. *Energies* **2018**, *11*, 50. [CrossRef]

40. Ni, Y.; van Heiningen, A. Swelling of pulp fibers derived from the ethanol-based organosolv process. *Tappi J.* **1997**, *80*, 211–213.
41. Teramura, H.; Sasaki, K.; Oshima, T.; Matsuda, F.; Okamoto, M.; Shirai, T.; Kawaguchi, H.; Ogino, C.; Hirano, K.; Sazuka, T.; et al. Organosolv pretreatment of sorghum bagasse using a low concentration of hydrophobic solvents such as 1-butanol or 1-pentanol. *Biotechnol. Biofuels* **2016**, *9*, 27. [CrossRef]
42. Mantanis, G.I.; Young, R.A.; Rowell, R.M. Swelling of compressed cellulose fiber webs in organic liquids. *Cellulose* **1995**, *2*, 1–22. [CrossRef]
43. Mosier, N.; Wyman, C.; Dale, B.; Elander, R.; Lee, Y.Y.; Holtzapple, M.; Ladisch, M. Features of promising technologies for pretreatment of lignocellulosic biomass. *Bioresour. Technol.* **2005**, *96*, 673–686. [CrossRef] [PubMed]
44. Wyman, C.E.; Balan, V.; Dale, B.E.; Elander, R.T.; Falls, M.; Hames, B.; Holtzapple, M.T.; Ladisch, M.R.; Lee, Y.Y.; Mosier, N.; et al. Comparative data on effects of leading pretreatments and enzyme loadings and formulations on sugar yields from different switchgrass sources. *Bioresour. Technol.* **2011**, *102*, 11052–11062. [CrossRef]

Publisher's Note: MDPI stays neutral with regard to jurisdictional claims in published maps and institutional affiliations.

© 2020 by the authors. Licensee MDPI, Basel, Switzerland. This article is an open access article distributed under the terms and conditions of the Creative Commons Attribution (CC BY) license (http://creativecommons.org/licenses/by/4.0/).

agronomy

MDPI

Article

Evaluating Feruloyl Esterase—Xylanase Synergism for Hydroxycinnamic Acid and Xylo-Oligosaccharide Production from Untreated, Hydrothermally Pre-Treated and Dilute-Acid Pre-Treated Corn Cobs

Lithalethu Mkabayi [1], Samkelo Malgas [1], Brendan S. Wilhelmi [2] and Brett I. Pletschke [1,*]

1 Enzyme Science Programme (ESP), Department of Biochemistry and Microbiology, Rhodes University, Grahamstown, Eastern Cape 6140, South Africa; lithalethum@gmail.com (L.M.); samkelomalgas@yahoo.com (S.M.)

2 Department of Biochemistry and Microbiology, Rhodes University, Grahamstown, Eastern Cape 6140, South Africa; b.wilhelmi@ru.ac.za

* Correspondence: b.pletschke@ru.ac.za; Tel.: +27-46-6038081

Received: 4 April 2020; Accepted: 30 April 2020; Published: 13 May 2020

check for updates

Abstract: Agricultural residues are considered the most promising option as a renewable feedstock for biofuel and high valued-added chemical production due to their availability and low cost. The efficient enzymatic hydrolysis of agricultural residues into value-added products such as sugars and hydroxycinnamic acids is a challenge because of the recalcitrant properties of the native biomass. Development of synergistic enzyme cocktails is required to overcome biomass residue recalcitrance, and achieve high yields of potential value-added products. In this study, the synergistic action of two termite metagenome-derived feruloyl esterases (FAE5 and FAE6), and an endo-xylanase (Xyn11) from *Thermomyces lanuginosus*, was optimized using 0.5% (*w*/*v*) insoluble wheat arabinoxylan (a model substrate) and then applied to 1% (*w*/*v*) corn cobs for the efficient production of xylo-oligosaccharides (XOS) and hydroxycinnamic acids. The enzyme combination of 66% Xyn11 and 33% FAE5 or FAE6 (protein loading) produced the highest amounts of XOS, ferulic acid, and *p*-coumaric acid from untreated, hydrothermal, and acid pre-treated corn cobs. The combination of 66% Xyn11 and 33% FAE6 displayed an improvement in reducing sugars of approximately 1.9-fold and 3.4-fold for hydrothermal and acid pre-treated corn cobs (compared to Xyn11 alone), respectively. The hydrolysis product profiles revealed that xylobiose was the dominant XOS produced from untreated and pre-treated corn cobs. These results demonstrated that the efficient production of hydroxycinnamic acids and XOS from agricultural residues for industrial applications can be achieved through the synergistic action of FAE5 or FAE6 and Xyn11.

Keywords: Ferulic acid; Feruloyl esterase; Xylanase; Synergy; Xylo-oligosaccharides

1. Introduction

Lignocellulosic biomass is widely considered one of the most promising low-cost feedstocks for the production of renewable energy and value-added chemicals. It is primarily composed of three major components, namely: cellulose (40–60%), hemicellulose (20–40%), and lignin (10–25%) [1]. Waste agricultural residues generated during crop harvesting and processing (e.g., rice straw, wheat straw, corn stover, corn cobs, sugarcane bagasse, sorghum bagasse, etc.) are renewable biomass resources that are readily available and inexpensive [2]. A significant amount of research has focused on developing environmentally friendly methods of utilising lignocellulosic biomass. In this regard, enzymatic conversion has emerged as a major technological platform that offers several advantages such as environmental benefits and lower energy costs, with no formation of undesirable by-products. Due to the complexity and heterogeneity in the structure of biomass, its enzymatic conversion requires the synergistic cooperation of several enzymes [3]. A detailed understanding of this synergistic cooperation is key in the development of enzyme cocktails for the optimal production of value-added chemicals from lignocellulosic biomass.

Feruloyl esterases (FAEs, EC 3.1.1.73) and endo-xylanases (EC 3.2.1.8) are essential enzymes for the degradation of xylan, the most abundant hemicellulose in biomass. FAEs catalyze the cleavage of covalent ester linkages between hydroxycinnamic acids and polysaccharides, releasing ferulic acid (FA) and *p*-coumaric acid (*p*-CA) from lignocellulosic biomass [4,5]. FA, the most abundant hydroxycinnamic acid, is usually esterified at the C-5 hydroxy group of the arabinofuranosyl units of arabinoxylan of commelinid plants [6]. Hydroxycinnamic acids are used in the food industry as precursors for vanillin, *p*-hydroxybenzoic acid and 4-vinylphenol production, and as food preservatives because of their antimicrobial properties [7], while in the pharmaceutical industry, they can be used for their antioxidant and anti-inflammatory properties [8–11]. Application of FAEs not only releases hydroxycinnamic acid, but also decreases biomass recalcitrance, making it more accessible for further hydrolysis by carbohydrate-active enzymes [11]. Endo-xylanases are essential in degrading xylan as they catalyze the random cleavage of β-1,4-D-xylosidic linkages, generating xylo-oligosaccharides (XOS) [12]. The enzymes have been classified into glycoside hydrolase (GH) families 5, 8, 10, 11, 30, 43, 62 and 98, with GH10 and 11 being the two families that have been extensively characterized (www.cazy.org). Endo-xylanase generated XOS from agricultural residues are used in food, feed, and pharmaceutical industries due to their prebiotic and antioxidant activity [13,14].

FAEs exhibit a synergistic interaction with xylanases during the hydrolysis of a range of lignocellulosic substrates, which is demonstrated by improved yields in the production of XOS and FA. Xylanases generate ferulated XOS, which become preferred substrates for FAEs to cleave ester bonds from, liberating FA as a product [15]. The removal of FA then increases the accessibility of xylanases to XOS for further hydrolysis into shorter XOS and/or xylose. Studies have reported significant increases in the amount of FA released from arabinoxylan after the enzymatic hydrolysis by FAEs in the presence of xylanases [16–18]. Although FAEs from different microorganisms have been used for the co-production of FA and XOS, some of these FAEs show limited hydrolysis efficiencies. Therefore, novel FAEs with exceptional catalytic properties are still required for the formulation of more efficient enzyme cocktails.

Rashamuse and co-workers [19] functionally screened novel FAEs from the hindgut prokaryotic symbionts of *Trinervitermes trinervoides* termite species, but the application of these enzymes in degrading agricultural residues has not yet been explored. In this study, the synergistic action of two new termite metagenome derived FAEs (FAE5 and FAE6) and a GH11 xylanase from *Thermomyces lanuginosus* was optimized on insoluble wheat arabinoxylan (model substrate) and then applied to corn cobs (a natural substrate) for the production of XOS and hydroxycinnamic acids. The study presented demonstrated that high quantities of XOS, *p*-CA and FA were generated from corn cobs as a result of synergistic interactions between Xyn11 and FAE5 or FAE6.

2. Materials and Methods

2.1. Chemicals, Substrates and Enzymes

Escherichia coli BL21 (DE3) was used for the expression of *fae5* and *fae6* genes harboured in pET28 plasmids. The cells were cultured in Luria-Bertani (LB) medium containing 50 μg/mL kanamycin at 37 °C. The induction was conducted by following the method described previously [20]. Enzymes were purified using 5 mL Ni-NTA Superflow Cartridges (QIAGEN®), purchased from QIAGEN (Germany), as per the manufacturer's instructions. Xylanase (Xyn11) from *Thermomyces lanuginosus* and ethyl ferulate (EFA) were purchased from Sigma Aldrich (South Africa). Insoluble wheat arabinoxylan (WAX) was purchased from Megazyme™ (Ireland).

2.2. Pre-Treatment of Corn Cobs

Milled corn cobs (CC) was treated using hydrothermal pre-treatment or dilute acid pre-treatment. A total of 10 g of biomass was suspended in Milli-Q water (for hydrothermal treatment) or in a 0.5% (*w/w*) sulphuric acid solution (for dilute acid treatment) (solid: liquid ratio of 1:10) and autoclaved for 20 min at 121 °C. The pre-treated CC slurry was filtered, followed by washing the solids repeatedly with Milli-Q water and then oven drying to constant weight at 50 °C for 48 h.

2.3. Chemical Characterization of CC

The total carbohydrates of CC were determined in triplicate by using a modified sulphuric acid method described previously [21]. Briefly, 300 mg of CC (untreated and pre-treated) was hydrolyzed with 72% (*v/v*) sulphuric acid at 30 °C for 1 h, diluted to 3% (*v/v*) sulphuric acid and autoclaved to solubilize the carbohydrate fraction. Following the hydrolysis, fractions were filtered to remove the insoluble lignin from the solution.

Determination of the alkali-extractable hydroxycinnamic acid content was carried out by treating 10 mg of biomass with 1 M NaOH solution for 24 h at room temperature and in the dark. The liquors obtained from alkaline treatments were separated from the solid fraction by centrifugation at 16,000× *g* for 5 min. The liquors were neutralized by 2 volumes of 1 M HCl and analyzed by HPLC as described in Section 2.4.

The morphological structure of untreated and pre-treated biomass was analyzed with a scanning electron microscope (SEM), JOEL JSM 840. CC samples were mounted on a metal stub with adhesive tape and coated with a thin layer of gold prior to SEM analysis.

In order to detect changes in functional groups, FTIR analysis of untreated and pre-treated CC was conducted by loading a few milligrams of pulverized samples in the universal ATR of a Spectrum 100 FT-IR spectrometer system (Perkin Elmer, Wellesley, MA). FT-IR spectra were recorded in quadruple at a range of 650–4000 cm^{-1} with a resolution of 4 cm^{-1}.

2.4. Determination of Enzyme Activities and Protein Concentration

For feruloyl esterase activity assay, ethyl ferulate (EFA) was used as substrate. The reaction (1 mL) was carried out in sodium phosphate buffer (50 mM, pH 7.4) that contained 1 mM EFA. The reaction was initiated by the addition of a diluted enzyme solution. After 15 min incubation at 40 °C, the enzymatic activity was terminated by incubation at 100 °C for 5 min. The released FA from the substrate was quantified using a Shimadzu HPLC system (Shimadzu Corp, Japan) equipped with a diode array detector (DAD), where the chromatographic separation was achieved using a Phenomenex® C18 5 µm (150 × 4.6 mm) LC column (Phenomenex, United States of America). Ambient conditions were used for analysis. The mobile phase A was 0.01 M phosphoric acid and the mobile phase B consisted of HPLC grade acetonitrile. The isocratic mobile phase consisted of A: 70% and B: 30% and ran for 10 min at a flow rate of 0.8 mL/min. The injected volume was 10 µL and the UV absorption of the effluent was monitored at 320 nm. One unit of enzyme activity was defined as the amount of enzyme that released 1 µmol of ferulic acid per min under standard conditions.

For xylanase activity assay, 10 mg/mL of insoluble wheat arabinoxylan (WAX) was used as a substrate. The reaction mixture consisted of 300 µL of 1.33% (*w/v*) substrate dissolved in 50 mM sodium phosphate buffer (pH 7.4) and 100 µL of diluted enzyme. After 15 min incubation at 40 °C, the enzymatic activity was terminated by incubation at 100 °C for 5 min, followed by centrifugation at 16,000× *g* for 5 min. The concentrations of reducing sugar were measured using 3,5-dinitrosalicylic acid (DNS) method described previously [22]. Briefly, 150 µL of the sample was mixed with 300 µL of DNS followed by boiling for 5 min. One unit of enzyme activity was defined as the amount of enzyme that released 1 µmol of reducing sugars per min. Reducing sugars were estimated using a xylose standard curve. The yield of reducing sugars was determined using the following equation:

$$\text{XOS yield (\%)} = ((\text{XOS released} \times 0.88) / \text{xylan content}) \times 100 \quad (1)$$

The Bradford method was used to determine the protein concentration of the enzymes [23]. A protein standard curve was constructed using bovine serum albumin as a suitable standard.

2.5. Synergy Studies

In order to study the synergistic interactions on agricultural residues, the FAE and Xyn11 combinations were first tested on WAX for the release of FA and XOS. Enzyme loadings for all single enzymes and all enzyme combinations were kept at a total protein loading of 4 mg protein per g of WAX and at 8 mg per g of CC. For combination experiments, an enzyme mixture consisting of Xyn11 and FAE in a 66:33% protein ratio was used. The enzymatic hydrolysis was performed at a substrate loading of 0.5% (*w*/*v*) for WAX and 1% (*w*/*v*) for CC in 50 mM phosphate buffer (pH 7.4) in a total volume of 400 μL. The reaction mixture was incubated at 40°C with agitation at 25 rpm for 24 h and terminated by boiling at 100 °C for 5 min. Hydrolysis controls included reactions without the addition of enzyme or substrate. All the experiments were performed in triplicate. The amount of FA and *p*-CA released was determined using the HPLC method described in Section 2.4, while reducing sugars were measured using the DNS method and XOS were quantified using the HPLC-RID method described in Section 2.7.

2.6. Determination of Xylo-Oligosaccharides Pattern Profiles

In order to determine the hydrolysis product profiles from synergy studies, XOS were analyzed by thin-layer chromatography (TLC). Five μL of hydrolysate sample and a mixture of XOS standards were applied on a silica gel 60 F_{254} plate (Merck, Darmstadt, Germany). The migration was repeated twice using a mobile phase consisting of 1-butanol, acetic acid and water in a 2:1:1 ratio, respectively. The plate was then submerged in Molisch's Reagent (0.3% (*w*/*v*) α-naphthol dissolved in methanol and sulphuric acid in a 95:5 ratio (*v*/*v*), respectively). The spots corresponding to the different XOS were visualized by heating the plate in the oven at 110 °C for 10 min.

The XOS were quantified by a Shimadzu HPLC system (Shimadzu Corp, Japan) equipped with a refractive index detector (RID) using a CarboSep CHO 411 column (Anatech, South Africa) with water as the mobile phase in isocratic mode. The column oven was set at 80 °C and separation was performed within 35 min at a flow rate of 0.3 mL per min. An injection volume of 20 μL was employed for all samples and XOS standards.

2.7. Statistical Analysis

All statistical analyses were performed on GraphPad Prism 6.0 software using the *t*-test. A *p*-value of less than 0.05 was considered to indicate statistically significant differences between compared data sets.

3. Results

3.1. Chemical Characterization of CC

The main hurdle in utilising lignocellulosic biomass lies in its recalcitrant nature due to the complexity of its biomass structure. Thus, to increase the enzymatic hydrolysis of biomass, various pre-treatment strategies are usually required. However, some of the pre-treatment methods are known for easily solubilizing some of the polysaccharides, mostly hemicellulose [24]. It is, therefore, critical to perform pre-treatment under conditions that lead to a recovery of the biomass components in a re-usable form while increasing the enzymatic digestibility. In this study, hydrothermal and dilute acid pre-treatment strategies were selected. An extremely low thermo-chemical pre-treatment severity was applied in order to preserve hemicellulose and hydroxycinnamic content in the solid fraction. In order to determine the effect and efficiency of the pre-treatments, the surface morphology of CC was visualized with SEM. Figure 1 shows the SEM micrographs of untreated and pre-treated CC.

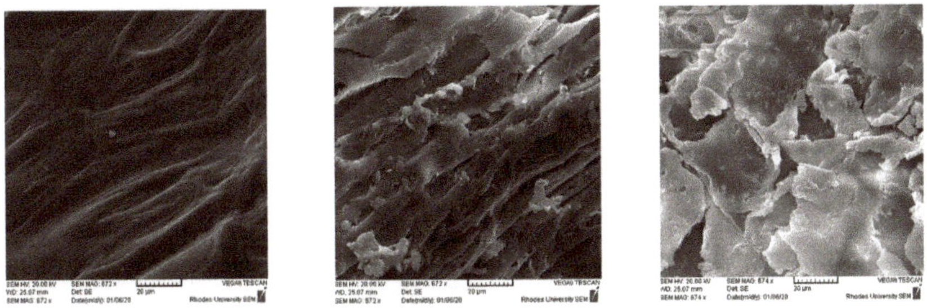

(a) (b) (c)

Figure 1. Morphological study of corn cobs (CC) by SEM. SEM micrographs of (**a**) untreated, (**b**) hydrothermal treated, and (**c**) acid-treated at a 2k× magnification.

Here it can be seen that the untreated sample appears to possess a compact structure. In contrast, the micrographs of pre-treated CC samples display distorted and fragmented structures on the surface. The structural changes due to pre-treatment might increase the surface area of pre-treated CC, which could lead to an enhanced enzymatic degradability.

To determine the recoverable sugars and hydroxycinnamic acids after pre-treatment, the chemical composition of the untreated and pre-treated CC was also analyzed (Table 1). For noting, the composition of other CC components such as lignin was omitted, and attention was focused on glucan, xylan, arabinan, and hydroxycinnamic acids. The content of glucan, xylan, and arabinan was slightly higher in the pre-treated biomass compared to untreated biomass samples. There were more recoverable sugars in the hydrothermal treated sample compared to the acid pre-treated sample. A slight increase in FA and *p*-CA content was observed in the acid-treated sample. The data on Table 1 confirmed that the relevant CC components were successfully retained upon pre-treatment.

Table 1. Chemical composition of untreated and pre-treated CC (on a percentage dry mass basis).

	Glucan [a]	Xylan [a]	Arabinan [a]	Reducing Sugars [b]	FA [c]	*p*-CA [c]
Untreated	30.86 ± 0.90	11.46 ± 0.48	7.79 ± 0.63	53.00 ± 0.49	0.61± 0.012	0.63 ± 0.026
Hydrothermal treated	36.03 ± 0.13	13.47 ± 0.91	11.24 ± 0.84	62.38 ± 0.29	0.68 ± 0.027	0.61 ± 0.024
Acid-treated	32.58 ± 0.85	12.22 ± 0.48	8.52 ± 0.55	59.20 ± 0.15	0.68 ± 0.027	0.67 ± 0.013

Analysis method: [a] Megazyme sugar kits, [b] DNS method, [c] HPLC. The data presented are averages ± standard deviations of triplicates.

To further investigate the chemical changes that took place during the pre-treatment of CC, FTIR analysis was conducted. The spectra of untreated, hydrothermal and acid-treated CC are shown in Figure 2. The absorption peaks at around 1730 cm^{-1} region are predominantly attributed to the C=O stretching vibration of the ester linkage of the carboxylic group of FA and *p*-CA of lignin and/or hemicellulose [25]. The spectra of all samples show this peak suggesting that changes due to pre-treatment (observed in SEM micrographs) did not lead to the removal of hemicellulose and hydroxycinnamic acids. The FTIR data is in agreement with composition analysis (Table 1), as the hydrothermal pre-treated sample displayed strong absorption bands around 1000 cm^{-1} and in the region 3500–3200 cm^{-1} (associated with β-glycosidic linkages and OH groups of glucose units, respectively) [26].

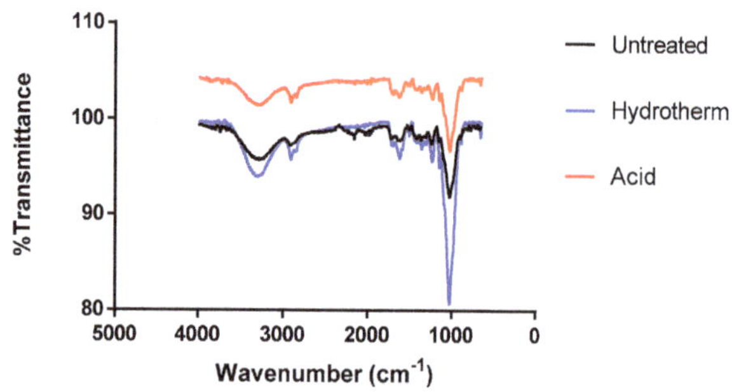

Figure 2. FTIR spectra of untreated, hydrothermal treated and acid-treated CC solids registered in the range of 450–4000 cm^{-1}.

3.2. Determination of Enzyme Specific Activities

The specific activities of the enzymes used in this study were determined as detailed in Section 2.4 under Materials and Methods. Xyn11 was the only enzyme active on WAX with a specific activity of 15.08 U/mg of xylanase, while FAE5 and FAE6 showed activity only on EFA with specific activities of 28.36 U/mg and 27.34 U/mg of feruloyl esterase, respectively. FAE5 and FAE6 demonstrated relatively similar feruloyl esterase activities.

3.3. Release of XOS from Substrates by Enzymatic Hydrolysis

In order to evaluate synergistic interactions between FAEs and Xyn11 on untreated and pre-treated CC (agricultural residue), the enzymes were tested for their ability to release XOS and FA from a model arabinoxylan substrate (WAX). A 0.5% WAX substrate loading was used for a better comparison with natural substrates since they have xylan contents of between 20–40%. Figure 3 shows the production of reducing sugars after enzymatic hydrolysis of WAX (a), untreated (b), hydrothermal treated (c) and acid-treated CC (d) by single or combinations of the enzymes. The trend in the production of reducing sugars was similar in all substrates, with the Xyn11 to FAE5/6 combinations releasing higher reducing sugars than those from the reactions containing individual enzymes. As expected, FAE5 or FAE6 alone were not able to release reducing sugars, this was also indicated in the specific activity determination study (Table 2). For hydrolysis of WAX (Figure 3a), both FAEs displayed synergy with Xyn11 by improving the release of reducing sugars. It was found that Xyn11 alone released 0.89 mg/mL, while co-incubation with FAE5 or FAE6 resulted in 1.18 mg/mL (25.61% increase) and 1.22 mg/mL reducing sugars (a 28.06% increase), respectively. The hydrolysis of untreated CC with xylanase alone and a combination of Xyn11 and FAE5 or FAE6 produced reducing sugars at concentrations of 0.94 mg/mL, 1.17 mg/mL, and 1.22 mg/mL, respectively (Figure 3b). The hydrothermal pre-treated sample exhibited a similar hydrolysis pattern to untreated CC, but the amounts of reducing sugars released in the co-incubation sets were higher (Figure 3c). However, the amount of reducing sugars was significantly less for acid pre-treatment when Xyn11 was applied alone (yield of 28.80%), but a combination of Xyn11 and FAE5 or FAE6 resulted in more reducing sugars (yield of 100% for both combinations) compared to untreated CC (Figure 3d). The hydrothermally pre-treated sample exhibited yields of 60.76% for Xyn11 alone, while co-incubation with FAE5 and FAE6 resulted in yields of 98.21% and 100%, respectively. These results indicate that FAEs play an important role in the synergistic hydrolysis of WAX, and the arabinoxylan contained in untreated and pre-treated CC samples.

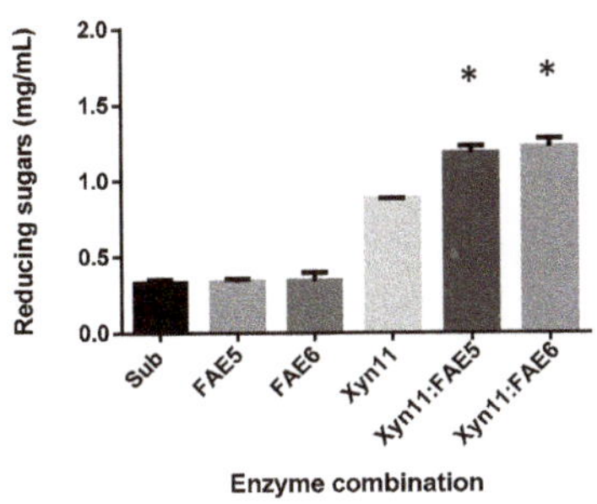

(**a**)

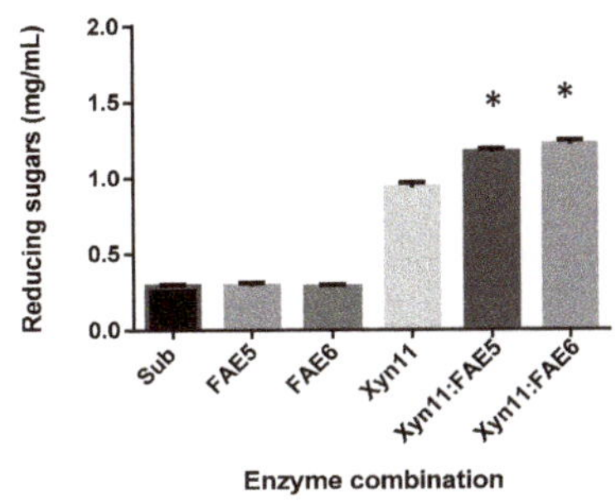

(**b**)

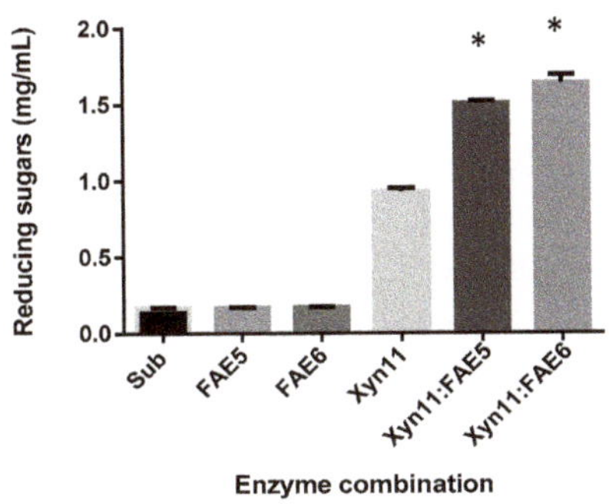

(**c**)

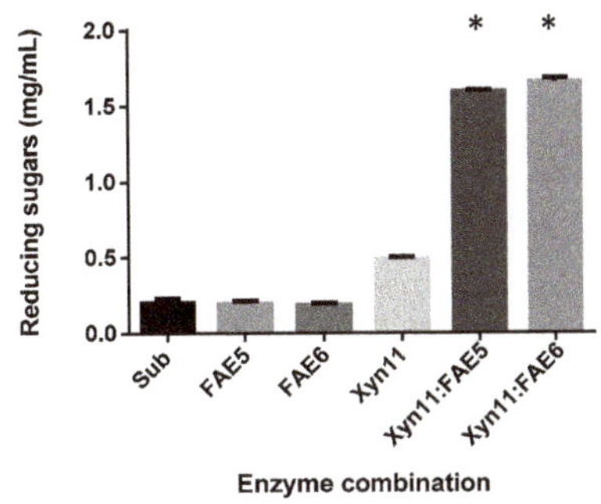

(**d**)

Figure 3. Release of reducing sugars during hydrolysis of 0.5% wheat arabinoxylan (WAX) (**a**), 1% untreated (**b**), hydrothermally pre-treated (**c**) and acid pre-treated CC (**d**) by individual enzymes or a combination of 66% Xyn11: 33% FAE5 or FAE6. Sub represents substrate control. Statistical analysis was conducted using *t*-test for improvement of hydrolysis with respect to reducing sugar by the enzyme combinations compared to single enzyme (Xyn11), key: * (p value < 0.05).

Table 2. Specific activities of the enzymes assessed in this study (U/mg protein).

Substrate	Enzyme Tested		
	Xyn11	**FAE5**	**FAE6**
WAX [a]	15.08	Nd	Nd
EFA [b]	Nd	28.36	27.34

[a] Analyzed by 3,5-dinitrosalicylic acid (DNS) method for xylanase activity and [b] analyzed by HPLC diode array detector (DAD) for feruloyl esterase activity. "Nd" = not detected.

3.4. Release of Hydroxycinnamic Acids from Substrates by Enzymatic Hydrolysis

To further understand the synergism between Xyn11 and FAEs, the release of hydroxycinnamic acids from substrates by individual enzymes and their combinations was evaluated. Figure 4 shows the release of FA from WAX (a), untreated (b), hydrothermal pre-treated (c), and acid pre-treated treated CC (d). In the case of WAX (Figure 4a), a slight increase in FA release by FAE5 or FAE6 alone was observed. The combination of Xyn11 and FAE5 or FAE6 significantly improved ($p < 0.05$) the release of FA from 0.64 µg/mL and 0.65 µg/mL (individual enzyme) to 1.92 µg/mL and 2.68 µg/mL, respectively. A similar pattern was observed for the hydrolysis of untreated CC, although a combination of Xyn11: FAE5 released more FA compared to Xyn11: FAE6 (Figure 4b). With respect to pre-treated CC, the amount of

FA released from the substrate without enzymatic hydrolysis was enhanced (Figure 4c,d). This could be attributed to the disruption of the close inter-component associations between major constituents of lignocellulose during the pre-treatment step. Only Xyn11: FAE6 could significantly improve ($p < 0.05$) the release of FA from both pre-treated samples. The release of *p*-CA from CC was also observed and is shown in Figure 5. Similar to the results presented in Figure 4c,d, the pre-treated samples already showed increased amounts of readily soluble *p*-CA before enzymatic hydrolysis. However, the combination of Xyn11 and FAE5 or FAE6 were able to release considerable quantities of *p*-CA compared to individual enzymes and those already present and soluble in substrate controls. Interestingly, contrary to the FA release from pre-treated samples, both Xyn11: FAE5 and Xyn11: FAE6 could release comparable quantities of *p*-CA (Figure 5b,c). However, the yields obtained (no more than 10%) were much less when compared to release efficiencies (more than 70%) reported in some studies in the literature [27,28]. These results indicate that FAEs acted synergistically with Xyn11 in the co-production of XOS, FA and *p*-CA from the CC substrates.

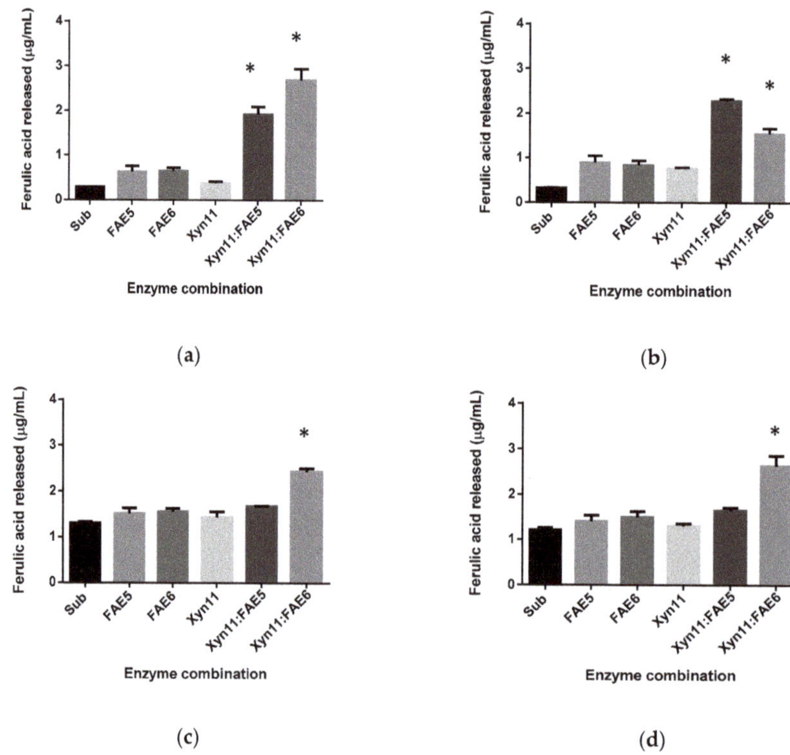

Figure 4. Release of ferulic acid during the degradation of 0.5% WAX (**a**), 1% untreated (**b**), hydrothermal pre-treated (**c**) and acid pre-treated CC (**d**) by individual enzymes or a combination of 66% Xyn11: 33% FAE5 or FAE6. Sub represents substrate control. Statistical analysis was conducted using *t*-test for improvement of Ferulic acid release by the enzyme combinations compared to single enzyme (FAE5/6), key: * (p value < 0.05).

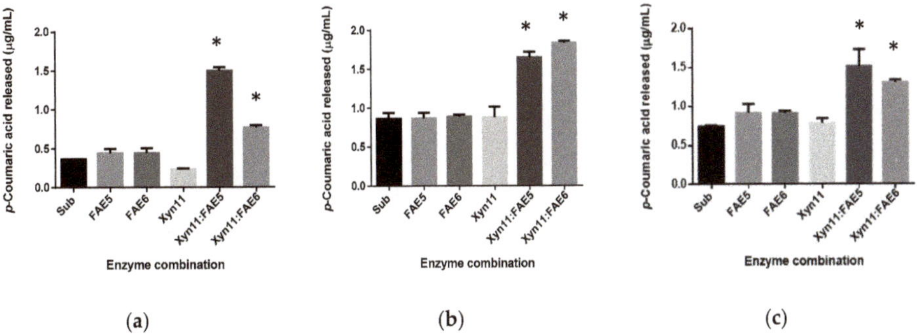

Figure 5. Release of *p*-coumaric acid during the degradation of 1% untreated (**a**), hydrothermal pre-treated (**b**) and acid pre-treated CC (**c**) by individual enzymes or a combination of 66% Xyn11: 33% FAE5 or FAE6. Note, WAX was not assessed as there was no *p*-coumaric acid detected in the substrate. Sub represents substrate control. Statistical analysis was conducted using *t*-test for improvement of *p*-coumaric acid release by the enzyme combinations compared to single enzyme (FAE5/6), key: * (p value < 0.05).

3.5. Determination of Hydrolysate Product Profiles

The data presented above indicates that FAE5 and FAE6 could release hydroxycinnamic acids in the presence of Xyn11 while improving the production of reducing sugars from WAX, untreated and pre-treated CC. To gain a deeper insight into these synergistic interactions, the product patterns of individual enzymes and their combinations were evaluated for the types of XOS generated. Figure 6 shows the TLC analysis of XOS released from the hydrolysis of WAX (a), untreated (b), hydrothermal treated (c) and acid-treated CC (d) by single and combinations of enzymes. It is important to note that the dark-yellow colored spots on the plates represent glycerol which was used as a stabilizer during storage of the purified FAE5 and FAE6. In the case of WAX hydrolysis (Figure 6a), individual enzymes (lane 2) and combinations (lane 3 and 4) generated xylobiose, xylotetraose, xylopentaose, and xylohexaose. The hydrolysis product patterns of CC (Figure 6b–d) appeared to consist of xylobiose (a dominant product) and small quantities of xylotetraose. The quantities of xylobiose seem to increase for the treated CC, most especially for enzyme combinations (lane 3 and 4). Also, for acid pre-treated CC (Figure 6d), Xyn11 alone (lane 2) produced very small quantities of XOS, this was also observed in Figure 3b during the quantification of reducing sugars.

We then further attempted to quantify the XOS generated using HPLC-RID. It is noteworthy that the HPLC-RID system used in this study couldn't detect XOS of less than 0.05 mg/mL. Figure 7 shows that the quantity of xylobiose produced by the combination of Xyn11 and FAE5 or FAE6 was enhanced compared to Xyn11 alone, this pattern was more pronounced on untreated and pre-treated CC. High quantities of xylotetraose (0.24 mg/mL) were observed for WAX (Figure 7a), but there was no significant increase between individual enzymes and their combinations. The highest quantities of xylobiose were produced by enzyme combinations for hydrothermal pre-treated CC (0.28 mg/mL for xyn11: FAE5 and 0.40 mg/mL for Xyn11: FAE6) and acid-treated CC (0.34 mg/mL for Xyn11: FAE5 and 0.43 mg/mL for Xyn11: FAE6). This improvement in xylobiose production was significant ($p < 0.05$) when compared to single enzyme incubations, which resulted in 0.14 mg/mL and 0.12 mg/mL for hydrothermal and acid-treated CC, respectively. From the results presented above, it appears that Xyn11 produced XOS which become substrates to FAEs for the removal of FA or *p*-CA, and this allows Xyn11 to further hydrolyze these long XOS into the shorter xylobiose.

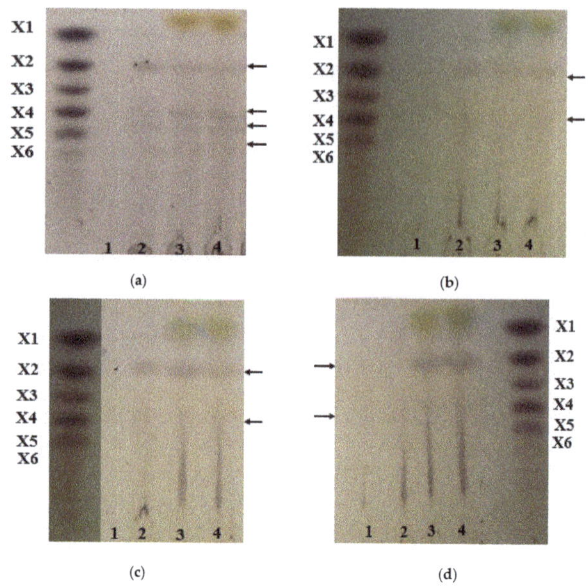

Figure 6. Thin-layer chromatography (TLC) analysis of hydrolysis of 0.5% WAX (**a**), 1% untreated (**b**), hydrothermal pre-treated (**c**) and acid pre-treated CC (**d**) by (2) Xyn11 alone, (3) a combination of 66% Xyn11: 33% FAE5, and (4) a combination of 66% Xyn11: 33% FAE6. Substrate without enzyme was used as a control (1). A mixture of xylo-oligosaccharides (X1-X6) was used as a standard. Arrows indicated observed bands, albeit feint in some instances.

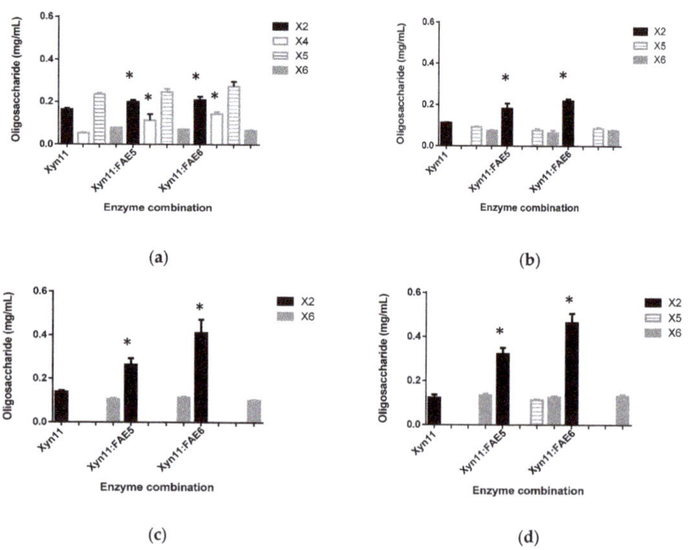

Figure 7. Xylo-oligosaccharide content measured from the hydrolysis of 0.5% WAX (**a**), 1% untreated (**b**), hydrothermal pre-treated (**c**) and acid pre-treated CC (**d**) after incubation with Xyn11 alone or a combination of 66% Xyn11: 33% FAE5 or FAE6. Statistical analysis was conducted using *t*-test for improvement of hydrolysis with respect to xylo-oligosaccharides (XOS) by the enzyme combinations compared to single enzyme (Xyn11), key: * (p value < 0.05).

4. Discussion

XOS and high value-added compounds generated from agricultural residues have great potential for application in pharmaceutical, food, and fine chemical industries. It has been reported that the total world production of corn, a key agricultural crop, was estimated to exceed 875,226,630 tons [29]. CC, a by-product of corn is an attractive agricultural residue due to its high xylan content and the fact that it's readily available. The production of XOS could be achieved by application of chemical and hydrothermal pre-treatments. However, higher severity pre-treatment conditions can lead to increased decomposition of XOS to xylose and generation of unwanted by-products [14]. The application of low severity pre-treatment conditions to increase accessible area for enzymes, followed by enzymatic hydrolysis, is an effective method for XOS production. In this study, the synergistic interactions between two termite metagenome-derived FAEs and a xylanase, Xyn11, for the production of XOS, *p*-CA and FA from untreated and pre-treated CC was evaluated. WAX was used as a model substrate to optimize the enzyme cocktail as it maintains the ferulic acid cross-linkages in the native arabinoxylan. CC was subjected to two pre-treatment strategies, hydrothermal and dilute sulfuric acid pre-treatment. The results for composition analysis (Table 1) showed that there was an increase in recoverable sugars and that hydroxycinnamic acid content was maintained in pre-treated samples. However, the xylan content of all samples was much less than the values reported in the literature (28%) [30]. Biomass (morphology) characterization indicated that these changes were associated with increased surface area due to pre-treatment (Figure 1). It can be speculated that the increased sugar content of pre-treated CC samples (Table 1) may be due to these structural modifications which may include changes in lignin content. An increase in the production of reducing sugars was observed during enzymatic hydrolysis of pre-treated CC (Figure 3), indicating the success of the pre-treatment strategies selected for this study.

Regarding the enzymatic release of hydroxycinnamic acids from the substrates, the data presented above suggests that FAE5 or FAE6 could release significant amounts of FA and *p*-CA only in the presence of Xyn11. The inability of FAEs to release high quantities of hydroxycinnamic acids when incubated alone could be attributed to the type of bonds available on the substrate. FA or *p*-CA is usually found esterified to polysaccharides such as arabinoxylan and could also ether-link with lignin or dimerize with other hydroxycinnamic acid-linked polysaccharides, forming cross-linkages between these polymers. FAEs are known for specifically catalyzing the cleavage of ester bonds between hydroxycinnamic acids and polysaccharides, but not the ether-linkages. It is, therefore, possible that FAE5 and FAE6 couldn't act on these cross-linked complex structures. Previous studies have indicated that FAEs could release FA from substrates when co-incubated with xylanase [27,31]. Zhang and co-workers [32] reported that the combination of a xylanase and AfFaeA increased the amount of FA released from steam exploded corn stalk 13-fold. It appears that FAE activity can be enhanced by the addition of a xylanase - it appears as if the xylanase generates feruloylated XOS from the xylan main chain which are easily hydrolyzed by FAE in comparison to feruloylated xylan. In turn, FAE action allows the xylanase to be able to further hydrolyze XOS with a high degree of polymerization into shorter chain XOS products. The patterns observed in the production of reducing sugars from acid pre-treated CC, presented in Figure 3d, are indicative of such a relationship between xylanase and FAE. The activity of Xyn11 alone was drastically reduced, while the enzyme combinations resulted in the highest reducing sugar release. It is likely acid pre-treatment of CC generated XOS esterified with FA or *p*-CA, which may pose limitations on their accessibility (steric hindrance) for the Xyn11 to hydrolyze them to shorter chain XOS products. Therefore, the addition of FAEs resulted in the improvement of Xyn11 activity on these long, feruloylated XOS. There have been reports on oligosaccharides resulting from the pre-treatment of several natural substrates [33–35].

The hydrolysis patterns of individual enzymes and enzyme combinations were then also investigated. Xyn11 and a combination of Xyn11 with FAEs released XOS with a low degree of polymerization (DP 6–4 and DP 2, respectively) from WAX and released mainly xylobiose from untreated and pre-treated CC. The variations in product patterns observed could be due to the differences in the xylan structure between WAX and CC. The yield of xylobiose from the pre-treated substrates was significantly increased by enzyme combinations compared to individual enzymes. It has been reported that xylobiose rich XOS are the preferred products for prebiotic activity [36]. The product patterns demonstrated that the Xyn11: FAEs enzyme combination (cocktail) has the potential of releasing XOS with a high xylobiose content from CC, most notably when combined with hydrothermal or dilute sulfuric acid pre-treatments.

5. Conclusions

This study reported, for the first time, the application of two termite metagenome-derived FAEs for the production of XOS and hydroxycinnamic acids from CC. The FA and/or *p*-CA released from WAX and CC was greatly enhanced when FAE5 and FAE6 were co-incubated with Xyn11 compared to when the enzymes were applied individually. The presence of FAEs played a major role in the production of XOS, mainly xylobiose, from untreated and pre-treated CC. This study suggests that FAEs act synergistically with Xyn11 during the enzymatic hydrolysis of CC to produce the industrially relevant products XOS and hydroxycinnamic acids.

Author Contributions: Conceptualization, L.M.; S.M.; B.S.W. and B.I.P.; methodology, L.M.; software, L.M.; validation, L.M.; S.M.; B.S.W. and B.I.P.; formal analysis, L.M.; S.M.; B.S.W. and B.I.P.; investigation, L.M.; resources, B.S.W. and B.I.P.; data curation, L.M.; writing—original draft preparation, L.M.; writing—review and editing, L.M.; S.M.; B.S.W. and B.I.P.; visualization, L.M.; supervision, B.I.P. and B.S.W.; project administration, B.I.P. and B.S.W.; funding acquisition, B.I.P. All authors have read and agree to the published version of the manuscript.

Funding: The work was funded by the South African Department of Science and Technology (DST)/Council for Scientific and Industrial Research (CSIR) National Biocatalysis Initiative, National Research Foundation (NRF) of South Africa and Rhodes University. Any opinion, findings and conclusions or recommendations expressed in this material are those of the author(s) and therefore the NRF does not accept liability in regard thereto.

Conflicts of Interest: The authors declare no conflict of interest.

References

1. Kang, Q.; Appels, L.; Tan, T.; Dewil, R. Bioethanol from lignocellulosic biomass: Current findings determine research priorities. *Sci. World J.* **2014**, *2014*, 1–13. [CrossRef] [PubMed]
2. Anwar, Z.; Gulfraz, M.; Irshad, M. Agro-industrial lignocellulosic biomass a key to unlock the future bio-energy: A brief review. *J. Radiat. Res. Appl. Sci.* **2014**, *7*, 163–173. [CrossRef]
3. Van Dyk, J.S.; Pletschke, B.I. A review of lignocellulose bioconversion using enzymatic hydrolysis and synergistic cooperation between enzymes—Factors affecting enzymes, conversion and synergy. *Biotechnol. Adv.* **2012**, *30*, 1458–1480. [CrossRef] [PubMed]
4. Wong, D.W.S. Feruloyl esterase: A key enzyme in biomass degradation. *Appl. Biochem. Biotechnol.* **2006**, *133*, 87–112. [CrossRef]
5. Faulds, C.B. What can feruloyl esterases do for us? *Phytochem. Rev.* **2009**, *9*, 121–132. [CrossRef]
6. Oliveira, D.M.; Finger-Teixeira, A.; Mota, T.; Salvador, V.H.; Moreira-Vilar, F.C.; Molinari, H.B.C.; Mitchell, R.; Marchiosi, R.; Ferrarese-Filho, O.; Dos Santos, W.D. Ferulic acid: A key component in grass lignocellulose recalcitrance to hydrolysis. *Plant Biotechnol. J.* **2014**, *13*, 1224–1232. [CrossRef]
7. Pei, K.; Ou, J.; Huang, J.; Ou, S. p-Coumaric acid and its conjugates: Dietary sources, pharmacokinetic properties and biological activities. *J. Sci. Food Agric.* **2016**, *96*, 2952–2962. [CrossRef]
8. Kumar, N.; Pruthi, V. Potential applications of ferulic acid from natural sources. *Appl. Biotechnol. Rep.* **2014**, *4*, 86–93. [CrossRef]
9. Shirai, A.; Matsuki, H.; Watanabe, T. Inactivation of foodborne pathogenic and spoilage micro-organisms using ultraviolet-A light in combination with ferulic acid. *Lett. Appl. Microbiol.* **2017**, *64*, 96–102. [CrossRef]
10. Chen, X.; Guo, Y.; Jia, G.; Zhao, H.; Liu, G.; Huang, Z. Ferulic acid regulates muscle fiber type formation through the Sirt1/AMPK signaling pathway. *Food Funct.* **2019**, *10*, 259–265. [CrossRef]
11. Oliveira, D.M.; Mota, T.R.; Oliva, B.; Segato, F.; Marchiosi, R.; Ferrarese-Filho, O.; Faulds, C.; Dos Santos, W.D. Feruloyl esterases: Biocatalysts to overcome biomass recalcitrance and for the production of bioactive compounds. *Bioresour. Technol.* **2019**, *278*, 408–423. [CrossRef] [PubMed]

12. Paës, G.; Berrin, J.-G.; Beaugrand, J. GH11 xylanases: Structure/function/properties relationships and applications. *Biotechnol. Adv.* **2012**, *30*, 564–592. [CrossRef] [PubMed]
13. Aachary, A.A.; Prapulla, S.G. Xylooligosaccharides (XOS) as an emerging prebiotic: Microbial synthesis, utilization, structural characterization, bioactive properties, and applications. *Compr. Rev. Food Sci. Food Saf.* **2010**, *10*, 2–16. [CrossRef]
14. Samanta, A.; Jayapal, N.; Jayaram, C.; Roy, S.; Kolte, A.; Senani, S.; Sridhar, M. Xylooligosaccharides as prebiotics from agricultural by-products: Production and applications. *Bioact. Carbohydr. Diet. Fibre* **2015**, *5*, 62–71. [CrossRef]
15. Malgas, S.; Mafa, M.S.; Mkabayi, L.; Pletschke, B. A mini review of xylanolytic enzymes with regards to their synergistic interactions during hetero-xylan degradation. *World J. Microbiol. Biotechnol.* **2019**, *35*, 187. [CrossRef] [PubMed]
16. Dilokpimol, A.; Mäkelä, M.R.; Mansouri, S.; Belova, O.; Waterstraat, M.; Bunzel, M.; De Vries, R.P.; Hildén, K. Expanding the feruloyl esterase gene family of Aspergillus niger by characterization of a feruloyl esterase, FaeC. *New Biotechnol.* **2017**, *37*, 200–209. [CrossRef]
17. Mäkelä, M.R.; Dilokpimol, A.; Koskela, S.; Kuuskeri, J.; De Vries, R.P.; Hildén, K. Characterization of a feruloyl esterase from Aspergillus terreus facilitates the division of fungal enzymes from Carbohydrate Esterase family 1 of the carbohydrate-active enzymes (CAZy) database. *Microb. Biotechnol.* **2018**, *11*, 869–880. [CrossRef]
18. Lau, T.; Harbourne, N.; Oruña-Concha, M.J. Optimization of enzyme-assisted extraction of ferulic acid from sweet corn cob by response surface methodology. *J. Sci. Food Agric.* **2019**, *100*, 1479–1485. [CrossRef]
19. Rashamuse, K.; Ronneburg, T.; Sanyika, W.; Mathiba, K.; Mmutlane, E.; Brady, D. Metagenomic mining of feruloyl esterases from termite enteric flora. *Appl. Microbiol. Biotechnol.* **2013**, *98*, 727–737. [CrossRef]
20. Beukes, N.; Pletschke, B.I. Effect of lime pre-treatment on the synergistic hydrolysis of sugarcane bagasse by hemicellulases. *Bioresour. Technol.* **2010**, *101*, 4472–4478. [CrossRef]
21. Sluiter, J.B.; Ruiz, R.O.; Scarlata, C.J.; Sluiter, A.D.; Templeton, D. Compositional Analysis of Lignocellulosic Feedstocks 1: Review and Description of Methods. *J. Agric. Food Chem.* **2010**, *58*, 9043–9053. [CrossRef]
22. Miller, G.L. Use of dinitrosalicylic acid reagent for the determination of reducing sugars. *Anal. Chem.* **1959**, *31*, 426–428. [CrossRef]
23. Bradford, M.M. A rapid and sensitive method for the quantitation of microgram quantities of protein utilizing the principle of protein-dye binding. *Anal. Biochem.* **1976**, *72*, 248–254. [CrossRef]
24. Rabemanolontsoa, H.; Saka, S. Various pretreatments of lignocellulosics. *Bioresour. Technol.* **2016**, *199*, 83–91. [CrossRef] [PubMed]
25. Gabhane, J.; William, S.P.; Vaidya, A.N.; Das, S.; Wate, S.R. Solar assisted alkali pretreatment of garden biomass: Effects on lignocellulose degradation, enzymatic hydrolysis, crystallinity and ultra-structural changes in lignocellulose. *Waste Manag.* **2015**, *40*, 92–99. [CrossRef] [PubMed]
26. Chandra, C.S.J.; George, N.; Narayanankutty, S.K. Isolation and characterization of cellulose nanofibrils from arecanut husk fibre. *Carbohydr. Polym.* **2016**, *142*, 158–166.
27. Wu, H.; Li, H.; Xue, Y.; Luo, G.; Gan, L.; Liu, J.; Mao, L.; Long, M. High efficiency co-production of ferulic acid and xylooligosaccharides from wheat bran by recombinant xylanase and feruloyl esterase. *Biochem. Eng. J.* **2017**, *120*, 41–48. [CrossRef]
28. Levasseur, A.; Navarro, D.; Punt, P.J.; Belaich, J.-P.; Asther, M.; Record, E. Construction of engineered bifunctional enzymes and their overproduction in aspergillus niger for improved enzymatic tools to degrade agricultural by-products. *Appl. Environ. Microbiol.* **2005**, *71*, 8132–8140. [CrossRef]
29. Ranum, P.; Peña-Rosas, J.P.; Garcia-Casal, M.N. Global maize production, utilization, and consumption. *Ann. N. Y. Acad. Sci.* **2014**, *1312*, 105–112. [CrossRef]
30. Da Silva, J.C.; De Oliveira, R.C.; Neto, A.D.S.; Pimentel, V.C.; Santos, A.D.A.D. Extraction, addition and characterization of hemicelluloses from corn cobs to development of paper properties. *Procedia Mater. Sci.* **2015**, *8*, 793–801. [CrossRef]
31. Nieter, A.; Kelle, S.; Linke, D.; Berger, R.G. Feruloyl esterases from Schizophyllum commune to treat food industry side-streams. *Bioresour. Technol.* **2016**, *220*, 38–46. [CrossRef] [PubMed]
32. Zhang, S.-B.; Zhai, H.-C.; Wang, L.; Yu, G. Expression, purification and characterization of a feruloyl esterase A from Aspergillus flavus. *Protein Expr. Purif.* **2013**, *92*, 36–40. [CrossRef] [PubMed]

33. Appeldoorn, M.M.; De Waard, P.; Kabel, M.A.; Gruppen, H.; Schols, H.A. Enzyme resistant feruloylated xylooligomer analogues from thermochemically treated corn fiber contain large side chains, ethyl glycosides and novel sites of acetylation. *Carbohydr. Res.* **2013**, *381*, 33–42. [CrossRef]
34. Jonathan, M.; DeMartini, J.; Thans, S.V.S.; Hommes, R.; Kabel, M.A. Characterisation of non-degraded oligosaccharides in enzymatically hydrolysed and fermented, dilute ammonia-pretreated corn stover for ethanol production. *Biotechnol. Biofuels* **2017**, *10*, 112. [CrossRef] [PubMed]
35. Bhatia, L.; Sharma, A.; Bachheti, R.K.; Chandel, A.K. Lignocellulose derived functional oligosaccharides: Production, properties, and health benefits. *Prep. Biochem. Biotechnol.* **2019**, *49*, 744–758. [CrossRef] [PubMed]
36. Singh, R.D.; Banerjee, J.; Sasmal, S.; Muir, J.; Arora, A. High xylan recovery using two stage alkali pre-treatment process from high lignin biomass and its valorization to xylooligosaccharides of low degree of polymerisation. *Bioresour. Technol.* **2018**, *256*, 110–117. [CrossRef]

© 2020 by the authors. Licensee MDPI, Basel, Switzerland. This article is an open access article distributed under the terms and conditions of the Creative Commons Attribution (CC BY) license (http://creativecommons.org/licenses/by/4.0/).

agronomy

MDPI

Article

The Effects of Alkaline Pretreatment on Agricultural Biomasses (Corn Cob and Sweet Sorghum Bagasse) and Their Hydrolysis by a Termite-Derived Enzyme Cocktail

Mpho. S. Mafa [1,2], Samkelo Malgas [1], Abhishek Bhattacharya [1,3], Konanani Rashamuse [4] and Brett I. Pletschke [1,*]

1 Enzyme Science Programme (ESP), Department of Biochemistry and Microbiology, Rhodes University, Grahamstown 6140, South Africa; mafams@ufs.ac.za (M.S.M.); samkelomalgas@yahoo.com (S.M.); abhishek.bhattacharya@biochemistry.lu.se (A.B.)

2 Department of Plant Sciences, University of the Free State P.O. Box 339, Bloemfontein 9300, South Africa

3 Department of Biochemistry and Structural Biology, Lund University, 22100 Lund, Sweden

4 CSIR Biosciences, Building 18, Meiring Naudé Road, Brummeria, PO Box 395, Pretoria 0001, South Africa; Krashamuse@gmail.com

* Correspondence: b.pletschke@ru.ac.za; Tel.: +27-46-603-8081

Received: 23 July 2020; Accepted: 12 August 2020; Published: 18 August 2020

check for updates

Abstract: Sweet sorghum bagasse (SSB) and corncob (CC) have been identified as promising feedstocks for the production of second-generation biofuels and other value-added chemicals. In this study, lime ($Ca(OH)_2$) and NaOH pretreatment efficacy for decreasing recalcitrance from SSB and CC was investigated, and subsequently, the pretreated biomass was subjected to the hydrolytic action of an in-house formulated holocellulolytic enzyme cocktail (HEC-H). Compositional analysis revealed that SSB contained 29.34% lignin, 17.75% cellulose and 16.28% hemicellulose, while CC consisted of 22.51% lignin, 23.58% cellulose and 33.34% hemicellulose. Alkaline pretreatment was more effective in pretreating CC biomass compared to the SSB biomass. Both $Ca(OH)_2$ and NaOH pretreatment removed lignin from the CC biomass, while only NaOH removed lignin from the SSB biomass. Biomass compositional analysis revealed that these agricultural feedstocks differed in their chemical composition because the CC biomass contained mainly hemicellulose (33–35%), while SSB biomass consisted mainly of cellulose (17–24%). The alkaline pretreated SSB and CC samples were subjected to the hydrolytic action of the holocellulolytic enzyme cocktail, formulated with termite derived multifunctional enzymes (referred to as MFE-5E, MFE-5H and MFE-45) and exoglucanase (Exg-D). The HEC-H hydrolysed NaOH pretreated SSB and CC more effectively than $Ca(OH)_2$ pretreated feedstocks, revealing that NaOH was a more effective pretreatment. In conclusion, the HEC-H cocktail efficiently hydrolysed alkaline pretreated agricultural feedstocks, particularly those which are hemicellulose- and amorphous cellulose-rich, such as CC, making it attractive for use in the bioconversion process in the biorefinery industry.

Keywords: alkaline pretreatment; enzyme cocktail; enzymatic hydrolysis; glycosyl hydrolase; termite metagenome derived enzymes

1. Introduction

Agricultural feedstocks such as sweet sorghum bagasse (SSB) and corncob (CC) have been identified as promising alternatives for the production of second-generation biofuels and other value-added chemicals [1–3]. These feedstocks consist of polysaccharides (cellulose and hemicellulose)

that can be hydrolysed to simple monosaccharides which can be fermented to produce ethanol. The cellulose consists of glucose moieties linked by β-1,4-glycosidic bonds, which forms crystalline cellulose that is recalcitrant to enzymatic hydrolytic activity, and also amorphous cellulose that is easily hydrolysed by cellulases. Mnich et al. [4] reported that in Poaceae (Gramineae), such as sorghum or maize plants, hemicellulose is mostly composed of a type of xylan called glucuronoarabinoxylans (GAX). Generally, xylans are composed of a backbone of β-1,4-glycosidic bonds linking xylose moieties, which can be O-acetylated or substituted at O-2 by α-arabinose, α-glucuronic acid or α-methyl-glucuronic acid side-chains that give rise to arabinoxylan, glucuronoxylan and GAX, respectively [4,5]. The cellulose and hemicellulose are generally covered and cross-linked by lignin which impedes glycoside hydrolase (GHs) hydrolytic activity [4,6]. As a result, to achieve efficient feedstock hydrolysis, the lignin is removed from the feedstock through chemical, physical, or biological pretreatments.

Agricultural biomasses have been pretreated with various acids such as phosphoric acid, sulfuric acid and acetic acid [6,7]. The acid pretreatment of agricultural feedstocks effectively removes lignin, however, it also removes a high amount of hemicellulose. In addition, some acids such as sulfuric acid are not environmentally friendly or require higher temperature for long periods (140 °C for 6 h) for the pretreatment to be effective [2]. Biological pretreatments such as the use of fungi or bacteria to degrade lignin were introduced as alternative environmentally friendly pretreatment technologies [8]. For instance, Mishra and Jana [2] pretreated SSB biomass with *Coriolus versicolor* using solid-state fermentation (SSF) in a bioreactor. *C. versicolor* removed more than 40% (w/w) of lignin from the SSB biomass, but this process takes many days to remove high amounts of lignin. An effective pretreatment process of the biomass should remove lignin within a few hours and must achieve the following; (1) a reduction of the cellulose crystallinity, (2) an increase in the surface area of the biomass, (3) an increase in porosity, (4) a reduction in the degree of polymerisation of cellulose, and (5) a reduction in the particle size of the biomass [6,7]. All these properties associated with the pretreatment of feedstocks improve the access of the enzymes to the biomass and significantly increases the enzymatic hydrolytic activity.

Alkaline pretreatment is among the most established techniques for the removal or redistribution of lignin and mercerisation of the biomass [9–12]. Alkaline pretreatments are relatively affordable compared to other methods (like acid pretreatments) because they generally use mild reaction conditions, alkaline chemicals can be recovered, reused and have a high selectivity for separation of lignin [13]. According to Kim et al. [13], alkaline pretreatment technologies can be divided into: (1) sodium hydroxide pretreatment, (2) ammonia pretreatment, (3) aqueous ammonia pretreatment, (4) anhydrous ammonia pretreatment, and (5) lime ($Ca(OH)_2$) pretreatment. In addition, da Silva Neto et al. [14] achieved more than 60% lignin removal from the SSB biomass using 6% (*w*/*v*) hydrogen peroxide within 4 h and solubilised 32% of the hemicellulose in the same process. A study showed that the alkaline pretreatment increased the swelling of the biomass, removed lignin, and converted the crystalline cellulose I into cellulose II which is less crystalline [9]. The crystallinity of the biomass is generally studied with X-ray diffraction (XRD) or Fourier-transform infrared spectroscopy (FTIR) [11,14,15]. These techniques show the changes of the biomass from a crystalline structure to a more amorphous one after pre-treatment, which is generally associated with improvements in the GH enzyme hydrolytic activity on the biomass. However, a few studies have investigated the performance of the activity of the GH enzymes on the biomass that was pretreated with $Ca(OH)_2$ and sodium hydroxide (NaOH/mercerised), and have determined which of the two pre-treatments is more effective at improving biomass saccharification.

A better understanding of lignocellulose digestion by termite-metagenome derived enzymes may help to uncover and eventually overcome challenges in the conversion of lignified plant cell walls into simple sugars [16]. The termite gut microflora acts as an important reservoir of novel GHs that are able to efficiently hydrolyse lignocellulosic biomass [16,17]. Functional metagenomic screening for hemicellulases in the termite, *Pseudacanthotermes militaris*, revealed a large number of hemicellulose degrading enzymes [18]. These enzymes included arabinofuranosidases, xylosidases and endo-xylanases. In addition, the metagenomic approach has been used previously to mine various

GHs and feruloyl esterase enzyme genes from bacterial symbionts in the hindgut of the *Trinervitermes trinervoides* termite species [19,20]. These studies demonstrated that termites and their symbionts possess hemicellulase enzymes which could be used to hydrolyse the hemicellulosic fraction of biomass in a synergistic manner.

In this study, we aimed to remove lignin from the SSB and CC biomasses by pretreating them with NaOH and $Ca(OH)_2$. Subsequently, the efficacy of each pretreatment on the biomass was evaluated with compositional analyses, morphological change assessments using a scanning electron microscope (SEM), and biomass chemical changes were analysed using FITR. Furthermore, the alkaline pretreated biomass was subjected to enzymatic hydrolysis using an in-house formulated holocellulolytic enzyme cocktail (HEC-H) consisting of multifunctional enzymes (MFEs) and an exoglucanase (Exg-D) derived from bacterial symbionts in the hindgut of *Trinervitermes trinervoides* (Sjöstedt).

2. Materials and Methods

2.1. Materials

The four GH enzymes [purified Exo-glucanase (Exg-D), MFE-5E, MFE-5H, and MFE-45] used in the current study were derived from hindgut bacterial symbionts of a termite (*T. trinervoides*) metagenome identified and isolated by Rashamuse et al. [20]. The model substrates; carboxymethylcellulose (CMC), Avicel PH101, locust bean gum and the *p*-nitrophenyl (*p*NP) based substrates (*p*NP-α-L-arabinofuranoside (A), *p*NP-β-D-glucopyranoside (G), *p*NP-β-D-mannopyranoside (M), *p*NP-β-D-cellobioside (C) and *p*NP-β-D-xylopyranoside (X)) were purchased from Sigma-Aldrich, South Africa. Beechwood xylan, soluble wheat flour xylan, insoluble wheat flour arabinoxylan and xyloglucan were purchased from Megazyme, Ireland. All chemicals used in this study were of analytical grade and were purchased from Sigma-Aldrich unless stated otherwise.

2.2. Lignocellulosic Biomass Pretreatment

The 5 g of dried and pulverised (>2 mm particles) sweet sorghum bagasse (SSB) and corncob (CC) samples were chemically pretreated with alkali (2 g of $Ca(OH)_2$ and 1.25 g of NaOH) according to protocols described by Panagiotopoulos et al. [21] and Beukes and Pletschke [10], respectively. After pretreatment, SSB and CC samples were filtered and washed with Milli-Q water until the filtrates had a pH of 7.0. The samples were air dried and stored in an airtight container until use.

2.3. Biomass Composition and Scanning Electron Microscope (SEM) Analysis

Biomass composition of untreated, $Ca(OH)_2$ and NaOH pretreated SSB and CC samples were determined according to protocols by the National Renewable Energy Lab (NREL) [22] at the Wood Sciences Faculty at Stellenbosch University, South Africa.

To determine the effect of each pretreatment ($Ca(OH)_2$ and NaOH) on biomass morphology, the dried SSB and CC samples were mounted on scanning electron microscope (SEM) stubs using double sided graphite tape, sputter coated with gold using a Balzers Union sputtering device. The gold coated SSB and CC samples were viewed under a Tescan Vega scanning electron microscope at 20 kV. Digital images were captured using the Vega imaging system.

2.4. Fourier-Transform Infrared Spectroscopy Analysis

The functional group analysis of the untreated or pretreated SSB and CC biomass samples was analysed with Fourier Transform Infrared (FTIR) Spectroscopy. The FTIR spectra of Avicel (cellulose control), untreated, $Ca(OH)_2$ and NaOH pretreated SSB and CC samples were recorded at room temperature using an UATR-FTIR instrument (PerkinElmer, USA). All FTIR spectra were collected at a spectrum resolution of 4 cm^{-1}, with 4 co-added scans per sample over the range of 4000 to 650 cm^{-1}. The Perkin-Elmer software (Spectrum version. 6.3.5) was used to perform spectra normalisation, baseline corrections, and peak integration. The spectra of the Avicel PH101 microcrystalline cellulose, untreated, pretreated SSB and CC samples were presented as absorbance values, and each value represented the means of four scans.

2.5. Formulation of the Holoenzyme Cocktail (HEC)

The substrate specificities of the MFE-5E, MFE-5H, and MFE-45 enzymes were performed according to Mafa et al. [23], and the released total reducing sugars were measured according to a modified 3,5-dinitrosalicylic acid (DNS) method [24]. An amount of 1% (*w*/*v*) beechwood xylan, wheat arabinoxylan, Avicel, xyloglucan, locust bean gum and CMC were used as substrates. For *p*NP based substrate specificity assays, 2 mM *p*NP-A, *p*NP-G, *p*NP-M, *p*NP-C and *p*NP-X were used. Four GH enzymes; exo-glucanase (Exg-D), MFE-5E, MFE-5H and MFE-45, derived from the termite bacterial hindgut metagenome were used to formulate the holocellulolytic enzyme cocktail HEC-H (60% MFE-5H, 20% MFE-5E, 10% MFE-45 and 10% Exg-D). HEC-H was then supplemented with 10% protein loading of each of the following enzymes, *Aspergillus niger* β-glucosidase (Novozyme 188) and *Selenomonas ruminantium* xylosidase, SXA, during the hydrolysis of pretreated feedstocks. The reactions were initiated by adding the HEC-H and auxiliary enzymes to 2% (*w*/*v*) of untreated, $Ca(OH)_2$ and NaOH pretreated SSB or CC biomass samples suspended in 50 mM sodium citrate buffer at pH 5.5. The reaction was carried out by incubating the samples at 37 °C in an incubation room. The release of the total reducing sugars was measured using the DNS assays.

2.6. Data Analysis

One-way analysis of variance (ANOVA) was used to analyse and detect significant increases in activity exhibited by the HEC-H. All pairwise comparison procedures were conducted using the data analysis feature in Microsoft Excel 3.

3. Results

3.1. Biomass Composition and SEM Analysis of Pre-Treated Feedstocks

Lignin in lignocellulosic biomass is an established GH enzyme inhibitor as it results in non-productive binding and enzyme stalling. Thus, it is important to pretreat the agricultural feedstocks before the application of GH enzymes for effective hydrolysis. In the current study, the untreated CC biomass composition showed that its lignin, cellulose and hemicellulose contents were about 22.51%, 23.58% and 33.34%, respectively (see Table 1). The $Ca(OH)_2$ pretreated CC composition showed that lignin content was significantly ($p < 0.01$) reduced by 7.7%, while cellulose content significantly ($p < 0.01$) increased by 3.76% compared to the untreated biomass. The NaOH pretreated CC lignin content was reduced by about 13.01%. Interestingly, when compared to the untreated biomass, the cellulose and hemicellulose contents were significantly ($p < 0.01$) increased by 9.74% and 1.69%, respectively. These results demonstrate that alkaline pretreatment effectively removed/modified lignin from the CC samples. The removal of lignin resulted in the increased cellulose and hemicellulose content of the CC feedstock. These results suggested that the carbohydrates were more exposed for hydrolytic activity because the pretreatments had removed significant amounts of lignin from these samples.

In addition, untreated SSB lignin, cellulose and hemicellulose contents were 29.34%, 17.75% and 16.28%, respectively (Table 1). The $Ca(OH)_2$ pretreated SSB lignin content was not significantly reduced compared to that of the untreated biomass. The hemicellulose content was reduced from 16.28 to 11.54% and the cellulose content significantly ($p < 0.01$) increased from 17.75% to 19.60% after the SSB was pretreated with $Ca(OH)_2$ (Table 1). The NaOH pretreated SSB led to an 18.45% reduction in the amount of lignin and a 6.06% increase in cellulose content compared to the untreated biomass. The NaOH pretreated SSB had a hemicellulose content that was significantly ($p < 0.01$) reduced from 16.28% to 13.05% compared to the untreated biomass. Also, the SSB pretreated with NaOH demonstrated that the pretreatment was effective in removing lignin and exposing the hemicellulose and cellulose.

Table 1. Alkaline pretreated feedstock compositional analysis.

Biomass Pre-Treatment	Biomass Composition (%)		
Corncob (CC)	**Lignin**	**Cellulose**	**Hemicellulose**
Untreated	22.51 ± 0.19	23.58 ± 0.47	33.34 ± 0.56
$Ca(OH)_2$	14.81 ± 0.10 #	27.34 ± 0.33 #	33.49 ± 0.11
NaOH	9.50 ± 0.04 #	33.32 ± 0.41 #	35.03 ± 0.63 *
Sweet sorghum bagasse (SSB)			
Untreated	29.34 ± 0.042	17.75 ± 0.36	16.28 ± 0.46
$Ca(OH)_2$	29.08 ± 0.092	19.60 ± 0.1 #	11.54 ± 0.47 *
NaOH	10.88 ± 0.031 #	23.81 ± 0.22 #	13.05 ± 0.25 #

Values represent the means ± SD and $n = 3$. * represents p-value < 0.05 and # represents p-value < 0.01.

SEM analysis demonstrated that $Ca(OH)_2$ and NaOH pretreatment resulted in different pretreatment severities in the CC and SSB feedstocks. Figure 1 shows that both untreated CC and SSB biomasses were covered with lignin, which formed a thick smooth whitish layer. After $Ca(OH)_2$ pretreatment, both biomasses had a morphology which showed that the lignin had broken down and condensed into droplets and tiny sheets. The hydrolysis of the lignin created a larger biomass surface area which was ideal for enzymatic activity. Table 1 shows that the amounts of the lignin for untreated SSB and $Ca(OH)_2$ pretreated were similar. SEM analysis also revealed that lignin was not totally removed, but rather modified to form droplets and sheets on the surface of the biomass. The SEM results for $Ca(OH)_2$ pretreatment also showed similar effects on CC biomass. However, the NaOH pretreatment was more effective in removing the lignin from CC and SSB biomass. Roughly more than 90% of lignin was removed from the biomass, leaving cellulose and hemicellulose threads exposed with pores on their surfaces (Figure 1). There was a high correlation between the biomass composition (Table 1) and the SEM results, which suggested that the NaOH pretreatment was more effective at removing lignin from the biomass compared to $Ca(OH)_2$ pretreatment.

3.2. FTIR Analysis of Pretreated Feedstock

FTIR is generally used to assess chemical functional group changes in biomass and this information is used to compare the structural changes of the biomass before and after chemical pretreatments. The alkaline pretreated CC and SSB biomasses were analysed with FTIR to validate the compositional analysis and SEM results. Figure 2 demonstrates that the alkaline pretreated CC and SSB biomasses possessed more cellulose II than crystalline cellulose I, which is an indication of reduced biomass crystallinity. The positive control for cellulose Iβ (Avicel PH101 microcrystalline cellulose) showed high intensity peaks at 1428.66, 1160.54 and 1104.67 cm^{-1}. These peaks represent the presence of predominately crystalline cellulose Iβ and small quantities of cellulose II in the biomass. Similar peaks (1428.06, 1157.46 and 1111.00 cm^{-1}) were present in the untreated SSB biomass, which suggest that there was a presence of crystalline cellulose I and cellulose II in this sample. However, only one peak was present in the $Ca(OH)_2$ (1158.11 cm^{-1}) and NaOH (1157.23 cm^{-1}) pretreated SSB biomass (Figure 2). These results demonstrated that after pretreatment, the SSB biomass had a reduced content of crystalline cellulose Iβ by producing cellulose II. In addition, the $Ca(OH)_2$ pretreatment shifted the 1428.66 cm^{-1} peak to 1410.45 cm^{-1}, while NaOH pretreatment shifted the 1428.66 cm^{-1} peak to 1421.90 cm^{-1}. This shift is characteristic to the formation of cellulose II and amorphous cellulose, and the disappearance of the crystalline cellulose I in the biomass. Figure 2B showed that the similar spectra were identified in the untreated and alkaline pretreated CC biomass, suggesting that alkaline pretreatment removed lignin and changed the CC biomass crystallinity in a similar fashion to the SSB biomass.

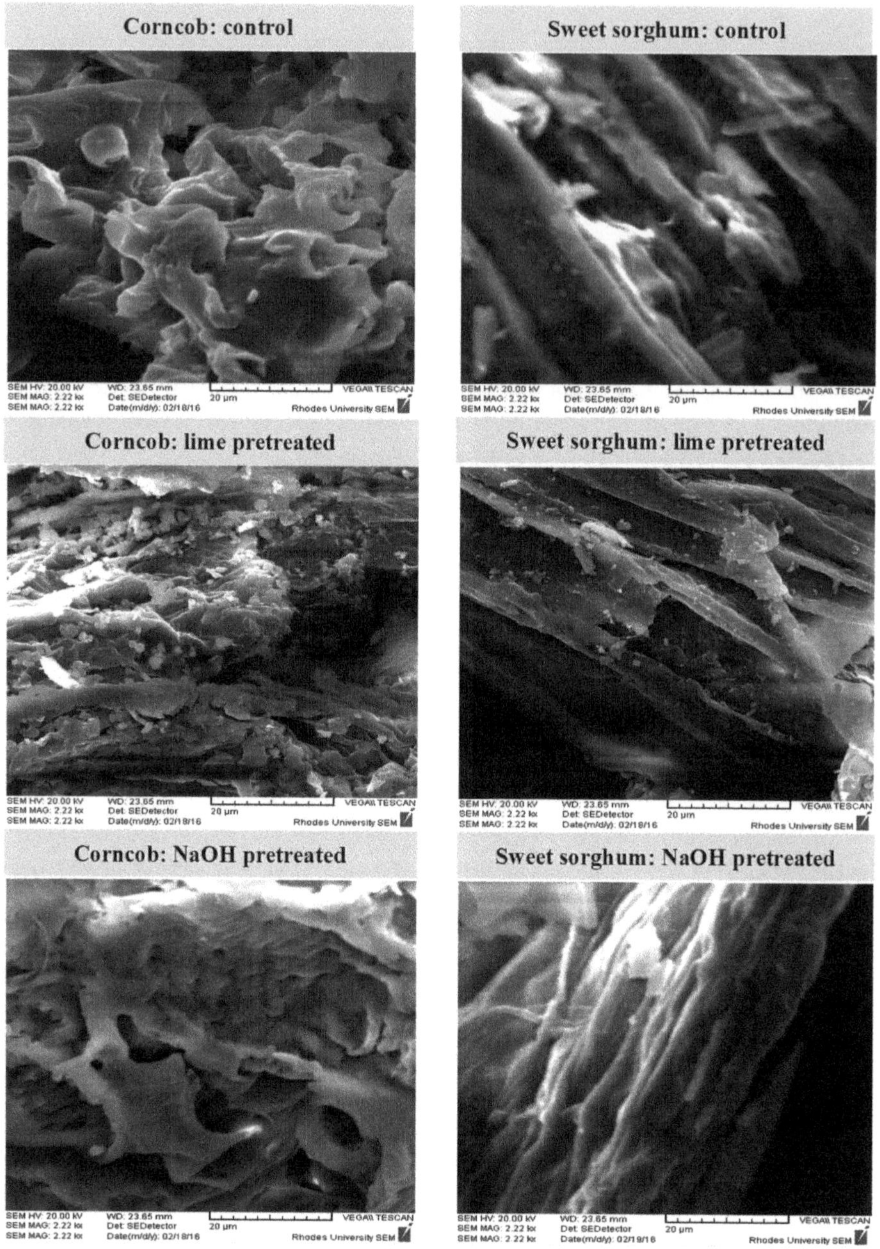

Figure 1. Topological analysis of pretreated sweet sorghum (SSB) and corncob (CC) feedstocks using scanning electron microscopy (SEM). The captions above the images indicate the control (untreated), type of treatment and the biomass in each SEM image. The scale bar was 20 μm and the magnification was 2.22 kx.

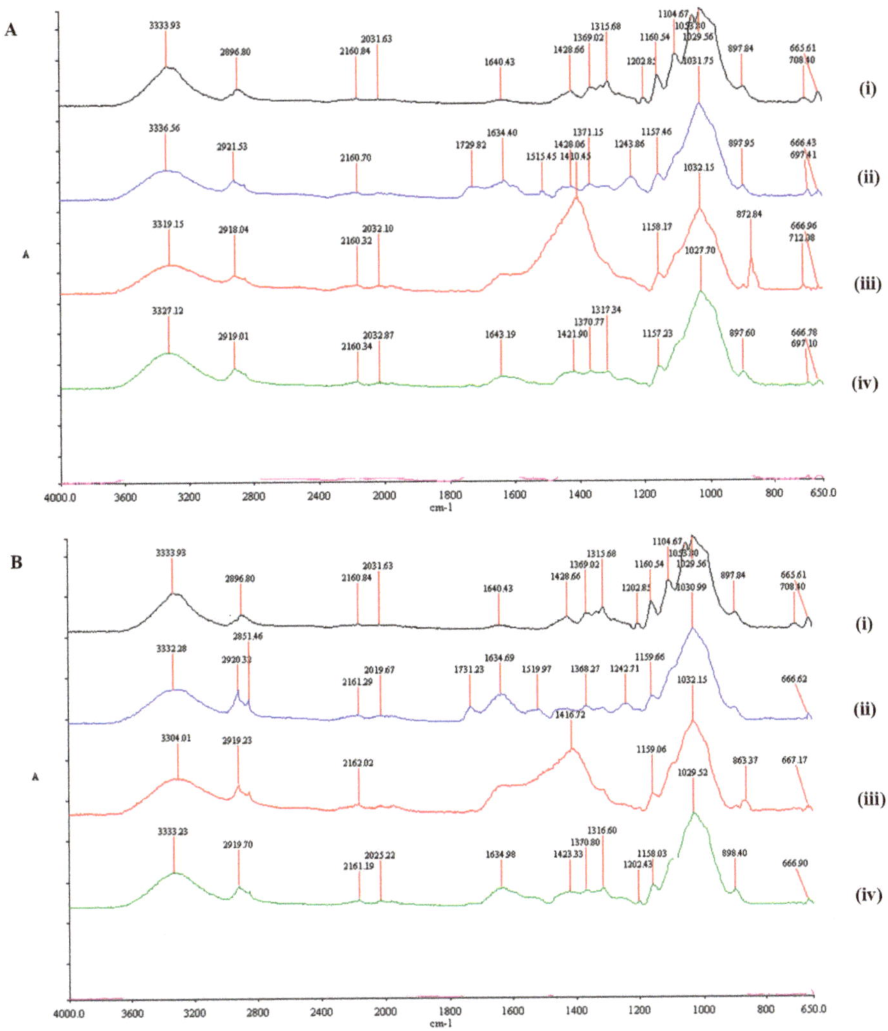

Figure 2. Comparative FTIR in the spectra range 4000–650 cm^{-1} corresponding to Avicel PH101 (i), untreated (ii), $Ca(OH)_2$ pretreated (iii) or sodium hydroxide pretreated biomass. (**A**) represents the SSB biomass and (**B**) represents the corncob biomass. Keys: (**A**) in the *y*-axis represents arbitrary units.

The peaks at 1515.45 cm^{-1} and 1519.97 cm^{-1} in the untreated SSB and CC biomasses, respectively, represent the C≡C vibration of the aromatic rings of the lignin. In contrast, these bands were absent in the IR spectrum of the positive control (Avicel) and NaOH pretreated SSB and CC biomasses (Figure 2). These results supported the biomass composition analysis and SEM findings which revealed that NaOH pretreatment removed the lignin from the biomass. However, the spectrum of the $Ca(OH)_2$ pretreated SSB and CC biomass samples had a broad and higher intensity peak starting from 1640 cm^{-1}, reaching a maximum at 1410 cm^{-1} and ending at 1370 cm^{-1} (Figure 2). This observation suggested that lignin was not completely removed by the $Ca(OH)_2$ pretreatment but was re-distributed as revealed by SEM (Figure 1). In addition, the untreated and the alkaline pretreated SSB and CC biomasses displayed one peak between 1033 and 1027, which represented the hemicellulose. These results revealed that the hemicellulose was still present in the SSB and CC after alkaline pretreatment.

3.3. Substrate Specificity

The enzyme substrate specificity results showed that MFE-5E was a multifunctional enzyme that displayed 17.189, 13.471, 7.483 and 6.831 U/mg protein specific activity on CMC, wheat flour xylan, beechwood xylan and xyloglucan, respectively. These findings demonstrate that MFE-5E had a higher propensity to hydrolyse amorphous cellulose followed by various xylan substrates and xyloglucan (Table 2). In addition, MFE-5H exhibited the highest specific activity on xyloglucan (84.975 U/mg protein), followed by wheat flour xylan, beechwood xylan and CMC with 67.365, 52.015 and 28.584 U/mg protein, respectively. Both MFE-5E and MFE-5H were classified under GH family 5 (with accession numbers AMO13175 and AMO13178, respectively), which consists of some of the enzymes displaying diverse activities on various carbohydrate substrates (http://www.cazy.org/GH5.html). This explains the broad range of MFE-5E and MFE-5H activity on various carbohydrate substrates. The results in Table 2 also shows that the MFE-45 displayed highest specific activity on xyloglucan (12.86 U/mg protein). MFE-45 showed similar activity during hydrolysis of beechwood xylan and CMC, which were 4.55 and 4.75 U/mg protein, respectively. MFE-45 is reported to belong to GH family 45 and its accession number is AMO13193. According to the CAZy Database (http://www.cazy.org/GH45_bacteria.html), enzymes in family 45 possess endoglucanase activity, xyloglucanase activity but not xylanase activity, which suggests that MFE-45 possesses a novel xylanase activity. The MFEs did not show any activity on Avicel, and the mannan substrate, locus bean gum, and several *p*-nitrophenyl-substrates (Table 2). However, MFEs presented in this study were true multifunctional enzymes that displayed specific activities on various substrates with their backbones linked by β-1,4-glycosic linkages, and possessed different sidechains.

Table 2. Substrate specificity of enzymes on polymeric and *p*-nitrophenyl substrates.

Substrates	Enzyme Specific Activity (U/mg Protein)		
	MFE-5E	**MFE-5H**	**MFE-45**
CMC	17.189 ± 0.049	28.584 ± 0.025	4.75 ± 0.053
Avicel	N/A	N/A	N/A
Beechwood xylan	7.483 ± 0.013	52.015 ± 0.071	4.55 ± 0.013
Wheat flour xylan	13.471 ± 0.161	67.365 ± 0.116	#
Xyloglucan	6.831 ± 0.008	84.975 ± 0.012	12.86 ± 0.19
Locus bean gum	N/A	N/A	N/A
*p*NP-A	N/A	N/A	NA
*p*NP-G	N/A	N/A	N/A
*p*NP-X	N/A	N/A	N/A
*p*NP-C	N/A	N/A	N/A
*p*NP-M	N/A	N/A	N/A

Values represent means ± SD and $n = 3$; N/A means there was no activity; # - means not tested on this substrate; pNP-A, G, -X, -C and M represents *p*-nitrophenyl-arabinofuranoside, -glucopyranoside, -xylopyranoside, -cellobioside and –mannopyranoside, respectively; Concentrations of polymeric and *p*-nitrophenyl substrates were 2% and 4 mM, respectively; U represents μmol/min.

3.4. Holocellulolytic Enzyme Cocktail Formulation and Its Application on Pretreated Biomass

The holocellulolytic enzyme cocktail (HEC-H) used to assess the effectiveness of the alkaline pretreatments in lowering biomass recalcitrance was formulated using a 60%: 20%: 10%: 10% combination of the MFE-5E, MFE-5H, MFE-45 and exoglucanase (Exg-D) enzymes. Exg-D was previously characterised by Mafa et al. [23] and its biochemical properties demonstrated that it can effectively be used in synergy with MFEs.

HEC-H hydrolysed the alkaline pretreated SSB and CC substrates better than their untreated biomasses (i.e., the controls) (Figure 3A). Also, HEC-H exhibited the highest activity on the CC biomass compared to SSB biomass. Higher activity (approximately 0.72 μmol/mL) was observed when HEC-H hydrolysed $Ca(OH)_2$ pretreated CC compared to SSB biomass, while an approximately 1.14 μmol/mL higher activity was recorded when the HEC-H hydrolysed NaOH-pretreated CC compared to SSB biomass. We propose that the HEC-H hydrolysed the alkaline pretreated CC better than the SSB due to differences in their biomass composition; CC had a higher hemicellulose content compared to SSB. Furthermore, our results also demonstrated that the alkaline pretreatment removed lignin from the biomass and improved the activity of the HEC-H, while there was no significant difference between the CC and SSB biomass controls (Figure 3A).

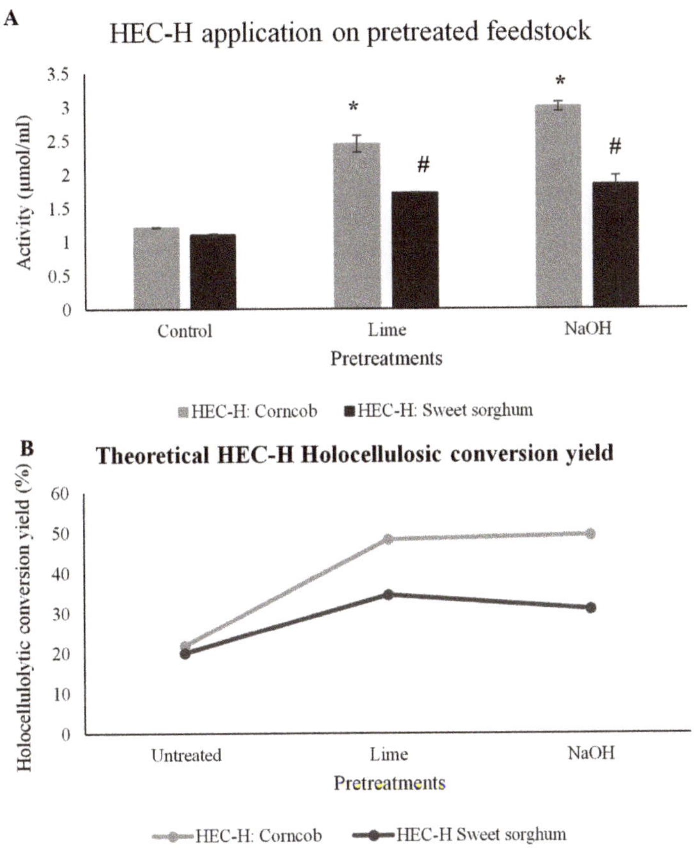

Figure 3. Application of the holocellulolytic enzyme cocktail (HEC-H) formulated with multifunctional enzymes (MFEs) and exoglucanase (Exg-D) derived from a termite bacterial hind-gut metagenome on pretreated sweet sorghum bagasse (SSB) and corncob (CC). (**A**) Activity in μmol/min, and (**B**) holocellulose conversion yields (total reducing sugars produced by HEC-H divided by the sum of cellulose and hemicellulose content) of SSB and CC samples. * Represents p-values < 0.01 and # represents p-values < 0.05. All experiments were performed in triplicate and the values represent means ± SD.

Figure 3B shows the theoretical holocellulosic hydrolytic yields of the HEC-H activity on the SSB and CC samples. The results suggest that the conversion rates of the $Ca(OH)_2$ pretreated CC samples was 48.30%, while the conversion rate of the NaOH pretreated CC was about 49.46%. The HEC-H only converted about 21.91% of the untreated CC samples, which illustrated that the enzyme cocktail displayed about 2.25-fold higher efficiency on the alkali pretreated biomass. In contrast, the HEC-H displayed an approximate 1.67-fold increase in the theoretical conversion rate of alkali pretreated SSB samples, compared to the untreated SSB (control). We suggest that the enzyme cocktail efficiency was higher on CC compared to SSB samples, because of the higher holocellulosic content in CC biomass (see Table 1).

4. Discussion

The SSB and CC biomass samples were pretreated before they were hydrolysed with the holocellulolytic enzyme cocktail formulated with MFEs and Exg-D to assess their hydrolysability. El-Naggar et al. [7] argued that one of the factors that hinders lignocellulosic hydrolysis are biomass structural or chemical factors, which include biomass crystallinity, cellulose degree of polymerisation, porosity, accessible surface area, particle size, lignin, hemicellulose and acetyl-group content. In addition, Álvarez et al. [25] reported that chemical pretreatments were essential to changes in the biomass composition. The pretreatment of the biomass results in the removal of lignin, which makes the biomass less recalcitrant to enzyme hydrolysis [6,25]. The current study demonstrated that NaOH pretreatment of the CC and SSB samples significantly removed the lignin that led to an increase in relative cellulose, hemicellulose and biomass porosity (Table 1; Figure 1). However, $Ca(OH)_2$ pretreatment did not remove all the lignin as shown in Table 1 and Figure 1. Beukes and Pletschke [10] also argued that the $Ca(OH)_2$ pretreatment of the sugar cane bagasse modified the lignin and did not remove most of it from the biomass. Alkaline pretreatments induce a saponification reaction, where free hydroxide ions break ester bonds that exist between the inner molecules of lignocellulose, which connect hemicellulose and other components (such as between lignin and other hemicelluloses) which results in an increase in the pore structures of lignocellulose due to disappearance of the connecting bonds [26].

The FTIR results validated the morphological and chemical changes of the biomass after pretreatment. FTIR results demonstrated that crystalline cellulose Iβ peaks were absent in the spectrum of the $Ca(OH)_2$ and NaOH pretreated SSB and CC biomass. The 1730 cm^{-1} peak corresponded to C-O stretching vibration for the acetyl and ester linkages in lignin and hemicellulose, while the 1250 cm^{-1} peak corresponded to C–O out of plane stretching due to the aryl group in lignin [27]. The presence of 1430, 1162 and 1111 cm^{-1} peaks indicates a prevalence of crystalline cellulose II [15]. The alkaline pretreated biomass showed improved cellulose II, hemicellulose and amorphous content. However, the SEM, chemical composition and FITR results revealed that the NaOH pretreatment was more effective in removing lignin from the biomass than $Ca(OH)_2$ pretreatment.

The MFEs displayed activity on the CMC, various xylan and xyloglucan substrates as shown in Table 2. Malgas et al. [5] reported that xylan is a complex substrate with different side chains that can hinder enzyme optimal function, if it possess a catalytic active site that is not suited to the side chains. Xyloglucan is also a complex substrate with various side chains as described by Fry et al. [28] and Rashmi and Siddalingamurthy [29]. These diverse and complex properties of the CMC, various xylan and xyloglucan substrates demonstrate that the MFE-5E, MFE-5H and MFE-45 were indeed multifucntional enzymes capable of hydrolysing the β-1,4-glucosidic bonds of the backbone chain. Several multifunctional (or promiscuous) enzymes have been reported previously and were defined as enzymes that play multiple physiological roles in a cell, or enzymes that display catalytic activities on a range of substrates which are structurally and chemically different [30–32]. In addition, mulifunctional/promiscous enzymes change their hydrolytic activities under different reaction conditions, which include various solvents, extreme temperature, various pHs and a range of substrates specificities. Aspeborg et al. [33] emphasised the idea that GH family 5 contains some of the multifunctional enzymes, which could explain the observed multifunctional nature of MFE-5E and MFE-5H.

HEC-H was applied to the alkaline pretreated biomass and the untreated biomass was used as a benchmark for hydrolysis improvement due to pretreatment. It was evident from Figure 3 that HEC-H demonstrated higher activity on pretreated biomass compared to the untreated biomass. In addition, the HEC-H performed much better on the CC biomass compared to the SSB biomass. The specific activity demonstrated that the MFEs used to formulate the HEC-H were highly active on the hemicellulose substrate and amorphous cellulose. These findings explain why the theoretical holocellulosic conversion yields for alkali pretreated biomass were between 35% and 50%. Takada et al. [3] also demonstrated that the chemical composition of CC biomass consisted mostly of hemicellulose. However, SSB biomass after alkaline pretreatment consisted mainly of cellulose as the major component. We believe that the HEC-H is the first enzyme cocktail formulated using GHs derived from a termite hindgut metagenome for the effective hydrolysis of hemicellulose and amorphous cellulosic components of agricultural feedstocks. Other studies have used two or three enzymes derived from termites, i.e., Feng et al. [34] reported on the synergy formed by two β-glucosidase (CfGlu1C and CfGlu1B) enzymes from a lower termite (*Coptotermes formosanus*). CfGlu1C and CfGlu 1B displayed synergy on lactose hydrolysis and significantly increased the rate of hydrolysis. The GH7 enzyme, endoglucanase and β-glucosidase from *Reticulitermes flavipes* worker termites also displayed synergism on pine sawdust [35]. The synergism between these enzymes significantly increased the amounts of glucose released during the hydrolysis of the pine sawdust.

5. Conclusions

Alkaline pretreatment was more effective in pretreating CC biomass compared to the SSB biomass. Both $Ca(OH)_2$ and NaOH pretreatment removed lignin from the CC biomass, while only NaOH removed lignin from the SSB biomass. Biomass compositional analysis revealed that these agricultural feedstocks differed in their chemical composition because the CC biomass contained hemicellulose (33–35%) as its major component, compared to SSB biomass that consisted mainly of cellulose (17–24%). Specific enzyme activity revealed that the termite metagenome derived enzymes MFE-5E, MFE-5H and MFE-45 were multifunctional enzymes, which hydrolysed substrates associated with hemicellulose (various xylan and xyloglucan) and amorphous cellulose (CMC). These observations explain why the HEC-H formulated with MFEs effectively hydrolysed the CC biomass preferentially to SSB biomass. We propose that the HEC-H cocktail can be used for the hydrolysis of hemicellulose- and amorphous cellulose-rich agricultural feedstocks during the bioconversion process in the biorefinery industry.

Author Contributions: M.S.M. carried out all the experiments and performed data analysis. K.R. purified and provided the termite derived enzymes. M.S.M., S.M. and A.B. participated in the experimental design of synergy studies and synergy data analysis. M.S.M., S.M. and A.B. designed the pretreatment and data analysis. M.S.M. and B.I.P. conceptualised the study and participated in its design coordination. M.S.M. drafted the manuscript. B.I.P. and K.R. supervised and co-supervised the study, respectively. M.S.M. prepared the manuscript while S.M., K.R. and B.I.P. contributed in editing and final preparation of the manuscript. All authors have read and agreed to the published version of the manuscript.

Funding: The current study was financially supported by the National Research Foundation (NRF) of South Africa, Department of Science and Technology (DST) and Council for Scientific and Industrial Research (CSIR) National Biocatalysis Initiative in South Africa, and Rhodes University.

Acknowledgments: We thank the Wood Sciences Faculty at Stellenbosch University, South Africa for biomass compositional charecterisation. Any opinion, findings and conclusions or recommendations expressed in this material are those of the author(s) and therefore the NRF does not accept liability in regard thereto.

Conflicts of Interest: The authors declare no conflict of interest.

References

1. Qi, W.; He, C.; Wang, Q.; Liu, S.; Yu, Q.; Wang, W.; Leksawasdi, N.; Wang, C.; Yuan, Z. Carbon-based solid acid pretreatment in corncob saccharification: Specific xylose production and efficient enzymatic hydrolysis. *ACS Sustain. Chem. Eng.* **2018**, *6*, 3640–3648. [CrossRef]
2. Mishra, V.; Jana, A.K. Sweet sorghum bagasse pretreatment by *Coriolus versicolor* in mesh tray bioreactor for selective delignification and improved saccharification. *Waste Biomass Valorization* **2019**, *10*, 2689–2702. [CrossRef]
3. Takada, M.; Niu, R.; Minami, E.; Saka, S. Characterization of three tissue fractions in corn (*Zea mays*) cob. *Biomass Bioenergy* **2018**, *115*, 130–135. [CrossRef]

4. Mnich, E.; Bjarnholt, N.; Eudes, A.; Harholt, J.; Holland, C.; Jørgensen, B.; Larsen, F.H.; Liu, M.; Manat, R.; Meyer, A.S.; et al. Phenolic cross-links: Building and de-constructing the plant cell wall. *Nat. Prod. Rep.* **2019**. [CrossRef] [PubMed]
5. Malgas, S.; Mafa, M.S.; Mkabayi, L.; Pletschke, B. A mini review of xylanolytic enzymes with regards to their synergistic interactions during hetero-xylan degradation. *World J. Microbiol. Biotechnol.* **2019**, *35*. [CrossRef] [PubMed]
6. Van Dyk, J.S.; Pletschke, B.I. A review of lignocellulose bioconversion using enzymatic hydrolysis and synergistic cooperation between enzymes-factors affecting enzymes, conversion and synergy. *Biotechnol. Adv.* **2012**, *30*, 1458–1480. [CrossRef]
7. El-Naggar, N.E.A.; Deraz, S.; Khalil, A. Bioethanol production from lignocellulosic feedstocks based on enzymatic hydrolysis: Current status and recent developments. *Biotechnology* **2014**, *13*, 1–21.
8. Sharma, H.K.; Xu, C.; Qin, W. Biological Pretreatment of Lignocellulosic Biomass for Biofuels and Bioproducts: An Overview. *Waste Biomass Valor* **2019**, *10*, 235–251. [CrossRef]
9. Dietrich, J.F.R.; Goring, D.A.I. Effect of mercerization on the crystallite size and crystallinity index in cellulose from different sources. *Can. J. Chem.* **1986**, *65*, 1724–1725.
10. Beukes, N.; Pletschke, B.I. Effect of lime pre-treatment on the synergistic hydrolysis of sugarcane bagasse by hemicellulases. *Bioresour. Technol.* **2010**, *101*, 4472–4478. [CrossRef]
11. Oka, D.; Kobayashi, K.; Isobe, N.; Ogawa, Y.; Yokoyama, T.; Kimura, S.; Kim, U.; Tokuyasu, K.; Wada, M. Enzymatic hydrolysis of wood with alkaline treatment. *J. Wood Sci.* **2013**, *59*, 484–488. [CrossRef]
12. Grimaldi, M.P.; Marques, M.P.; Laluce, C.; Cilli, E.M.; Pombeiro Sponchiado, S.R.P. Evaluation of lime and hydrothermal pretreatments for efficient enzymatic hydrolysis of raw sugarcane bagasse. *Biotechnol. Biofuels* **2015**, *8*, 1–14.
13. Kim, J.S.; Lee, Y.Y.; Kim, T.H. A review on alkaline pretreatment technology for bioconversion of lignocellulosic biomass. *Bioresour. Technol.* **2016**, *199*, 42–48. [CrossRef] [PubMed]
14. da Silva Neto, J.M.; Oliveira, C.; da Silva, L.S.H.; Tabosa, F.L.; Pacheco, J.N.; da Silva, M.J.V. Use of sweet sorghum bagasse (*Sorghum bicolor* (L.) Moench) for cellulose acetate synthesis. *BioRes* **2019**, *14*, 3534–3553.
15. Carrillo, F.; Colom, X.; Sunol, J.J.; Saurina, J. Structural FTIR analysis and thermal characterization of lyocell and viscose-type fibres. *Eur. Polym. J.* **2004**, *40*, 2229–2234. [CrossRef]
16. Brune, A. Symbiotic digestion of lignocellulose in termite guts. *Nat. Rev. Microbiol.* **2014**, *12*, 168–180. [CrossRef]
17. Ni., J.; Tokuda, G. Lignocellulose-degrading enzymes from termites and their symbiotic microbiota. *Biotechnol. Adv.* **2013**, *31*, 838–850. [CrossRef]
18. Bastien, G.; Arnal, G.; Bozonnet, S.; Laguerre, S.; Ferreira, F.; Fauré, R.; Henrissat, B.; Lefèvre, F.; Robe, P.; Bouchez, O.; et al. Mining for hemicellulases in the fungus-growing termite *Pseudacanthotermes militaris* using functional metagenomics. *Biotechnol. Biofuels* **2013**, *6*, 1–15.
19. Rashamuse, K.; Ronneburg, T.; Sanyika, W.; Mathiba, K.; Mmutlane, E.; Brady, D. Metagenomic mining of feruloyl esterases from termite enteric flora. *Appl. Microbiol. Biotechnol.* **2013**, *98*, 727–737. [CrossRef]
20. Rashamuse, K.; Sanyika, T.W.; Mathiba, K.; Ngcobo, T.; Mtimka, S.; Brady, D. Metagenomic mining of glycoside hydrolases from the hindgut bacterial symbionts of a termite., *Trinervitermise trinervoides* and the characterisation of a multimodular β-1, 4-Xylanase (GH11). *Biotechnol. Appl. Biochem.* **2016**, *64*. [CrossRef]
21. Panagiotopoulos, I.A.; Bakker, R.R.; De Vrije, T.; Koukios, E.G.; Claassen, P.A.M. Pretreatment of sweet sorghum bagasse for hydrogen production by *Caldicellulosiruptor saccharolyticus*. *Int. J. Hydrogen. Energy* **2010**, *35*, 7738–7747. [CrossRef]
22. Sluiter, J.B.; Ruiz, R.O.; Scarlata, C.J.; Sluiter, A.D.; Templeton, D.W. Compositional analysis of lignocellulosic feedstocks. 1. Review and description of methods. *J. Agric. Food Chem.* **2010**, *58*, 9043–9053. [CrossRef] [PubMed]
23. Mafa, M.S.; Dirr, H.W.; Malgas, S.; Krause, R.W.M.; Rashamuse, K.; Pletschke, B.I. A novel dimeric exoglucanase (GH5_38): Biochemical and structural characterisation towards its application in alkyl cellobioside synthesis. *Molecules* **2020**, *25*, 746. [CrossRef] [PubMed]
24. Miller, L.G. Use of dinitrosalicylic acid reagent for determination of reducing sugar. *Anal. Chem.* **1959**, *31*, 426–428. [CrossRef]
25. Álvarez, C.; Reyes-Sosa, F.M.; Díez, B. Enzymatic hydrolysis of biomass from wood. *Microb. Biotechnol.* **2016**, *9*, 149–156. [CrossRef]

26. Chen, H.; Liu, J.; Chang, X.; Xue, Y.; Liu, P.; Lin, H.; Han, S. A review on the pretreatment of lignocellulose for high-value chemicals. *Fuel Proc. Technol.* **2017**, *160*, 196–206. [CrossRef]
27. Kumar, A.; Negi, Y.S.; Choudhary, V.; Bhardwaj, N.K. Characterization of cellulose nanocrystals produced by acid-hydrolysis from sugarcane bagasse as agro-waste. *J. Mater. Phys. Chem.* **2014**, *2*, 1–8.
28. Fry, S.C.; York, W.S.; Albersheim, P.; Darvill, A.; Hayashi, T.; Joseleau, J.-P.; Kato, Y.; Lorences, P.E.; Maclachlan, G.A.; McNeil, M.; et al. An unambiguous nomenclature for xyloglucan-derived oligosaccharides. *Physiol. Plant.* **1993**, *89*, 1–3. [CrossRef]
29. Rashmi, R.; Siddalingamurthy, K.R. Microbial xyloglucanases: A comprehensive review. *Biocatal. Biotransform.* **2018**, *36*, 280–295. [CrossRef]
30. Cheng, X.-Y.; Huang, W.-J.; Hu, S.-C.; Zhang, H.-L.; Wang, H.; Zhang, J.-X.; Lin, H.-H.; Zou, Y.-Z.Q.; Ji, Z.-L. Global Characterization and Identification of Multifunctional Enzymes. *PLoS ONE* **2012**, *7*, e38979. [CrossRef]
31. Huberts, D.H.E.W.; van der Klei, I.J. Moonlighting proteins: An intriguing mode of multitasking. *Biochim. Biophys. Acta* **2010**, *1803*, 520–525. [CrossRef] [PubMed]
32. Jeffery, J.C. Moonlighting proteins: Old proteins learning new tricks. *Trends Genet.* **2003**, *19*, 415–417. [CrossRef]
33. Aspeborg, H.; Coutinho, P.M.; Wang, Y.; Brumer, H.; Henrissat, B. Evolution, substrate specificity and subfamily classification of glycoside hydrolase family 5 (GH5). *BMC Evol. Biol.* **2012**, *12*, 1–16. [CrossRef] [PubMed]
34. Feng, T.; Liu, H.; Xu, Q.; Sun, J.; Shi, H. Identification and characterization of two endogenous β-glucosidases from the termite *Coptotermes formosanus*. *Appl. Biochem. Biotechnol.* **2015**, *176*, 2039–2052. [CrossRef] [PubMed]
35. Sethi, A.; Kovaleva, E.S.; Slack, J.M.; Brown, S.; Buchman, G.W.; Scharf, M.E. A GHF7 cellulase from the protist symbiont community of *Reticulitermes flavipes* enables more efficient lignocellulose processing. *Arch. Insect Biochem. Physiol.* **2013**, *84*, 175–193. [CrossRef]

© 2020 by the authors. Licensee MDPI, Basel, Switzerland. This article is an open access article distributed under the terms and conditions of the Creative Commons Attribution (CC BY) license (http://creativecommons.org/licenses/by/4.0/).

agronomy

Article

Optimized Bioconversion of Xylose Derived from Pre-Treated Crop Residues into Xylitol by Using *Candida boidinii*

Soma Bedő [1], Anikó Fehér [1], Panwana Khunnonkwao [2], Kaemwich Jantama [2] and Csaba Fehér [1,*]

[1] Biorefinery Research Group, Department of Applied Biotechnology and Food Science, Budapest University of Technology and Economics, 1111 Budapest, Hungary; bedo.soma@mail.bme.hu (S.B.); feher.aniko@mail.bme.hu (A.F.)
[2] Metabolic Engineering Research Unit, School of Biotechnology, Institute of Agricultural Technology, Suranaree University of Technology, 111 University Avenue, Suranaree Sub-District, Muang District, Nakhon Ratchasima 30000, Thailand; panwana@outlook.com (P.K.); kaemwich@g.sut.ac.th (K.J.)
* Correspondence: feher.csaba@vbk.bme.hu

Abstract: Crop residues can serve as low-cost feedstocks for microbial production of xylitol, which offers many advantages over the commonly used chemical process. However, enhancing the efficiency of xylitol fermentation is still a barrier to industrial implementation. In this study, the effects of oxygen transfer rate (OTR) (1.1, 2.1, 3.1 mmol $O_2/(L \times h)$) and initial xylose concentration (30, 55, 80 g/L) on xylitol production of *Candida boidinii* NCAIM Y.01308 on xylose medium were investigated and optimised by response surface methodology, and xylitol fermentations were performed on xylose-rich hydrolysates of wheat bran and rice straw. High values of maximum xylitol yields (58–63%) were achieved at low initial xylose concentration (20–30 g/L) and OTR values (1.1–1.5 mmol $O_2/(L \times h)$). The highest value for maximum xylitol productivity (0.96 g/(L × h)) was predicted at 71 g/L initial xylose and 2.7 mmol $O_2/(L \times h)$ OTR. Maximum xylitol yield and productivity obtained on wheat bran hydrolysate were 60% and 0.58 g/(L × h), respectively. On detoxified and supplemented hydrolysate of rice straw, maximum xylitol yield and productivity of 30% and 0.19 g/(L × h) were achieved. This study revealed the terms affecting the xylitol production by *C. boidinii* and provided validated models to predict the achievable xylitol yields and productivities under different conditions. Efficient pre-treatments for xylose-rich hydrolysates from rice straw and wheat bran were selected. Fermentation using wheat bran hydrolysate and *C. boidinii* under optimized condition is proved as a promising method for biotechnological xylitol production.

Keywords: wheat bran; rice straw; acidic hydrolysis; fermentation; aeration; detoxification

check for updates

Citation: Bedő, S.; Fehér, A.; Khunnonkwao, P.; Jantama, K.; Fehér, C. Optimized Bioconversion of Xylose Derived from Pre-Treated Crop Residues into Xylitol by Using *Candida boidinii*. *Agronomy* **2021**, *11*, 79. https://doi.org/10.3390/agronomy11010079

Received: 27 November 2020
Accepted: 29 December 2020
Published: 1 January 2021

Publisher's Note: MDPI stays neutral with regard to jurisdictional claims in published maps and institutional affiliations.

Copyright: © 2021 by the authors. Licensee MDPI, Basel, Switzerland. This article is an open access article distributed under the terms and conditions of the Creative Commons Attribution (CC BY) license (https://creativecommons.org/licenses/by/4.0/).

1. Introduction

Biotechnological valorization of lignocellulosic residues derived from the agro-industrial sector is of great importance in many countries with intense agriculture in order to deal with the increasing demand of the society for energy and materials, the mitigation of greenhouse gas emissions and waste production and the development of a bio-based circular economy [1]. For such a sustainable bio-based economy, biorefineries that provide integrated facilities to produce a wide range of bio-products and bioenergy from biomass residues within a zero waste approach are considered as main pillars [2]. Around 4.6 billion tonnes of lignocellulosic biomass are produced annually as agricultural residues worldwide, of which only about 25% are used intensively [3,4]. However, they could serve as cheap, renewable and widely available raw materials for biorefineries [5].

Wheat (*Triticum aestivum*) and rice (*Oryza sativa*) are considered the most important crops in the human diet by contributing about 20% and 19% of the average calorie intake at global level, respectively [6]. The current worldwide production of wheat and rice are estimated at around 760 and 500 million tonnes, respectively, by FAO [7,8]. Moreover, increased demands for wheat and rice in the near future are usually predicted [9,10]. Hence, the intense cultivation and processing of these crops generate huge amount of lignocellulosic by-products having the potential to be valorized in biorefinery processes [10,11]. Wheat bran and rice straw are the main lignocellulosic by-products of wheat-milling and rice-harvesting processes, with an estimated global production of 150 and 730 million tonnes per year, respectively [12,13]. They contain a considerable amount of xylan-type hemicellulose fractions besides cellulose and lignin [14,15]. In an appropriate biorefinery process, the main constituents of the lignocellulosic raw material (cellulose, hemicellulose, lignin, protein etc.) can be sharply separated and converted into high-value platform chemicals, bio-products and energy [16]. Due to the relatively high xylan content of wheat bran and rice straw, one of the most promising platform chemicals that can be produced from them in a biorefinery is xylitol [17,18].

Xylitol is a five-carbon polyol that is mainly used in the food industry as an alternative sweetener. It has an equivalent sweetness to sucrose but with lower caloric value and glycemic index. It is an ideal sweetener for diabetics because its metabolism is independent of insulin [19,20]. Moreover, due to its other specific properties, it is extensively used in personal health products such as mouthwash and toothpaste [21]. It is also used in cosmetics and by the pharmaceutical industry as therapeutic or coating agents [22]. Xylitol is currently produced by the chemical reduction of xylose on an industrial scale. However, a high purity of xylose is required for the chemical reduction process to avoid the formation of by-products (e.g., arabitol). Thus an extensive xylose purification step is inevitable prior to the chemical reduction, which contributes to the high production cost of xylitol [23,24]. On the other hand, xylitol can be produced by a microbial process. The microbial reduction takes place under mild conditions (in terms of pressure and temperature), and it does not require highly pure xylose as a carbon substrate, thus allowing the direct utilization of hemicellulosic hydrolysates that contain a mixture of sugars [25]. There are several microorganisms among bacteria, yeasts, and fungi with the capability of producing xylitol, though, the most efficient species belong to yeasts of *Candida*, such as *C. tropicalis*, *C. guilliermondi*, and *C. boidinii* [26]. However, there are several factors (e.g., pH, temperature, aeration, initial substrate concentration, inoculum level, medium composition etc.) affecting the xylitol fermentation. Oxygen supply and initial substrate concentration are considered as critical variables [27,28]. For the efficient production of xylitol, yeasts usually require a so-called micro-aerobic condition, providing oxygen limitation during the xylitol formation phase of the fermentation. The aeration conditions are usually characterized by the oxygen transfer rate (OTR) or oxygen mass transfer coefficient (k_La) of the fermentation system applied. The optimal OTR for xylitol production strongly depends on the yeast strain and fermentation medium applied, so it has to be determined in every case. Moreover, it can be affected by other fermentation parameters as well. Interestingly, the possible interactions between the fermentation parameters affecting the xylitol production are poorly investigated in the literature on this topic. Many studies reported that high xylose concentration can strongly inhibit xylitol production [29,30]; however, possible explanations were usually only hypothesized. Further investigations on the effects of the initial substrate concentration and its interaction with other fermentation parameters are still required in order to provide a proper explanation of the phenomena. Xylose-rich, hemicellulosic hydrolysates derived from the pre-treatment or fractionation processes of lignocellulosic residues are promising raw materials for biotechnological xylitol production [31].

In this study, the effects of initial xylose concentration and aeration on the xylitol production of *Candida boidinii* were investigated and the xylitol production was optimised. In addition, xylose-rich hydrolysates derived from acidic pre-treatments of wheat bran and rice straw were characterized and tested in xylitol fermentation by *C. boidinii*. This study revealed the main factors affecting the xylitol production of *C. boidinii* and provided validated models to predict xylitol yield and productivity depending on the initial xylose concentration and OTR. Moreover, promising methods to obtain xylose-rich hydrolysates from rice straw and wheat bran were selected and xylitol production under the most favorable conditions was successfully performed.

2. Materials and Methods

2.1. Microorganism

Candida boidinii NCAIM Y.01308 was purchased from the National Collection of Agricultural and Industrial Microorganisms (Budapest, Hungary). The strain was maintained at 4 °C on malt agar slants: 2 *w*/*w*% malt extract and 2 *w*/*w*% agar.

2.2. Lignocellulosic Materials

Wheat bran was provided from Gyermelyi Ltd. (Gyermely, Hungary). Ground rice straw (average particle size of 1.42 mm) was provided from Suranaree University of Technology (Nakhon Ratchasima, Thailand). Wheat bran and rice straw had dry matter contents of 91.6% and 91.2%, respectively, and they were stored at room temperature.

2.3. Compositional Analysis

The amounts of structural carbohydrates (glucan, xylan and arabinan) and lignin in rice straw and wheat bran regarding their dry matter content (*w*/*w*%) were determined according to the method published by the National Renewable Energy Laboratory [32]. Determination of the protein content of the raw materials was performed by using the Dumas method [33]. The inorganic content was determined gravimetrically after incineration of the raw materials in muffle furnace equipped with a ramping program [34].

The ground rice straw contained 38.8% glucan, 19.4% xylan, 3.8% arabinan, 14.0% lignin, 3.7% protein, and 13.2% inorganic compounds (*w*/*w*, on dry basis). Wheat bran contained 26.6% glucan, 16.8% xylan, 9.5% arabinan, 6.4% lignin 17.4% protein, and 5.7% inorganic compounds (*w*/*w*, on dry basis).

2.4. Preparation of Xylose-Rich Hydrolysates from Wheat Bran and Rice Straw

Xylose-rich hydrolysates from ground and fine ground rice straw samples were prepared by using phosphoric and sulfuric acids. Ground rice straw (average particle size: 1.42 mm) was treated in a blade grinder for 10 min to obtain the fine ground rice straw (average particle size 0.67 mm). The phosphoric acid treatment of ground and fine ground rice straws was performed according to the method of Jampatesh et al. [35]. Briefly, ground and fine ground rice straw samples were soaked in 2N phosphoric acid for 4 h at room temperature and 220 rpm. The dry matter (DM) content of the reaction mixtures was 10 *w*/*v*%. (Thus, the obtained reaction mixtures contained 5.9 *w*/*w*% phosphoric acid.) After the soaking, the reaction mixtures were treated in autoclave at 121 °C for 30 min and then filtered through a nylon filter with 50 µm pore size to separate the obtained xylose-rich hydrolysates and the remaining solid fractions. The xylose-rich hydrolysates derived from the phosphoric acid treatments of ground and fine ground rice straws are referred to as GRS/P and FGRS/P, respectively. The sulfuric acid pre-treatments of ground and fine ground rice straws were performed in reaction mixtures containing 1.5 *w*/*w*% sulfuric acid and 10 *w*/*w*% DM content. The reaction mixtures were treated in autoclave at 121 °C for 30 min. Then the xylose-rich supernatants were separated from the solid fractions by using a nylon filter with 50 µm pore size. The xylose-rich hydrolysates derived from the sulfuric acid treatment of ground and fine ground rice straws are referred to as GRS/S and FGRS/S, respectively. Xylose-rich hydrolysates from wheat bran were prepared following a two-step acidic hydrolysis process detailed in our previous study [17]. Briefly, wheat bran (10 *w*/*w*% DM) was treated in the first step in a 90 °C water bath by using 1.16 *w*/*w*% sulfuric acid for 50 min or 1.61 *w*/*w*% sulfuric acid for 47 min, resulting in solid residues referred to as WBI and WBII, respectively. In the second step, WBI and WBII (7.5% *w*/*w* DM) were treated by using 1 *w*/*w*% sulfuric acid solution in an autoclave for 30 min at 121 °C in order to obtain xylose-rich hydrolysates referred to as WB1/S and WB2/S, respectively. WB1/S and WB2/S were separated from the solid residues by using a nylon filter with 50 µm pore size. The conditions of the acidic hydrolyses of ground and fine ground rice straws and wheat bran are summarized in Table 1. Xylose-rich hydrolysates of GRS/S and WB1/S were used for xylitol fermentation experiments.

Table 1. Different pre-treatment methods of rice straw and wheat bran for xylose-rich hydrolysates.

Type of Acid	Raw Material	Treatments		Xylose-Rich Hydrolysate
		First Step	Second Step	
H_3PO_4	ground rice straw	soaking (10% DM, 2N acid, 4 h, 25 °C)	in autoclave (soaked slurry, 121 °C, 30 min)	GRS/P
	fine ground rice straw	soaking (10% DM, 2N acid, 4 h, 25 °C)	in autoclave (soaked slurry, 121 °C, 30 min)	FGRS/P
H_2SO_4	ground rice straw	in autoclave (10% DM, 1.5% acid, 121 °C, 30 min)	-	GRS/S
	fine ground rice straw	in autoclave (10% DM, 1.5% acid, 121 °C, 30 min)	-	FGRS/S
	wheat bran	in water bath (10% DM, 90 °C, 1.16% acid, 50 min)	in autoclave (first solid residue, 7.5% DM, 1% acid sol., 121 °C, 30 min)	WB1/S
	wheat bran	in water bath (10% DM, 90 °C, 1.61% acid, 47 min)	in autoclave (first solid residue, 7.5% DM, 1% acid sol., 121 °C, 30 min)	WB2/S

2.5. Treatments and Supplementations of the Xylose-Rich Hydrolysates before Xylitol Fermentation

The pHs of GRS/S and WB1/S were adjusted to 6 by adding $Ca(OH)_2$, and the precipitated gypsum was removed by filtration (filter paper). In certain cases, the rice straw hydrolysate (GRS/S) was supplemented with 2 g/L ammonium-sulphate or peptone. A combined treatment of GRS/S including a clarification by activated carbon and a subsequent supplementation with 2 g/L peptone was also tested. The clarification was performed by using 5 *w/w*% activated carbon (Norit DX ULTRA 8005.3) for 30 min with continuous stirring. The activated carbon was then separated by filtration through paper filter. After the treatments described above, all of the hydrolysates were sterilized in autoclave at 121 °C for 20 min.

2.6. Inoculum Preparation

A single colony of *Candida boidinii* NCAIM Y.01308 was transferred from the malt agar slants into glucose agar (1 *w/v*% glucose, 1 *w/v*% peptone, 0.3 *w/v*% yeast extract, and 2 *w/v*% agar) and propagated for three days at room temperature. Then the cells were transferred into the inoculum medium (pH 6) containing 10 g/L yeast extract, 15 g/L KH_2PO_4, 1 g/L $MgSO_4.7H_2O$, 3 g/L $(NH_4)_2HPO_4$, and 30 g/L xylose. The cells were propagated in the inoculum medium for 72 h at 220 rpm and 30 °C.

2.7. Xylitol Fermentation

Xylitol fermentation experiments were performed using *Candida boidinii* NCAIM Y.01540 in shake flasks. Xylitol fermentation on a semi-defined xylose medium was carried out in 100 mL-shake flasks containing 35, 50, or 65 mL medium (pH 6) at 30 °C in a rotary shaker (125 rpm) for 96 h. The semi-defined xylose medium contained 10 g/L yeast extract, 15 g/L KH_2PO_4, 1 g/L $MgSO_4.7H_2O$, 3 g/L $(NH_4)_2HPO_4$ and 30, 55, or 80 g/L xylose. Before inoculation, the solutions containing the xylose and other components were sterilized in an autoclave (121 °C, 20 min) separately to avoid the Maillard reaction. Initial cell concentrations were 5 g (dry cell mass)/L. The experiments were carried out according to designed experiments (Table 2). The experiments for model validation were carried out in duplicates.

Table 2. Results of xylitol fermentation during the designed experiments (3^2) by using *Candida boidinii* NCAIM Y.01308. (OTR: oxygen transfer rate, IXC: initial xylose concentration).

OTR	IXC (Nominal Values)	Max. Xylitol Yield	Time of Max. Xylitol Yield	Max. Xylitol Productivity	24-h Xylitol Yield	Xylose at Max. Xylitol Yield	Spec. Xylitol Yield at Max. Xylitol Yield	Max. Ethanol
mmol $O_2/(L \times h)$	g/L	%	h	$g/(L \times h)$	%	g/L	g/g	g/L
3.1	30	49	24	0.61	49 *	3	0.55	4
2.1	30	58	24	0.66	58 *	4.5	0.71	2.7
1.1	30	58	48	0.49	41	1.4	0.61	2
3.1	55	42	72	0.88	35	0.2	0.43	2
2.1	55	50	72	0.88	39	0.5	0.51	6.1
2.1	55	48	72	0.91	40	0.4	0.49	5.9
2.1	55	49	72	0.92	40	0.6	0.5	6
1.1	55	52	72	0.59	26	5.3	0.58	5.7
3.1	80	34	96	0.93	27	9.7	0.39	3.1
2.1	80	36	96	0.87	26	19.8	0.49	5.9
1.1	80	39	96	0.6	18	30.3	0.64	2.8

* Equal to the maximum xylitol yield.

Xylitol fermentations on xylose-rich hydrolysates of wheat bran and rice straw (WB1/S, GRS/S, and GRS/S supplemented with ammonium-sulphate; GRS/S supplemented with peptone, and GRS/S clarified by activated carbon and supplemented with peptone) were performed in 100-mL shake flasks filled with 50 mL hydrolysates at 30 °C in a rotary shaker (125 rpm) for 96 h. The initial cell concentrations were 5 g (dry cell mass)/L.

All the fermentations were monitored by taking and analyzing samples every 24 h. The samples were analyzed by spectrophotometer for optical density and high-performance liquid chromatography (HPLC) for sugars, alcohols and organic acids.

2.8. Analytical Methods

2.8.1. Concentration of Sugars, Alcohols, Organic Acids, Phenols, and Total Protein

Concentrations of glucose, xylose, arabinose, xylitol, ethanol, acetic acid, formic acid, furfural, and hydroxymethylfurfural (HMF) were determined by HPLC using BioRad (Hercules, CA, USA) Aminex HPX-87H (300 × 7.8 mm) column equipped with Micro-Guard Cation H+ Refill Cartridge (30 × 4.6 mm) pre-column at 65 °C, and a refractive index detector. The eluent was 5 mmol/L sulfuric acid at a flow rate of 0.5 mL/min. The injection volume was 40 μL. Due to the overlapping peaks of xylose and galactose, those sugars were determined as one component in the xylose-rich hydrolysates of wheat bran and rice straw. It was referred to as xylose throughout the text because of the low concentration of galactose expected in the hydrolysates [14,36]. The total phenol content was determined by the Folin-Ciocalteu (FC) reagent according to the method described by Guo et al. [37]. The total protein content of the hydrolysates was determined by the Bradford method [38].

2.8.2. Cell Concentration

The cell concentration in fermentation samples was determined by measuring the optical density at a wavelength of 600 nm by spectrophotometer (Ultrospec III, Pharmacia LKB, Uppsala, Sweden). The cell concentration was calculated by using a calibration curve based on the relationship of optical density and cell dry weight [39].

2.8.3. Thin-Layer Chromatography

Thin-layer chromatography (TLC) was performed to qualitatively analyse the xylo-oligomeric composition of liquid samples. The samples (5 μL) and the standard mixture (5 μL) containing xylose and xylo-oligosaccharides (DP2-DP6) (Megazyme, Bray, Ireland) were dotted on a silica gel plate. The analysis was performed using butanol/acetic acid/water (2:1:1) as running solvent and revealed by water/ethanol/sulfuric acid (20:70:3) solution with 1% *v/v* orcinol over flame [40].

2.8.4. Determination of Oxygen Transfer Rate

Oxygen transfer rate (OTR) in shake-flask fermentation systems was determined by multiplying the maximum level of dissolved oxygen concentration (C*) achievable and the gas–liquid mass transfer coefficients (k_La). The C* was determined by optical oxygen sensor (VisiFerm DO 120, HAMILTON Bonaduz AG, Switzerland) in shake flasks containing semi-defined xylose medium or xylose-rich hydrolysates (WB1/S, GRS/S) at 30 °C. The k_La was determined by a non-fermentative gassing-out method [41]. Dissolved oxygen was removed from the fermentation medium by bubbling it with nitrogen gas. After that, the headspace of the flasks was washed out with air, and the shaking was restarted. Then the dissolved oxygen concentration was continuously increased until reaching a constant value (maximum level of dissolved oxygen concentration (C*)). The expression of -ln(1-(C/C*)) was plotted as a function of time, where C is the actual dissolved oxygen concentration at that given point in time. A linear curve was fitted to the plotted data. The slope of the fitted linear curve was equal to the value of k_La [42]. The OTR in 100-mL shake flasks filled with 35, 50, and 65 mL semi-defined xylose medium were 1.1 (0.1), 2.1 (0.1), and 3.1 (0.1) mmol $O_2/(L \times h)$, respectively, with the standard deviations indicated in parenthesis. The OTR values of the wheat bran and rice straw hydrolysates were 1.6 (0.2) and 2.1 (0.2) mmol $O_2/(L \times h)$, respectively. OTR values of 1.1, 2.1, and 3.1 mmol $O_2/(L \times h)$ in semi-defined medium correspond to k_La values of 5, 10, and 15 (1/h), respectively. OTR values of 1.6 mmol $O_2/(L \times h)$ in WB1/S and 2.1 mmol $O_2/(L \times h)$ in GRS/S correspond to k_La values of 8 and 10 (1/h), respectively.

2.9. Calculation of Xylose Yield, Xylitol Yield, Xylitol Volumetric Productivity, Specific Xylitol Yield and Combined Severity Factor

Xylose yield obtained during the hydrolysis of wheat bran and rice straw was expressed as percentage of theoretical. Theoretical xylose yield was calculated from the composition of the raw material used by assuming a complete hydrolysis of its xylan content into xylose.

Xylitol yield achieved in xylitol fermentation experiments was also expressed as percentage of theoretical. Theoretical xylitol yield was calculated from the initial xylose concentration by assuming a complete (stoichiometric) conversion into xylitol. Xylitol volumetric productivity ($g/(L \times h)$) was calculated by dividing the xylitol concentration by the elapsed fermentation time. The specific xylitol yield was calculated as the amount of xylitol produced divided by the amount of xylose consumed and expressed as g/g.

In order to compare different pre-treatment methods, combined severity factor was calculated according to Wyman et al. [43].

2.10. Statistical Evaluations and Optimisation

A full factorial, orthogonal design of experiments (3^2) with triplicates in the center point was performed in order to investigate the effects of two independent variables (OTR, initial xylose concentration) and their interactions on the maximum xylitol yield, maximum xylitol volumetric productivity and the xylitol yield after 24 h of fermentation. Maximum xylitol yield and maximum xylitol volumetric productivity refers to the highest values achieved during the given experiment. The results were evaluated by StatisticaTM v.13 (TIBCO Software, Palo Alto, USA) software. The settings of the two factors of initial xylose concentration and OTR value were the following: 30, 55, and 80 g/L initial xylose concentration and 1.1, 2.1, and 3.1 mmol $O_2/(L \times h)$ OTR (Table 2). A quadratic polynomial model was fitted to the measured data. The adequacy of the model was tested with an F-test ($p = 0.05$). The fitted model is described by Equation (1).

$$Y = \beta_0 + \beta_1 X_1 + \beta_2 X_2 + \beta_{12} X_1 X_2 + \beta_{11} X_1^2 + \beta_{22} X_2^2 \quad (1)$$

where Y represents the response variable, β_0 is the interception coefficient, b_1 and b_2 are the linear terms, β_{11} and β_{22} are the quadratic terms and X_1 and X_2 represent the independent variables studied [44]. The model was reduced by non-significant terms, where it was possible. In the model equations, the original numerical values of OTR and initial xylose concentration were used without their units, and initial xylose concentration was referred to as IXC. The Pareto chart was also used to investigate the effects of the terms and interactions of the independent variables. Critical values of the fitted models were determined within the experimental range in order to find the optimum condition of xylitol fermentation.

To investigate the effect of the particle size of rice straw on the acidic hydrolysis, a grinding process was applied prior to acidic treatment, and the results were evaluated by one-way ANOVA analysis at a significance level of 5%. It was performed by StatisticaTM v.13 (TIBCO Software, Palo Alto, USA) software.

3. Results

3.1. Investigating the Effects of OTR and Initial Xylose Concentration on Xylitol Production by Candida boidinii

In order to investigate the effects of initial xylose concentration and OTR on xylitol production, statistical analysis of the results of the designed experiments was performed (Table 2).

Three dependent parameters of the fermentation were evaluated: the maximum xylitol yield, the maximum xylitol volumetric productivity and the xylitol yield after 24 h (Table 2). Maximum xylitol yields were between 34–58, and the fitted model showed that the maximum xylitol yield was affected by the linear and quadratic terms of initial xylose concentration and by the linear term of OTR (Equation (2)). The maximum xylitol yields were reached at different times depending on the initial xylose concentration (Table 2). In general, they were obtained after 24, 72, and 96 h in the cases of 30, 55, and 80 g/L initial xylose concentrations, respectively, independent of the OTR applied. There was only one exception (1.1 mmol $O_2/(L \times h)$ of OTR and 30 g/L initial xylose concentration) where the maximum xylitol yield was reached in 48 h. The highest maximum xylitol yield (58%) was obtained in the case of 30 g/L initial xylose concentration with 2.1 mmol $O_2/(L \times h)$ OTR after 24 h. The fitted surface (Figure 1A) clearly shows that high xylitol yields (58–63%) can be achieved at low initial xylose concentrations (20–30 g/L) and OTR values (1.1–1.5 mmol $O_2/(L \times h)$). When 30 and 55 g/L initial xylose concentrations were applied, only small amounts of xylose were remaining at the points of the maximum xylitol yields (Table 2). Maximum xylitol volumetric productivities were obtained after 24 h in all of the cases, and the values varied between 0.49 and 0.93 g/$(L \times h)$ (Table 2). The statistical analysis showed that all of the terms (linear and quadratic terms of initial xylose concentration, linear and quadratic terms of OTR, and the interaction between the linear terms of initial xylose concentration and OTR) had a significant effect on the maximum xylitol volumetric productivity (Equation (3)). The fitted model predicted a highest value (0.96 g/L $\times$ h) for maximum xylitol volumetric productivity in the case of 71.1 g/L initial xylose concentration and 2.7 mmol $O_2/(L \times h)$ OTR (Figure 1B). Xylitol yields after 24 h of fermentation were also included in the evaluation in order to investigate the effects of the two independent factors at the time points of the maximum xylitol productivities. The maximum xylitol yield and the xylitol yield after 24 h were equal when the maximum xylitol yield was reached at 24 h. However, this occurred in only two cases (initial xylose concentration 30 g/L and OTR 3.1 or 2.1 mmol $O_2/(L \times h)$). All of the terms of the independent factors had significant effect on xylitol yield after 24 h, except the term of the interaction between the linear terms of initial xylose concentration and OTR (Equation (4)). The fitted surface (Figure 1C) clearly shows that the optimal OTR range in terms of the xylitol yield after 24 h is between 2.0–2.5 mmol $O_2/(L \times h)$. Hence, it is different than the optimal OTR range for the maximum xylitol yield (1.0–1.5 mmol $O_2/(L \times h)$). The fitted surface area (Figure 1C) also shows that the lower initial xylose concentration resulted in higher xylitol yield after 24 h which was similar to that experienced in the case of maximum xylitol yields.

At the end of the fermentations, xylose was completely consumed in most of the cases, except in the cases of 80 g/L initial xylose concentration, where 30.3, 19.8, and 9.7 g/L xylose remained at OTR 1.1, 2.1 and 3.1 mmol $O_2/(L \times h)$, respectively (Figure S1). A small amount of xylose (5.3 g/L) also remained in the case of 60 g/L initial xylose concentration at 1.1 mmol $O_2/(L \times h)$ OTR (Figure S1). The cell concentrations increased by 0.5–1.9 g/L throughout the fermentations, and they showed an increasing tendency by increasing the OTR (Figure S1). At the end of the fermentations, different ethanol concentrations were observed depending on the initial xylose concentration and OTR values applied (Table 2). The ethanol production was increased by increasing the initial xylose concentration and OTR. The highest ethanol concentration was 7.8 g/L at 80 g/L initial xylose concentration and 3.1 mmol $O_2/(L \times h)$ OTR (Table 2).

$$\text{Maximum xylitol yield} = 65.80 + 6.37 \times \text{OTR} - 2.44 \times \text{OTR}^2 - 0.37 \times \text{IXC} \quad (2)$$

$$\text{Maximum xylitol volumetric productivity} = -0.48 + 0.59 \times \text{OTR} - 0.14 \times \text{OTR}^2 + 0.02 \times \text{IXC} - 0.0002 \times \text{IXC}^2 + 0.002 \times \text{OTR} \times \text{IXC} \quad (3)$$

$$\text{Xylitol yield after 24 h} = 34.04 + 40.84 \times \text{OTR} - 8.65 \times \text{OTR}^2 - 1.02 \times \text{IXC} + 0.005 \times \text{IXC}^2 \quad (4)$$

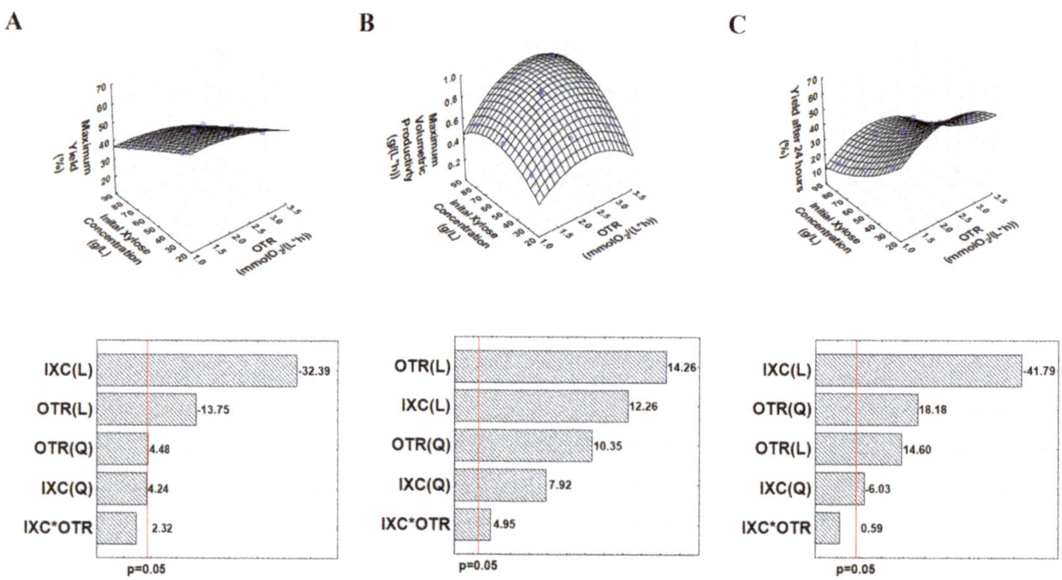

Figure 1. The fitted surface areas and Pareto charts of the evaluation of designed xylitol fermentation experiments using *Candia boidinii* NCAIM Y.01308. (**A**): maximum xylitol yield, (**B**): maximum xylitol volumetric productivity, (**C**): xylitol yield after 24 h. (OTR: oxygen transfer rate; IXC: initial xylose concentration).

In order to check the reproducibility of the fermentations and validate the fitted models, experiments were performed under two different conditions. In the first case, fermentations were performed under the conditions where the highest maximum xylitol yield was obtained during the designed experiments (30 g/L initial xylose concentration, 2.1 mmol $O_2/(L \times h)$ OTR). In the second case, a previously unmeasured point was selected, namely, 71 g/L initial xylose concentration and 2.1 mmol $O_2/(L \times h)$ OTR. This was near to the point where the highest maximum xylitol volumetric productivity was predicted by the model (71.1 g/L initial xylose concentration, 2.7 mmol $O_2/(L \times h)$). Fermentation profiles of the validation experiments are shown in Figure 2. In the first case, 55% maximum xylitol yield and 0.75 g/(L × h) maximum xylitol productivity were achieved. The xylitol concentration reached its maximum value at 24 h, and then, it slightly decreased until the end of the fermentation (Figure 2A). A small amount of ethanol was produced in the first 48 h (4.3 g/L); however, that was completely consumed by the cells until the end of the fermentation (Figure 2). The maximum productivity was also obtained after 24 h. A small amount of xylose (4.7 g/L) remained in the broth at this point, but it was completely depleted after 48 h (Figure 2A). The achieved xylitol yields and volumetric productivity are in accord with the ones predicted by the fitted model and previously obtained during the designed experiments (Tables 2 and 3), indicating the good reproducibility of the fermentations. All the results fitted into the predicted intervals of the models (Table 3). In the second case, 46% maximum xylitol yield, 32% 24-h xylitol yield, and 1.01 g/(L × h) maximum xylitol productivity were achieved. These values are similar to that predicted by the fitted models, and all of them are in the prediction intervals (Table 3). The fermentation profile observed was quite different to that of the first case (Figure 2B). The xylitol concentration continuously increased, meanwhile the xylose concentration decreased until the end of the fermentation. In addition, a significant amount of ethanol was produced by the end of the fermentation (13.8 g/L). In both cases, a small increase in the cell concentration (0.8–1 g/L) was also observed. Based on the results of the verification experiment, it can be concluded that the fitted models are adequate and suitable for good predictions within the experimental range.

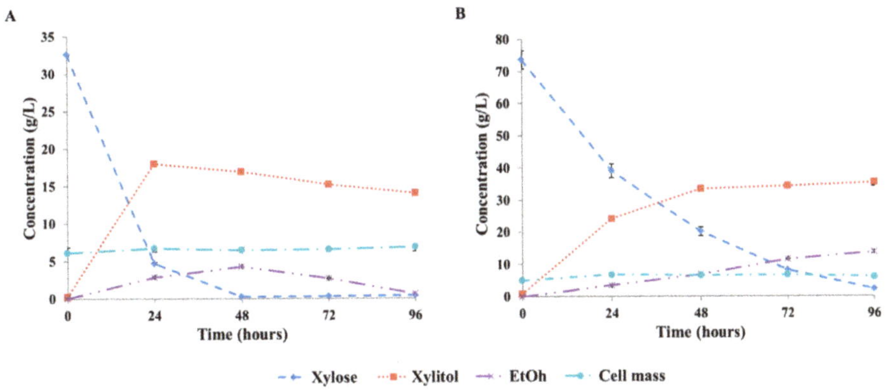

Figure 2. Profiles of fermentations performed at 2.1 mmol O_2/(L × h) OTR on semi-defined medium containing an initial xylose concentration of 30 g/L (**A**) and 71 g/L (**B**) using *Candida boidinii* Y.01308. Standard deviations are calculated from duplicates.

Table 3. Maximum xylitol yield, maximum xylitol volumetric productivity, and xylitol yield after 24 h of model predictions and experiments. (OTR: oxygen transfer rate, IXC: initial xylose concentration.)

Conditions	**30 g/L IXC, 2.1 mmol O_2/(L × h) OTR, Semi-Defined Medium**			**71 g/L IXC, 2.1 mmol O_2/(L × h) OTR, Semi-Defined Medium**		
Parameters	**Max.. Xylitol Yield**	**Max. Xylitol Productivity**	**24-h Xylitol Yield**	**Max. Xylitol Yield**	**Max. Xylitol Productivity**	**24-h Xylitol Yield**
	%	**g/(L × h)**	%	%	**g/(L × h)**	%
Predicted	57	0.68	55	42	0.92	32
95% prediction interval	54–61	0.57–0.79	51–59	39–46	0.81–1.02	29–36
Experimentally obtained	55 (0.18)	0.75 (0.00)	55 (0.18)	46 (1.67)	1.01 (0.01)	32 (1.43)

Standard deviations are calculated from duplicates and presented in parentheses.

3.2. Xylose-Rich Hydrolysates from Rice Straw and Wheat Bran

In order to produce xylose-rich hydrolysates, acidic pre-treatments of wheat bran and rice straw were investigated. In the case of wheat bran, two-step sulfuric acid hydrolyses, suggested in our previous study [17], were performed, resulting in two types of xylose-rich hydrolysates (Table 1). In the case of rice straw, a phosphoric acid hydrolysis, proposed by Jampatesh et al. [35] as an optimal pre-treatment for succinate production and with high hemicellulose saccharification yield, was used. The effects of the particle size on the efficiency of phosphoric acid pre-treatment were evaluated by using ground (average particle size of 1.42 mm) and fine ground (average particle size of 0.67 mm) rice straw samples (Table 1). In addition, sulfuric acid pre-treatments of ground and fine ground rice straws under mild conditions were also examined (Table 1). In order to predict the applicability of those hydrolysates in xylitol fermentation experiments, the sugar compositions, amount of inhibitory compounds and protein content were evaluated. The sugar composition of the wheat bran hydrolysates were quite similar to that obtained previously [17] (Table 4); however, a higher amount of acetic acid (0.7–1 g/L (Table 4) compared to 0.3 g/L [17]) was observed in this study. Interestingly, only phenolic substances were observed as inhibitory compounds beside acetic acid. Furfural, HMF and formic acid were not detected in the hydrolysates (Table 4). In addition, wheat bran hydrolysates contained relatively high amount of solubilized proteins (7 g/L), which is advantageous in terms of their fermentability. WB1/S contained a slightly higher amount of xylose (22.6 g/L) compared to WB2/S (21.1 g/L) (Table 4); hence, WB1/S was selected for xylitol fermentation experiments. In the case of rice straw, sulfuric acid and phosphoric acid treatments by using ground and fine ground raw material were investigated, resulting in four different types of xylose-rich hydrolysate (Table 1). In the case of using phosphoric acid, glucose concentration of the hydrolysates was not quantified, because of the overlapping peaks of phosphoric acid and glucose in our HPLC analyses. The phosphoric acid treatment of ground and fine ground rice straws yielded relatively low xylose concentrations of 11.1 g/L and 11.0 g/L, respectively. These concentrations correspond to the xylose yields of 69% and 68%, respectively. The total concentration of furfural, formic acid, HMF and acetic acid was around 3 g/L in both cases, which is similar to that obtained by Jampatesh et al. [35] under the same conditions. Besides those inhibitors, GRS/P and FGRS/P also contained considerable amounts of phenolic substances, resulting in total inhibitor concentrations of 4.5 g/L and 4.2 g/L, respectively (Table 4). Comparing the xylose and different inhibitor concentrations of GRS/P and FGRS/P at 0.05 significance level, significant differences were not observed. Thus, it could be concluded that particle size of rice straw does not have a significant effect on the performance of the phosphoric acid hydrolysis within the investigated range. Sulfuric acid treatment of ground and fine ground rice straws resulted in 20.6 g/L and 19.9 g/L xylose, respectively. These values correspond to the xylose yields of 94% and 91%, respectively, which are considerably higher than the xylose yields achieved during the phosphoric acid treatment. That could be partly because of an incomplete degradation of the solubilized xylan part in the case of the phosphoric acid treatment. Thin-layer chromatography analysis showed sharper spots for the xylo-oligosaccharides (DP 2–6) when phosphoric acid treatment was performed, suggesting an incomplete hydrolysis of xylan (Figure S2). When the severity factors of the two treatments were compared, a value of 1.3 was obtained for both treatments, suggesting that sulfuric acid is more efficient in decomposing hemicellulose completely. The total inhibitor concentrations of GRS/P and FGRS/P were 4.5 g/L and 4.2 g/L, respectively (Table 4). Significant differences were not observed when the xylose and different inhibitor concentrations of GRS/S and FGRS/S were compared at 0.05 significance level. Thus, particle size had no significant effect on the quality of the sulfuric acid hydrolysates within the investigated range. The protein content of both the sulfuric acid- and phosphoric acid-catalyzed hydrolysates was very low (below 1 g/L) (Table 4). The hydrolysates from wheat bran seemed to be a better medium for xylitol production due to their higher xylose but lower inhibitor content, compared to the hydrolysates from rice straw. However, in terms of the concentration of other sugars, rice straw hydrolysates were more advantageous. They contained much less arabinose beside the xylose, which can enhance the purity of the fermented broth. GRS/P contained quite low xylose concentration, which would probably result in decreased xylitol production, thus only GRS/S was tested in xylitol fermentation beside WB1/S.

Table 4. Composition of different lignocellulosic hydrolysates.

	Composition (g/L)					
	WB1/S	**WB2/S**	**GRS/S**	**FGRS/S**	**GRS/P**	**FGRS/P**
Glucose	0.8 (0.1)	1.0 (0.0)	5.5 (0.1)	5.3 (0.3)	n.m.	n.m.
Xylose	22.6 (1.3)	21.1 (0.4)	20.6 (0.7)	19.9 (0.5)	11.1 (0.5)	11.0 (0.2)
Arabinose	12.4 (0.6)	9.1 (1.3)	3.9 (0.2)	3.7 (0.2)	3.3 (0.1)	3.2 (0.1)
Protein	7.3 (0.3)	7.0 (0.4)	0.3 (0.0)	n.m.	0.7 (0.0)	n.m.
Acetic acid	1.0 (0.1)	0.8 (0.0)	2.4 (0.2)	2.4 (0.1)	1.6 (0.1)	1.3 (0.0)
Formic acid	n.d.	n.d.	1.1 (0.2)	1.1 (0.0)	1.0 (0.1)	1.0 (0.0)
HMF	n.d.	n.d.	0.5 (0.1)	0.5 (0.0)	0.6 (0.0)	0.5 (0.0)
Furfural	n.d.	n.d.	n.d.	n.d.	n.d.	n.d.
Phenols	1.4 (0.1)	1.4 (0.1)	1.4 (0.3)	1.7 (0.1)	1.3 (0.1)	1.4 (0.0)
Total inhibitors	2.5	2.4	5.4	5.7	4.5	4.2

n.m.: not measured; n.d.: not detected; Standard deviations are calculated from triplicates and presented in parentheses.

3.3. *Xylitol Fermentation on Xylose-Rich Hydrolysates of Rice Straw and Wheat Bran*

In Section 3.1, the effects of OTR and initial xylose concentration on the xylitol production of *Candida boidinii* were investigated. The applicability and adequacy of the models for maximum xylitol yield (Equation (2)), maximum xylitol volumetric productivity (Equation (3)), and xylitol yield after 24 h (Equation (4)), developed by using semi-defined medium, were also tested in the case of using xylose-rich hydrolysates derived from rice straw (GRS/S) and wheat bran (WB1/S). Xylitol fermentation experiments on the xylose-rich hydrolysates were performed under the aeration condition that provided 2.1 mmol $O_2/(L \times h)$ OTR value in the case of semi-defined medium. However, these conditions resulted in 1.6 mmol $O_2/(L \times h)$ OTR when WB1/S was used, probably due to the different chemical composition. Xylitol fermentation on WB1/S showed similar profile than that of the fermentation obtained by using semi-defined medium (30 g/L initial xylose, 2.1 mmol $O_2/(L \times h)$ OTR) (Figures 2A and 3A); however, both the initial xylose concentration (22.1 g/L) and the OTR (1.6 mmol $O_2/(L \times h)$) were a bit lower in the case of WB1/S. Maximum xylitol concentration (14.2 g/L) was obtained at 24 h, resulting in the maximum xylitol yield of 60% (spec. xylitol yield: 0.72 g/g) (Figure 3A). The maximum xylitol productivity was also reached after 24 h, and it was 0.58 g/(L × h). Predictions for the maximum xylitol yield, xylitol yield after 24 h, and maximum xylitol volumetric productivity were calculated by the previously fitted models (Equations (2)–(4)) using 22.1 g/L initial xylose concentration and 1.6 mmol $O_2/(L \times h)$ OTR value as input parameters. The maximum xylitol yield was achieved also in 24 h, thus it was equal to the xylitol yield after 24 h. Due to the fact that different models were fitted to the maximum and 24-h xylitol yields, the models provided slightly different predictions for those response variables but with overlapping predicted intervals. The models predicted 61% maximum xylitol yield with a prediction interval of 58–65%, and 57% xylitol yield after 24 h with a prediction interval of 52–61%. Thus, the experimentally measured xylitol yield (60%) fitted in with the predictions of both models. Maximum xylitol productivity was estimated to be 0.49 g/(L × h) by the model with a prediction interval of 0.35–0.63 g/(L × h), showing a good agreement with the experimentally obtained value (0.58 g/(L × h)). Therefore, the models developed by using semi-defined medium were found appropriate to provide adequate predictions for xylitol fermentations on WB1/S, indicating that wheat bran hydrolysate is a suitable raw material to produce xylitol by using *Candida boidinii* NCAIM Y.01308. Ethanol production was observed in both cases of using a semi-defined medium or WB1/S. The ethanol concentration reached its maximum in 48 h, and then, ethanol was totally consumed by the end of the fermentation (Figures 2A and 3A). The growth of the cells was also similar in both cases. The cell concentrations increased by about 1 g/L during the fermentations. The profiles of xylose decrease were also similar, except that a small amount of xylose seemed to remain in wheat bran hydrolysate (Figure 3A). It could be explained by the presence of small amount of galactose in the wheat bran hydrolysate [36], which could be detected as xylose due to the overlapping peaks in the HPLC analysis used in this study. Residual galactose during xylitol fermentation on corn fiber hydrolysate by *Candida boidinii* was observed in our previous study [39]. Xylitol fermentations were also performed on GRS/S under the same aeration condition, resulting in similar OTR value (2.1 mmol $O_2/(L \times h)$) to that measured in the semi-defined medium. The profile of the xylitol fermentation on GRS/S (Figure 3B) was different from that observed on the semi-defined medium or WB1/S. Although the initial xylose concentration was similar to that in WB1/S, the achieved maximum xylitol yield was significantly lower (20%, spec. xylitol yield: 0.26 g/g), and it was reached latter (72 h). The remaining xylose concentration (3.4 g/L) at this point was similar to that observed in the cases of a semi-defined medium and WB1/S (Figures 2A and 3A,B). However, it could contain small amount of galactose as well, which was not analyzed in this study. The xylitol yield after 24 h (19%) was similar to the maximum xylitol yield (20%), but at 24 h, the remaining xylose concentration was much higher (8 g/L). This is almost half of the initial xylose concentration and nearly twice of the xylose concentration measured in semi-defined medium and WB1/S after 24 h of fermentation (Figures 2A and 3A,B). The maximum xylitol volumetric productivity obtained in 24 h was also very low (0.14 $g/(L \times h)$) compared to the previous fermentations. In the case of GRS/S, a higher amount of ethanol was produced (8.4 g/L after 72 h) compared to the previous cases. Moreover, ethanol production exceeded xylitol production (Figure 3B).

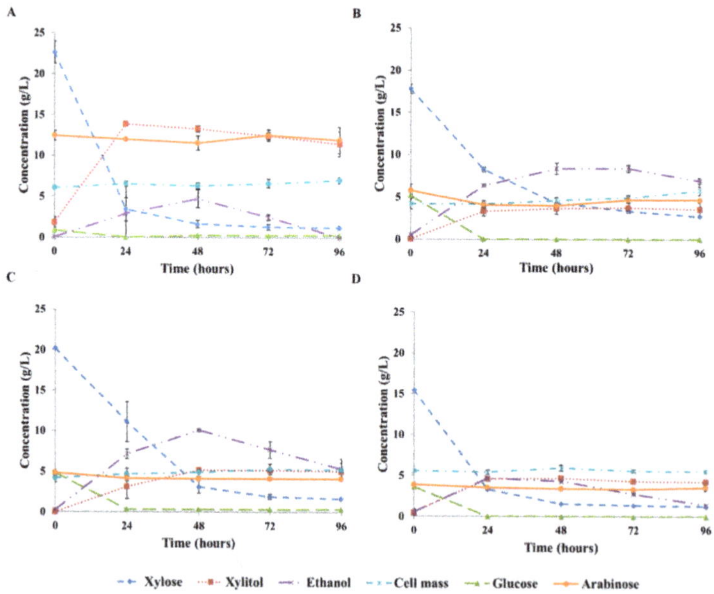

Figure 3. Profiles of the xylitol fermentations on WB1/S (**A**) GRS/S (**B**) GRS/S supplemented with 2 g/L peptone (**C**) and GRS/S treated by activated carbon and supplemented with 2 g/L peptone (**D**) using *Candida boidinii* NCAIM Y.01308. Standard deviations are calculated from duplicates.

Since GRS/S hydrolysate had a low protein content (Table 4), supplementations with 2 g/L of inorganic (ammonium sulphate) or organic (peptone) nitrogen sources were investigated with the aim of enhancing xylitol production. In the case of supplementation with ammonium-sulphate, the maximum xylitol yield was 15% (spec. xylitol yield: 0.22 g/g at this point), and it was achieved after 48 h (data not shown). The xylitol yield after 24 h and the maximum xylitol productivity were 10% and 0.08 g/(L × h), respectively (data not shown). All of these results are lower than that achieved without supplementation. In contrast, the addition of peptone resulted in a small increase in maximum xylitol yield (25%, (spec. xylitol yield: 0.30 g/g at this point)), xylitol yield after 24 h (15%) and maximum xylitol volumetric productivity (0.15 g/(L × h)) (Figure 3C) as well. Hence, xylitol fermentation was slightly enhanced by adding peptone; however, ethanol concentration was also considerably increased. The ethanol concentration was 10 g/L at 48 h beside only 5 g/L of xylitol (Figure 3C).

To improve xylitol fermentation on GRS/S, activated carbon treatment was tested in addition to the supplementation with peptone. After the activated carbon treatment, detectable amount of phenols and HMF were not present in GRS/S. On the other hand, the amount of organic acids (0.9 g/L formic acid and 2.4 g/L acetic acid) were not reduced significantly compared to the GRS/S without detoxification (1.1 g/L formic acid and 2.4 g/L acetic acid). Moreover, the activated carbon treatment caused significant xylose loss, resulting in 15.4 g/L initial xylose concentration. After the detoxification step, GRS/S was also supplemented by 2 g/L peptone. Due to the activated carbon treatment, the maximum xylitol yield increased to 30% (spec. xylitol yield: 0.33 g/g), and it was reached after 48 h. The xylitol yield after 24 h and the maximum xylitol productivity were 29% and 0.19 g/(L × h), respectively. All of these results are lower compared to those achieved on WB1/S or a semi-defined medium under similar aeration conditions. The ethanol production was reduced due to the detoxification step, resulting in only 4.7 g/L ethanol concentration at the point of the maximum xylitol yield. This was a half of the ethanol produced in the hydrolysate without detoxification. The maximum specific xylitol yields were reached after 24 h in all fermentations on lignocellulosic hydrolysates. The specific xylitol yields were 0.35 g/g, 0.24 g/g, 0.35 g/g, and 0.38 g/g in the cases of GRS/S, GRS/S supplemented with ammonium-sulphate, GRS/S supplemented with peptone, GRS/S clarified by activated carbon and supplemented with peptone, respectively. The cell concentrations increased by 1–1.8 g/L during the fermentations, except in the cases of GRS/S supplemented with peptone and detoxified GRS/S with peptone supplementation, where the cell mass did not change. Small amount of glucose (3.5–5 g/L) was also present in the rice straw hydrolysates, which was totally consumed after 24 h (Figure 3B–D).

4. Discussion

The effects of initial xylose concentration and aeration (OTR) on maximum xylitol yield, maximum xylitol volumetric productivity, and xylitol yield after 24 h were investigated first in this study, by using *Candida boidinii* NCAIM Y.01308 in a semi-defined xylose medium. Vandeska et al. [45] investigated the effect of the initial xylose concentration on the achievable xylitol yield during fermentation on a model xylose medium by *C. boidinii* NRRL Y17213. They varied the initial xylose concentration between 20 and 200 g/L under the same aeration conditions (50 mL medium in 125 mL-flask, 150 rpm shaking), and concluded that the xylitol yield of 14 days fermentation continuously increased from 6% to the maximum value of 52% by increasing the initial xylose concentration from 20 g/L to 150 g/L. Interestingly, initial xylose concentration of 200 g/L significantly inhibited the xylitol fermentation. Vongsuvanlert and Tani [46] reported that a xylose concentration of 150 g/L resulted in lower xylitol production than that of 100 g/L, when *C. boidinii* no. 2201 was used on xylose basal medium. They performed the fermentations in 500 mL-flasks filled with 100 mL medium at 100 rpm shaking. Those studies showed that under given conditions of aeration, increasing xylose concentration increased the xylitol production until a certain value above which it had a negative effect. Osmotic stress on the cells is often hypothesized as a possible reason for the negative effect of high xylose concentration [46,47]. However, in our study, a decreasing tendency in the xylitol yield was observed by increasing the xylose concentration under the investigated range of xylose concentration and aeration. Oxygen availability during xylitol fermentation is a key factor due to its influence on the intracellular redox balance. The key enzymes for assimilation of D-xylose in yeasts are the xylose reductase (XR), which catalyses the reduction of D-xylose to xylitol requiring NAD(P)H as cofactor, and the NAD-dependent xylitol dehydrogenase (XDH) catalysing the xylitol conversion to xylulose [48]. Xylitol accumulation under microaerobic conditions is the result of a deficient NAD regeneration by the respiratory chain which leads to diminished XDH activity [45,48]. In addition, formation of certain by-products during the xylitol fermentation can also contribute to maintaining the intracellular cofactor balance [49]. Winkelhausen et al. [50] investigated the effect of different k_La values on the xylitol production of *C. boidinni* NRRL Y17213 in a xylose model medium using shake flasks (50 g/L initial xylose concentration) and fermenter (130 g/L initial xylose concentration). The k_La values were varied between 0 and 46 1/h in shake flasks, and it was concluded that increasing k_La resulted in decreasing xylitol and increasing cell mass production. Interestingly, the highest specific xylitol yield (0.3 g/g) was achieved under anaerobic conditions (0 1/h k_La). As a comparison, Winkelhasuen et al. [50] achieved a specific xylitol yield of 0.16 g/g at 50 g/L initial xylose concentration and 16 1/h k_La after 96 h, which is lower than that of obtained in our study (0.43 g/g) at an initial xylose concentration of 55 g/L and k_La of 15 1/h (3.1 mmol O_2/(L × h) OTR) after 72 h. The lower initial cell concentration (1.3 g/L) applied by Winkelhasuen et al. [50] compared to that of used in our study (5 g/L) could be one of the reasons for that. However, it is clear that xylitol producing capability can significantly differ with the subspecies of *Candida boidinii* also. Subspecies isolated from different environmental conditions might have XR and XDH enzymes with different characteristics, and different metabolic pathways for co-factors regeneration might have been activated in them, resulting in variable capability in xylitol fermentation under certain fermentation conditions. During the fermentations in a bioreactor using 130 g/L initial xylose and 5 g/L initial cell concentrations, Winkelhasuen et al. [50] achieved the highest specific xylitol yield (around 0.45 g/g) and xylitol volumetric productivity (around 0.26 g/(L × h)) at k_La of 47 1/h. It seemed that the increase in xylose and aeration together caused an increase in xylitol yield and productivity, however the increased initial cell mass could also have a positive effect on that. In our study, it was concluded that both initial xylose concentration and OTR had significant effects on the maximum xylitol yield and xylitol yield achieved after one day. However the interaction between xylose concentration and OTR had no significant effect within the investigated experimental range. It is worth it to note that the extent of their effects and their significant terms were different when the achievable maximum xylitol yield and the xylitol yield after a certain fermentation time (e.g., one day) was examined. In contrast, when the maximum xylitol volumetric productivity was evaluated, a clear interaction between OTR and initial xylose concentration was observed.

Xylitol fermentation on lignocellulosic hydrolysates by using *Candida boidinii* was previously tested by other studies. Santana et al. [51] investigated *Candida boidinii* XM02G (4 g/L initial cell mass) on cocoa pod husk hemicellulose hydrolysate detoxified by activated carbon, and a specific xylitol yield of 0.52 g/g was achieved after 372 h of fermentation. Fehér et al. [39] published a xylitol yield of 53% of theoretical and a xylitol volumetric productivity of 0.14 g/(L × h) reached after 72 h of fermentation by using *C. boidinii* NCAIM Y.01308 (5 g/L initial cell mass) on corn fibre hydrolysate detoxified by activated carbon. Lopez-Linares et al. [52] investigated the xylitol production of *C. boidinii* NCAIM Y.01308. (5 g/L initial mass) on exhausted olive pomace hydrolysate detoxified by ion-exchange resin and achieved 0.43 g/g specific xylitol yield and 0.07 g/(L × h) volumetric productivity after 96 h. In this study, the xylitol yield, the specific xylitol yield, and the volumetric productivity were 60%, 0.72 g/g, and 0.58 g/(L × h), respectively, on WB1/S after 24 h. Those results exceeded the ones mentioned before. It is also important to note that in the case of WB1/S, no detoxification step was required prior to the fermentation. In our previous study, xylitol fermentation was performed by using *Ogataea zsoltii* NCAIM Y.01540 on xylose-rich wheat bran hydrolysate [17]. Comparing the maximum xylitol yields and volumetric productivities achieved under the same conditions by using

O. zsoltii NCAIM Y.01540 (56% and 0.24 g/(L × h)) and *C. boidinii* NCAIM Y.01308 (60% and 0.58 g/(L × h)), xylitol production of *C. boidinii* NCAIM Y.01308 was found to be more advantageous. Based on these results, this study confirms that the xylose-rich hydrolysate of wheat bran is a suitable medium for xylitol fermentation without detoxification.

Mayerhoff et al. [53] investigated different yeast strains to ferment xylitol on sulfuric acid treated rice straw hydrolysate. In their work, the initial xylose concentration was 53.9 g/L and 50 mL medium was used in 125 mL-flasks at 200 rpm shaking. High specific xylitol yields (>0.6 g/g) were achieved after 75 h by several *Candida* strains such as *C. guilliermondii* FTI-20037, *C. mogii* NRRL Y-17032, *C. parapsilosis* IZ-1710, and *C. veronae* IZ-945. However, it was only 0.17 g/g by using *C. boidinii* NRRL Y-17213. Compared to that, a higher specific xylitol yield was achieved by *C. boidinii* NCAIM Y.01308 in our study (0.26 g/g after 72 h). Nitrogen source in the fermentation media is also an important factor influencing the xylitol production. Since rice straw hydrolysates contained very low amount of proteins, supplementation by ammonium-sulphate and peptone was tested in this study. Generally, organic nitrogen sources (e.g., yeast extract and urea) result in higher xylitol yield compared to the inorganic ones [54]. In accord with that, higher maximum xylitol yield was achieved on peptone-supplemented GRS/S (25%) compared to GRS/S (20%). Silvia and Roberto [55] also investigated the effect of the nutrient supplementation (2 g/L $(NH_4)_2SO_4$, 0.1 g/L $CaCl_2$*$2H_2O$ and 10 g/L rice bran extract) of rice straw hydrolysate in order to improve the xylitol production of *C. guilliermondii* FTI 20037. They found that the nitrogen supplementation had no effect on the specific xylitol yield (0.36–0.37 g/g) achieved. In this study, a similar result was obtained, supplementation of GRS/S with peptone did not improve the specific xylitol yield (0.35 g/g after 24 h); however it increased the maximum xylitol yield from 20% to 25% of theoretical. Zeid et al. [56] investigated the effects of activated carbon treatment on xylitol production by *C. tropicalis* and *C. guilliermondii* using rice straw hydrolysate. After the activated carbon treatment, specific xylitol yields were increased from 0.25 g/g and 0.47 g/g to 0.61 g/g and 0.69 g/g in the cases of *C. tropicalis* and *C. guilliermondii*, respectively. Similarly, as a result of the activated carbon treatment of GRS/S, the maximum xylitol yield was increased from 25% to 30% in this study. Lopez Linares et al. [52] published a specific xylitol yield of 0.36 g/g achieved on exhausted olive pomace hydrolysate treated by activated carbon after 72 h of fermentation by *C. boidinii* NCAIM Y.01308 (5 g/L initial cell mass, 50 mL medium in 100 mL-flask, 150 rpm shaking). That is slightly lower than that obtained in our study (0.38 g/g) using peptone-supplemented GRS/S treated by activated carbon and the same yeast strain. One of the reasons for the low xylitol yields obtained in our study on rice straw hydrolysates is probably the presence of considerable amount of glucose (glucose/xylose ratio of 0.29 in GRS/S). A glucose/xylose ratio that is higher than 0.1 could negatively affect xylitol fermentation [52]. Moreover, the activated carbon treatment was not effective in removing the organic acids from GRS/S, which could also contribute to the insufficient bioconversion of xylose into xylitol. Bio-purification processes selectively removing organic acids and glucose from lignocellulosic hydrolysates [57,58] or appropriate genetic modifications of the xylose-fermenting microorganisms [59,60] are promising ways to overcome these kinds of obstacles.

5. Conclusions

Xylitol fermentation by *C. boidinii* NCAIM Y.01308 was optimised by investigating the effects of oxygen transfer rate and initial xylose concentration applied. The fitted models of maximum xylitol yield, maximum xylitol volumetric productivity, and xylitol yield after 24 h were verified and they were successfully used to predict xylitol production on wheat bran hydrolysate. Wheat bran hydrolysate was found to be an outstanding substrate for *C. boidinii* NCAIM Y.01308 to produce xylitol. Meanwhile, xylitol fermentation on rice straw hydrolysates by *C. boidinii* NCAIM Y.01308 requires further investigations to be improved.

Supplementary Materials: The following are available online at https://www.mdpi.com/2073-4395/11/1/79/s1. Figure S1: Fermentation profiles obtained during the designed experiments of xylitol production on semi-defined medium by *Candida boidinii* NCAIM Y.01308. Figure S2: Thin-layer chromatography analysis of rice straw hydrolysates. STD: standard (X1: xylose, X2: xylobiose, X3: xylotriose, X4: xylotetraose, X5: xylopentaose, and X6: xylohexaose).

Author Contributions: S.B. conducted experiments, analyzed the data and wrote the manuscript. A.F. conducted experiments. C.F. and K.J. conceived and designed the research, were responsible for funding acquisition and corrected the manuscript. P.K. revised the manuscript. All authors have read and agreed to the published version of the manuscript.

Funding: The authors gratefully acknowledge the projects of TNN_16-123305 of the National Research, Development and Innovation Fund of Hungary, the project of VEKOP-2.3.2-16-2017-00013 supported by the European Union, the State of Hungary and the European Regional Development Fund for their financial support. This article is based on work from COST Action (CA18229, Yeast4Bio), supported by COST (European Cooperation in Science and Technology, www.cost.eu). The research was financed by the Higher Education Institutional Excellence Programme (NKFIH-1150-6/2019) of the Ministry of Innovation and Technology of Hungary, within the framework of the Biotechnology thematic programme of the University of Debrecen.

Institutional Review Board Statement: Not applicable.

Informed Consent Statement: Not applicable.

Data Availability Statement: The data presented in this study are available on request from the corresponding author. The data are not publicly available due to privacy restrictions.

Acknowledgments: The authors are grateful to Gyermelyi Ltd. (Gyermely, Hungary), Suranaree University of Technology (Thailand), and National Collection of Agricultural and Industrial Microorganisms (Budapest, Hungary) for kindly providing research material; to Kinga Kóder for technical support and to Julie Johnson for linguistics editing.

Conflicts of Interest: The authors declare no conflict of interests.

References

1. Ubando, A.T.; Felix, C.B.; Chen, W.H. Biorefineries in circular bioeconomy: A comprehensive review. *Bioresour. Technol.* **2020**, *299*, 122585. [CrossRef] [PubMed]
2. Fehér, C.; Barta, Z.; Réczey, K. Process considerations of a biorefinery producing value-added products from corn fibre. *Period. Polytech. Chem. Eng.* **2012**, *56*, 9–19. [CrossRef]
3. Dahmen, N.; Lewandowski, I.; Zibek, S.; Weidtmann, A. Integrated lignocellulosic value chains in a growing bioeconomy: Status quo and perspectives. *GCB Bioenergy* **2019**, *11*, 107–117. [CrossRef]
4. Piotrowski, S.; Essel, R.; Carus, M.; Dammer, L.; Engel, L. Nachhaltig nutzbare Potenziale für Biokraftstoffe in Nutzungskonkurrenz zur Lebens-und Futtermittelproduktion, Bioenergie sowie zur stofflichen Nutzung in Deutschland, Europa und der Welt. *Eur. Welt Nova-Institut* **2015**, *22501112*, 1–252.
5. Isikgor, F.H.; Becer, C.R. Lignocellulosic biomass: A sustainable platform for the production of bio-based chemicals and polymers. *Polym. Chem.* **2015**, *6*, 4497–4559. [CrossRef]
6. Shiferaw, B.; Smale, M.; Braun, H.J.; Duveiller, E.; Reynolds, M.; Muricho, G. Crops that feed the world 10. Past successes and future challenges to the role played by wheat in global food security. *Food Secur.* **2013**, *5*, 291–317. [CrossRef]
7. Alexandri, M.; López-Gómez, J.P.; Olszewska-Widdrat, A.; Venus, J. Valorising agro-industrial wastes within the circular bioeconomy concept: The case of defatted rice bran with emphasis on bioconversion strategies. *Fermentation* **2020**, *6*, 42. [CrossRef]
8. FAO. Cereal Supply and Demand Brief. Available online: http://www.fao.org/worldfoodsituation/csdb/en/ (accessed on 27 November 2020).
9. Bandumula, N. Rice Production in Asia: Key to Global Food Security. *Proc. Natl. Acad. Sci. India Sect. B Biol. Sci.* **2018**, *88*, 1323–1328. [CrossRef]
10. Sharma, I.; Tyagi, B.S.; Singh, G.; Venkatesh, K.; Gupta, O.P. Enhancing wheat production—A global perspective. *Indian J. Agric. Sci.* **2015**, *85*, 3–13.
11. Muthayya, S.; Sugimoto, J.D.; Montgomery, S.; Maberly, G.F. An overview of global rice production, supply, trade, and consumption. *Ann. N. Y. Acad. Sci.* **2014**, *1324*, 7–14. [CrossRef]
12. Prückler, M.; Siebenhandl-Ehn, S.; Apprich, S.; Höltinger, S.; Haas, C.; Schmid, E.; Kneifel, W. Wheat bran-based biorefinery 1: Composition of wheat bran and strategies of functionalization. *LWT Food Sci. Technol.* **2014**, *56*, 211–221. [CrossRef]
13. Kim, S.; Dale, B.E. Global potential bioethanol production from wasted crops and crop residues. *Biomass Bioenergy* **2004**, *26*, 361–375. [CrossRef]

14. Balan, V.; Da Costa Sousa, L.; Chundawat, S.P.S.; Vismeh, R.; Jones, A.D.; Dale, B.E. Mushroom spent straw: A potential substrate for an ethanol-based biorefinery. *J. Ind. Microbiol. Biotechnol.* **2008**, *35*, 293–301. [CrossRef] [PubMed]
15. Koegelenberg, D.; Chimphango, A.F.A. Effects of wheat-bran arabinoxylan as partial flour replacer on bread properties. *Food Chem.* **2017**, *221*, 1606–1613. [CrossRef] [PubMed]
16. Kumar, A.; Gautam, A.; Dutt, D. Biotechnological Transformation of Lignocellulosic Biomass in to Industrial Products: An Overview. *Adv. Biosci. Biotechnol.* **2016**, *7*, 149–168. [CrossRef]
17. Bedő, S.; Antal, B.; Rozbach, M.; Fehér, A.; Fehér, C. Optimised fractionation of wheat bran for arabinose biopurification and xylitol fermentation by *Ogataea zsoltii* within a biorefinery process. *Ind. Crops Prod.* **2019**, *139*, 111504. [CrossRef]
18. Liaw, W.-C.; Chen, C.-S.; Chang, W.-S.; Chen, K.-P. Xylitol Production from Rice Straw Hemicellulose Hydrolyzate by Polyacrylic Hydrogel Thin Films with Immobilized *Candida subtropicalis* WF79. *J. Biosci. Bioeng.* **2008**, *105*, 97–105. [CrossRef]
19. Ylikahri, R. Metabolic and Nutritional Aspects of Xylitol. In *Advances in Food Research*; Chichester, C.O., Ed.; Academic Press: Cambridge, MA, USA, 1979; Volume 25, pp. 159–180.
20. Mussatto, S.I. Application of Xylitol in Food Formulations and Benefits for Health. In *D-Xylitol*; Springer: Berlin/Heidelberg, Germany, 2012; pp. 309–323.
21. Ravichandran, K.; Jain, J.; Sadhu, B.J.; Gunasekaran, S.; Poojitha, M.C.; Majumder, P.; Isaac, N. The efficacy of xylitol based oral hygiene products on salivary parameters—An invivo study. *Int. J. Res. Pharm. Sci.* **2020**, *11*, 953–959. [CrossRef]
22. Delgado Arcaño, Y.; Valmaña García, O.D.; Mandelli, D.; Carvalho, W.A.; Magalhães Pontes, L.A. Xylitol: A review on the progress and challenges of its production by chemical route. *Catal. Today* **2018**, *344*, 2–14. [CrossRef]
23. Mikkola, J.P.; Sjöholm, R.; Salmi, T.; Mäki-Arvela, P. Xylose hydrogenation: Kinetic and NMR studies of the reaction mechanisms. *Catal. Today* **1999**, *48*, 73–81. [CrossRef]
24. Yadav, M.; Mishra, D.K.; Hwang, J.S. Catalytic hydrogenation of xylose to xylitol using ruthenium catalyst on NiO modified TiO_2 support. *Appl. Catal. A Gen.* **2012**, *425–426*, 110–116. [CrossRef]
25. Tran, L.H.; Yogo, M.; Ojima, H.; Idota, O.; Kawai, K.; Suzuki, T.; Takamizawa, K. The production of xylitol by enzymatic hydrolysis of agricultural wastes. *Biotechnol. Bioprocess Eng.* **2004**, *9*, 223–228. [CrossRef]
26. Espinoza-Acosta, J.L. Biotechnological Production of Xylitol from Agricultural Waste. *Biotecnia* **2020**, *XXII*, 126–134.
27. Winkelhausen, E.; Pittman, P.; Kuzmanova, S.; Jeffries, T.W. Xylitol formation by *Candida boidinii* in oxygen limited chemostat culture. *Biotechnol. Lett.* **1996**, *18*, 753–758. [CrossRef]
28. Mussatto, S.I.; Roberto, I.C. Xylitol production from high xylose concentration: Evaluation of the fermentation in bioreactor under different stirring rates. *J. Appl. Microbiol.* **2003**, *95*, 331–337. [CrossRef]
29. Tamburini, E.; Costa, S.; Marchetti, M.G.; Pedrini, P. Optimized production of xylitol from xylose using a hyper-acidophilic *Candida tropicalis*. *Biomolecules* **2015**, *5*, 1979–1989. [CrossRef]
30. Nolleau, V.; Preziosi-Belloy, L.; Delgenes, J.P.; Navarro, J.M. Xylitol production from xylose by two yeast strains: Sugar tolerance. *Curr. Microbiol.* **1993**, *27*, 191–197. [CrossRef]
31. Felipe Hernández-Pérez, A.; de Arruda, P.V.; Sene, L.; da Silva, S.S.; Kumar Chandel, A.; de Almeida Felipe, M. das G. Xylitol bioproduction: State-of-the-art, industrial paradigm shift, and opportunities for integrated biorefineries. *Crit. Rev. Biotechnol.* **2019**, *39*, 924–943. [CrossRef]
32. Sluiter, A.; Hames, B.; Ruiz, R.; Scarlata, C.; Sluiter, J.; Templeton, D. Determination of Sugars, Byproducts, and Degradation Products in Liquid Fraction Process. *Natl. Renew. Energy Lab.* **2008**, *NREL/TP-510-42623*, 1–11.
33. Hames, B.; Scarlata, C.; Sluiter, A. Determination of protein content in biomass. *Natl. Renew. Energy Lab.* **2008**, *NREL/TP-510-42625*, 1–5.
34. Sluiter, A.; Hames, B.; Ruiz, R.; Scarlata, C.; Sluiter, J.; Templeton, D. Determination of ash in biomass. *Natl. Renew. Energy Lab.* **2008**, *NREL/TP-510-42622*, 1–5.
35. Jampatesh, S.; Sawisit, A.; Wong, N.; Jantama, S.S.; Jantama, K. Evaluation of inhibitory effect and feasible utilization of dilute acid-pretreated rice straws on succinate production by metabolically engineered *Escherichia coli* AS1600a. *Bioresour. Technol.* **2019**, *273*, 93–102. [CrossRef] [PubMed]
36. Palmarola-Adrados, B.; Chotěborská, P.; Galbe, M.; Zacchi, G. Ethanol production from non-starch carbohydrates of wheat bran. *Bioresour. Technol.* **2005**, *96*, 843–850. [CrossRef] [PubMed]
37. Guo, X.; Cavka, A.; Jönsson, L.J.; Hong, F. Comparison of methods for detoxification of spruce hydrolysate for bacterial cellulose production. *Microb. Cell Fact.* **2013**, *12*, 93. [CrossRef] [PubMed]
38. Bradford, M.M. A rapid and sensitive method for the quantitation of microgram quantities of protein utilizing the principle of protein-dye binding. *Anal. Biochem.* **1976**, *72*, 248–254. [CrossRef]
39. Fehér, C.; Gazsó, Z.; Gál, B.; Kontra, A.; Barta, Z.; Réczey, K. Integrated Process of Arabinose Biopurification and Xylitol Fermentation Based on the Diverse Action of *Candida boidinii*. *Chem. Biochem. Eng. Q.* **2015**, *29*, 587–597. [CrossRef]
40. Ghio, S.; Ontañon, O.; Piccinni, F.E.; Marrero Díaz de Villegas, R.; Talia, P.; Grasso, D.H.; Campos, E. *Paenibacillus sp.* A59 GH10 and GH11 Extracellular Endoxylanases: Application in Biomass Bioconversion. *Bioenergy Res.* **2018**, *11*, 174–190. [CrossRef]
41. WISE, W.S. The measurement of the aeration of biological culture media. *J. Gen. Microbiol.* **1950**, *4*, 167–177.
42. Nishikawa, M.; Nakamura, M.; Yagi, H.; Hashimoto, K.; Nishikawa, M.; Yagi, H. Gas absorption in aerated mixing vessels. *J. Chem. Eng. Jpn.* **1981**, *14*, 219–226. [CrossRef]

43. Wyman, C.E.; Yang, B. Combined Severity Factor for Predicting Sugar Recovery in Acid-Catalyzed Pretreatment Followed by Enzymatic Hydrolysis. In *Hydrothermal Processing in Biorefineries: Production of Bioethanol and High Added-Value Compounds of Second and Third Generation Biomass*; Ruiz, H.A., Hedegaard Thomsen, M., Trajano, H.L., Eds.; Springer: Cham, Switzerland, 2017; pp. 161–180. ISBN 978-3-319-56457-9.
44. Fehér, C.; Gazsó, Z.; Tatijarern, P.; Molnár, M.; Barta, Z.; Réczey, K. Investigation of selective arabinose release from corn fibre by acid hydrolysis under mild conditions. *J. Chem. Technol. Biotechnol.* **2015**, *90*, 896–906. [CrossRef]
45. Vandeska, E.; Amartey, S.; Kuzmanova, S.; Jeffries, T. Effects of environmental conditions on production of xylitol by *Candida boidinii*. *World J. Microbiol. Biotechnol.* **1995**, *11*, 213–218. [CrossRef] [PubMed]
46. Vongsuvanlert, V.; Tani, Y. Xylitol production by a methanol yeast, *Candida boidinii* (*Kloeckera sp.*) No. 2201. *J. Ferment. Bioeng.* **1989**, *67*, 35–39. [CrossRef]
47. Kim, J.H.; Han, K.C.; Koh, Y.H.; Ryu, Y.W.; Seo, J.H. Optimization of fed-batch fermentation for xylitol production by *Candida tropicalis*. *J. Ind. Microbiol. Biotechnol.* **2002**, *29*, 16–19. [CrossRef] [PubMed]
48. Aranda-Barradas, J.S.; Garibay-Orijel, C.; Badillo-Corona, J.A.; Salgado-Manjarrez, E. A stoichiometric analysis of biological xylitol production. *Biochem. Eng. J.* **2010**, *50*, 1–9. [CrossRef]
49. Granström, T.B.; Izumori, K.; Leisola, M. A rare sugar xylitol. Part I: The biochemistry and biosynthesis of xylitol. *Appl. Microbiol. Biotechnol.* **2007**, *74*, 277–281. [CrossRef]
50. Winkelhausen, E.; Amartey, S.A.; Kuzmanova, S. Xylitol production from D-xylose at different oxygen transfer coefficients in a batch bioreactor. *Eng. Life Sci.* **2004**, *4*, 150–154. [CrossRef]
51. Santana, N.B.; Teixeira Dias, J.C.; Rezende, R.P.; Franco, M.; Silva Oliveira, L.K.; Souza, L.O. Production of xylitol and bio-detoxification of cocoa pod husk hemicellulose hydrolysate by *Candida boidinii* XM02G. *PLoS ONE* **2018**, *13*, e0195206. [CrossRef]
52. López-Linares, J.C.; Ruiz, E.; Romero, I.; Castro, E.; Manzanares, P. Xylitol production from exhausted olive pomace by *Candida boidinii*. *Appl. Sci.* **2020**, *10*, 6966.
53. Mayerhoff, Z.D.V.L.; Roberto, I.C.; Silva, S.S. Xylitol production from rice straw hemicellulose hydrolysate using different yeast strains. *Biotechnol. Lett.* **1997**, *19*, 407–409. [CrossRef]
54. Dasgupta, D.; Bandhu, S.; Adhikari, D.K.; Ghosh, D. Challenges and prospects of xylitol production with whole cell bio-catalysis: A review. *Microbiol. Res.* **2017**, *197*, 9–21. [CrossRef]
55. Silva, C.J.S.M.; Roberto, I.C. Optimization of xylitol production by *Candida guilliermondii* FTI 20037 using response surface methodology. *Process Biochem.* **2001**, *36*, 1119–1124. [CrossRef]
56. Zeid, A.A.A.; El-Fouly, M.Z.; El-Zawahry, Y.A.; El-Mongy, T.M.; El-Aziz, A.B.A. Bioconversion of rice straw xylose to xylitol by a local strain of Candida Tropicalis. *J. Appl. Sci. Res.* **2008**, *4*, 975–986.
57. Schneider, H. Selective removal of acetic acid from hardwood-spent sulfite liquor using a mutant yeast. *Enzyme Microb. Technol.* **1996**, *19*, 94–98. [CrossRef]
58. Fehér, A.; Fehér, C.; Rozbach, M.; Rácz, G.; Fekete, M.; Hegedűs, L.; Barta, Z. Treatments of Lignocellulosic Hydrolysates and Continuous-Flow Hydrogenation of Xylose to Xylitol. *Chem. Eng. Technol.* **2018**, *41*, 496–503. [CrossRef]
59. Hua, Y.; Wang, J.; Zhu, Y.; Zhang, B.; Kong, X.; Li, W.; Wang, D.; Hong, J. Release of glucose repression on xylose utilization in *Kluyveromyces marxianus* to enhance glucose-xylose co-utilization and xylitol production from corncob hydrolysate. *Microb. Cell Fact.* **2019**, *18*, 24. [CrossRef] [PubMed]
60. Kogje, A.; Ghosalkar, A. Xylitol production by *Saccharomyces cerevisiae* overexpressing different xylose reductases using non-detoxified hemicellulosic hydrolysate of corncob. *3 Biotech* **2016**, *6*, 127. [CrossRef]

agronomy

MDPI

Article

Bioconversion Process of Barley Crop Residues into Biogas—Energetic-Environmental Potential in Spain

Carlos Morales-Polo [1,2,3,*], María del Mar Cledera-Castro [1,2,3], Marta Revuelta-Aramburu [2] and Katia Hueso-Kortekaas [2]

1 Institute for Research in Technology, Comillas Pontifical University, 28015 Madrid, Spain; mcledera@comillas.edu
2 Department of Mechanical Engineering, ICAI School of Engineering, Comillas Pontifical University, 28015 Madrid, Spain; mrevuara@comillas.edu (M.R.-A.); khueso@comillas.edu (K.H.-K.)
3 Rafael Mariño Chair for New Energy Technology, Comillas Pontifical University, 28015 Madrid, Spain
* Correspondence: cmorales@comillas.edu; Tel.: 34-91-542-28-00

Abstract: Barley fields reach 1.7 million hectares in Spain, of which 320,000 are used to produce malt, generating 450,000 tons of crop residue from barley intended for malt production. One way to treat this waste in an environmentally sound, energy-sustainable and economically cost-effective manner is anaerobic digestion. The biogas generated can be used as fuel and as a renewable source of energy (providing a solution to the energy supply problem from an environmental point of view). It has been shown that, when treated along with sludge from a Upflow Anaerobic Sludge Blanket (UASB) reactor, the crop malt residue produces about 1604 NmL of biogas per 100 g; with a content in methane of 27.486%. The development of the process has been studied with a novel indicator, hydrogen generation, and it has been determined that the process takes place in two phases. It has been demonstrated that this solution is beginning to be energy-efficient and therefore to produce energy for external uses in regions that have at least 6000 hectares of planted barley. At best, it can be considered, in a given region, the equivalent of a 115 MW power plant. It could supply energy to 10 thousand homes per year. Therefore, it is considered an energy-efficient solution that com-plies with the Sustainable Development Goals #1, #7, #10, #12 and #13. It guarantees access to energy in isolated areas or with supply problems, and results in a 55.4% reduction in emissions of equivalent-CO_2 (which equals 38,060 tons of equivalent-CO_2 in Spain).

Keywords: barley crop residue; biochemical methane potential; material degradability; anaerobic indicators; biogas feasibility; biogas emissions

check for updates

Citation: Morales-Polo, C.; Cledera-Castro, M.d.M.; Revuelta-Aramburu, M.; Hueso-Kortekaas, K. Bioconversion Process of Barley Crop Residues into Biogas—Energetic-Environmental Potential in Spain. *Agronomy* **2021**, *11*, 640. https://doi.org/10.3390/agronomy11040640

Academic Editor: Carlos Martín

Received: 5 February 2021
Accepted: 23 March 2021
Published: 26 March 2021

Publisher's Note: MDPI stays neutral with regard to jurisdictional claims in published maps and institutional affiliations.

Copyright: © 2021 by the authors. Licensee MDPI, Basel, Switzerland. This article is an open access article distributed under the terms and conditions of the Creative Commons Attribution (CC BY) license (https://creativecommons.org/licenses/by/4.0/).

1. Introduction

Agri-food trade in the European Union (EU) is one of the most important in the world economy [1]. In 2019, the EU positioned itself as the world's largest exporter and the second largest importer of agri-food products. The value of exports increased to 14.7 billion euros in 2019 compared to 2018, while import values increased to 10.7 billion euros [2]. Both imports and exports have been growing since 2002, contributing to a monthly trade surplus in the agri-food sector of 4.0 billion euros.

According to EUROSTAT data, the demographic situation in the EU reflects an upward growth; since 2008, the population has increased by 13 million inhabitants [3]. Alongside this population growth, an increase in needs and consumption is associated, especially in the agri-food sector, given the basic need for population feeding, but also in the energy field, as discussed below.

The agri-food industry comprises activities from all economic sectors [1]. The food supply chain (FSC) begins with stages of the primary sector (agriculture and livestock), which generates by-products (i.e., manure, waffle, cornstalk) and food waste and food loss in the form of low-quality products, damaged production, or products with no commercial value [4,5]. In response to the first part of FSC, and primary sector activities, the main crop in the EU is cereal (including rice), which in 2018 was 295.1 million tons, corresponding to 11.3% of global production [6]. That is why cereal production is particularly influential at European level. In Spain, one of the main producers of the agricultural sector in Europe, an average of 6 million hectares of cereals are grown These are distributed in 38.7% corresponding to barley, 28.9% soft wheat, 15.8% maize, 6.2% oat, 5.7% for durum wheat, 2.9% triticale, 1.7% rye and 0.1% sorghum [7].

Agricultural activities, as well as livestock and forestry activities, generate in their various stages variable quantities of by-products and waste whose storage, disposal or disposal represents an additional task and source of costs for the producer [8], as well as constituting an environmental problem by the increasing generation of waste associated with human consumption and population growth. These wastes are considered an under-utilized source of resources as they are produced continuously and renewably [9]. These residues are, in a very small part, partially valued at different levels (production of biofuel, food and animal bed, composting or building materials) [10]; however, a significant volume of them is not reused and constitutes a serious problem that negatively affects the overall sustainability of the agricultural sector [10]. The agricultural waste known as industrial, those that must be eliminated because they are not usable in the area in which they are generated, are in the case of cereal and grain crops, straw [11]. In Spain, 35.7% of all agricultural waste produced consists of residues from cereal crops (mostly barley, followed by wheat and maize), with annual quantities exceeding 9 million tons [12].

Within the production of barley in Spain, an important part is intended for its processing into malt. Malting is a process applied to cereal grains, by which they germinate by submerging them in water, then being quickly dried by injection of hot air. During this process, malted grains develop enzymes that convert grain starch into sugar. Barley is the most commonly used cereal for malting because of this reason [13].

According to the results of the authors, in Spain 320,000 hectares of barley are allocated to the production of malt. This means that 18.850% of barley production (See Section 3.10.1) is destined for this purpose and is therefore a potential source of straw waste generation, which can be harnessed in some way.

Taking up the problem of population growth previously introduced, it is also associated with an increase in energy consumption (Figure 1). This growth rate is expected to continue over time. In fact, the International Energy Outlook [14] predicts a 28% increase in energy consumption for the period 2015–2040. Other agencies such as the International Energy Agency estimate this growth by 35% for the period 2010–2035 [15]. Of all the energy consumed in Europe in 2017, approximately 80% came from fossil sources [16] such as coal, oil, natural gas and derivatives. Finding new forms of energy to reduce this dependence on foreign and fossil fuels is strategic. Moreover, given that fossil fuels are known emitters of greenhouse gases, the reduction in their use is not only strategic, but also necessary. In fact, of the 4.66 gigatons of equivalent CO_2 emitted by the EU, 82.8% of emissions came from the energy sector [16]. These two reasons mark the roadmap for a more renewable energy model [17].

Both growth in cereal consumption and production, as well as growth in energy consumption, lead to several environmental problems such as increased waste generation, and the possible increase in greenhouse emissions when fossil fuels are used for energy production [18]. A joint solution for both problems can be provided by Anaerobic Digestion (AD) [19]. Anaerobic digestion allows the bioconversion of the organic matter present in organic substrates, such as a residue, into biogas (a renewable energy source) via microbiological degradation [20]. AD also generates a digestate that can be used as fertilizer. In this way, an energy source is obtained by reducing the waste generated and it also creates a circular economy in that process, solving both environmental problems [21].

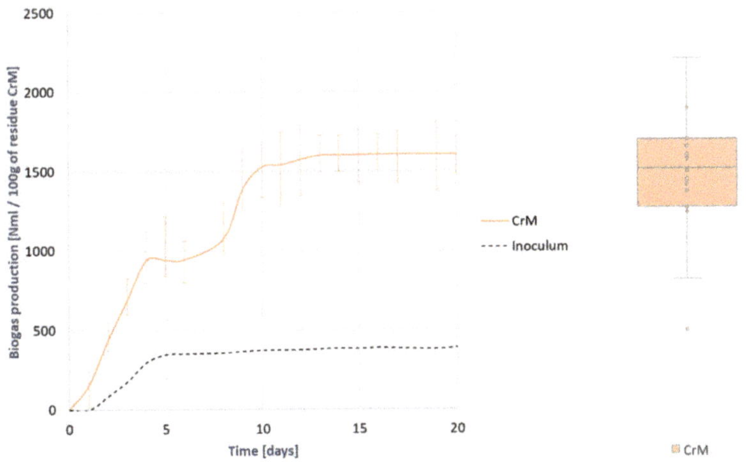

Figure 1. Gross biogas production mean curve from the anaerobic digestion of 100 g of residue (Crop Barley Residue) CrM, and box diagram of its variation.

AD is a very complex process in which different stages occur simultaneously with different microorganisms [22], each producing the substrates necessary for the next stage to take place, depending on the substrate composition. It is then important to perfectly know the composition of the substrate to be digested, so that the process development can be determined. In addition, because of the complexity of the AD process, in which different stages occur simultaneously with different microorganisms, a medium in which all influential parameters are controlled is required to ensure a stable environment [23]. One of the novelties of this study is the use as development of the process indicator, typical variables such as pH combined with the study of the evolution of the hydrogen content generated and consumed during the process. H_2 is intermediate gas that marks with its appearance the beginning of the stages of hydrolysis, acidogenesis and acetogenesis, and with its reduction the development of methanogenesis since it is transformed into methane via hydrogenotrophic methanogenesis, so that monitoring its evolution can also monitor the development of the process.

For all of the above, the main objective of this research is to propose and study the feasibility and process development of treatment by anaerobic digestion of crop residues generated during harvest of barley that is intended for malt production (barley straw).

To this end, the substrate is completely characterized, the anaerobic digestion process is analyzed through all phases and processes, focusing on the inhibitions and alterations of the process and the potential for biogas and methane generation is determined. Finally, the amount of energy that can be obtained through this route is estimated, against with the reduction of equivalent CO_2 emissions that comes when using this renewable energy source to replace a conventional energy source.

2. Materials and Methods

2.1. Test Samples

The studied substrate is the residue from barley intended for malt production, hereinafter CrM. To avoid variability in the composition of the substrate, a laboratory-generated residue was used. For this purpose, barley of the *Hordeum distichon* type was used, which undergoes a malting process, keeping the grains moist for 5 days, dried at 40 °C for 24 h and then at 60 °C for one night.

As an inoculum, and therefore the source that provides methane-based and anaerobic biomass to trigger the biomethanization process, sewage sludge from a Wastewater Treatment

Plant (WWTP) from Madrid, Spain. To perform the Biochemical Methane Potential (BMP) tests it is recommended to use a stable and easily accessible inoculum [24]. Fundamental standards such as UNE-EN ISO 11734 and VDI-4630 [25] recommend using sludge from WWTP. Several authors advise it because of its accessibility and permanence of biomass [26,27], including pioneers in conducting Biochemical Methane Potential (BMP) test in 1979 [28], the following pioneers in the conduct of BMP assays [29]. In particular, in this study it is a granular sludge from an Upflow Anaerobic Sludge Blanket (UASB) reactor, from an agri-food industry sewage treatment plant. This type of sludge, along with its granule agglomeration characteristic makes it resistant to internal or process alterations [30].

2.2. Analytical Methods for Compositional Characterization of Samples, Inoculum and Mixture Introduced in the Reactor

One of the sub-objectives of this research is to characterize the substrates in depth to know, draw conclusions and assist in the determination of the development of the process. For this purpose, the CrM substrate and the inoculum are characterized. In addition, the same compositional study is also performed to the mixture of substrate + inoculum introduced into the reactor, before the digestion test and after completion of the BMP test. In this way it can be compared and determined what changes in the composition has caused anaerobic digestion, such as how much organic matter has been degraded, whether the mixture has developed accumulation of (Volatile Fatty Acid) VFA, or how much volatile solids have been degraded.

In particular, the following compositional determinations according to the standardized methods were carried out (Table 1).

2.3. Biochemical Methane Potential Tests to Determine de Anaerobic Degradation

The Biochemical Methane Potential (BMP) test procedure has been developed in accordance with UNE-EN ISO 11734 [31], including a gas chromatograph for the measurement of biogas composition. In order to determine the amount of methane produced the UNE-EN ISO 11734 [31] standard uses manometric methods. The VDI-4630 [25] procedure has been used to transform the measured pressure inside the reactor into the gas flow generated.

The test conditions were the same as those used on the full scale. Digestion is performed under mesophilic conditions, 37 ± 1 °C, with a ratio between substrate and inoculum of 1:3 m/v, that is, for 300 mL of UASB sludge, 100 g of CrM residue are added for digestion [25].

Pressure changes inside the reactor are measured daily. Also, a sample of the gas generated in the reactors is extracted daily and analyzed in a gas chromatograph to determine the proportion of methane and hydrogen in the biogas. An Agilent (California, USA) 7820A gas chromatograph (GC) was used with a thermal conductivity detector (TCD), equipped with a Molsieve 5A-CP molecular sieve, and a PoraPLOT Q capillary column. N_2 is used as a carrier gas to detect H_2 more easily. In addition, by using N_2 as a gas to ensure air displacement and anaerobic conditions inside the bottles.

Twenty-one repetitions are made with the aim of obtaining reliable results with the least dispersion possible. Since the substrate is a lab prepared sample, there is no dispersion in the results due to it. The only dispersion variable is the inoculum, which, as discussed, is obtained from a UASB reactor from a WWTP. However, attempts have been made to minimize this dispersion by taking inoculum samples on the same days of the week, when the level of production was similar and the characterization of the inoculum too.

2.4. Statistical Analysis of Results

As the tests were repeated 21 times, a statistical analysis was necessary to determine the dispersion between them is so large that there is no relationship, an ANOVA statistical analysis with a 95% confidence level is performed accompanied by their respective DMS and Tukey contrasts. The starting hypotheses are H0 = Equality between curves, and H1 = Inequality between curves.

Table 1. Standardized procedures used for compositional characterization of substrates and mixtures.

	Method Procedure
Physical parameters	
Humidity—Hum [% $_{wet\ weight}$] Total Solids—TS [% $_{wet\ weight}$] Volatile Solids—vs. [% $_{wet\ weight}$]	APHA 2540-G
Macronutritional Analysis (LPCH Content)	
Lipids (L) [% $_{wet\ weight}$] Proteins (P) [% $_{wet\ weight}$] Carbohydrates (CH) [% $_{wet\ weight}$]	UNE-EN-ISO 13804:2013
Organic Content Analysis (Chemical Oxygen Demand COD)	
Total COD—CODt [mg O_2/g—mL] Filtered COD—CODf [mg O_2/g—mL] Solubility [%]	APHA 5220-B
Nitrogen Content Analysis	
Total Kjeldahl Nitrogen—TKN [mg N/g—mL] Amoniacal Nitrogen—AN [mg N/g—mL] Organic Nitrogen—ON [mg N/g—mL]	APHA 4500-N APHA 4500-NH_3 APHA 4500-N_{org}
pH and Alkalinity Analysis	
pH Total Alkalinity—TA [mg $CaCO_3$/g—mL] Partial Alkalinity—PA [mg $CaCO_3$/g—mL] Intermediate Alkalinity—IA [mg $CaCO_3$/g—mL]	APHA 2320-B
Elemental Analysis	
C [% $_{dry\ weight}$] H [% $_{dry\ weight}$] N [% $_{dry\ weight}$] S [% $_{dry\ weight}$]	UNE-EN-ISO 15104:2011

Some descriptive statistics are also calculated to help determine the level of dispersion between the results. For this purpose, a visual method is used using box diagrams and a quantitative method through Analysis of the Coefficient of Variation (CV).

2.5. Mathematical Adjustment and Determinations of Results

Once the results are obtained these are treated mathematically. Methane generation curves are treated as first-order kinetics, according to the described process by Veeken and Hamlers [32]. The disintegration constant (k_{dis}) can be obtained directly by considering the complete anaerobic digestion process, giving information about the depth and speed of the hydrolysis process. Also, maximum amount of actual methane obtained in the process (CH_{4max}) can be determined, adjusting the generation data by means of least squares, to a first order kinetics reaction:

$$CH_4(t) = CH_{4\ max} \cdot \left[1 - e^{-k_{dis} \cdot t}\right], \quad (1)$$

where $CH_4(t)$ represents the production of methane on the t-day; CH_{4max} the maximum generation of methane recorded (which can be assumed by the latest generation); and k_{dis} the average disintegration constant measured in days^{-1}.

The theoretical generation of methane that was to be expected after anaerobic degradation of the substrate can be obtained from the reduction of COD experienced before and after the BMP test.

$$V_{\text{theoretical } CH4_{\text{accumulated}}}[\text{Nml}] = \frac{(COD_0 - COD_f)\cdot V_{\text{test}}\cdot 340}{1000}\cdot\frac{1}{0.9869}, \quad (2)$$

where COD_0 and COD_f represents the COD levels measured at baseline at the end of the BMP test expressed in mg/l; V_{test} the test volume occupied by the mixture substrate + sludge expressed in litters; 340 the conversion factor of COD in methane; and 1/0.9869 the conversion factor from standard conditions (0 °C and 1 bar) to normal conditions (0 °C and 1 atm).

The level of degradation of the substrate or residue can be calculated analytically through the COD reduction degradation levels and provides information about the level of the degradation of the substrate, regardless of degradation inoculum.

$$BD_{\text{residue}}[\%] = \frac{\left(COD_{\text{mixture}_0} - COD_{\text{sludge}_0}\right) - \left(COD_{\text{mixture}_f} - COD_{\text{sludge}_f}\right)}{\left(COD_{\text{mixture}_0} - COD_{\text{sludge}_0}\right)}\cdot 100\,. \quad (3)$$

2.6. Energy Suitability of the Proposed Process: Energy Balance Analysis

One of the objectives to be covered by this research is to determine whether the proposed solution (anaerobic treatment of barley crop residues) is highly cost-effective. To do this, it analyses its potential, taking into account the energy generated in form of biogas, the needs that must be met in the process, and whether there is energy available for external uses.

The procedure has been developed according to the one described by the authors in the article [33], which in turn is based on the one described in [19]. Firstly, analyses and calculates the energy needs of the anaerobic process. Subsequently, with the laboratory data obtained in this research calculates the energy available in the biogas generated, depending on the amount of biogas that is produced and its methane content. Finally, it is determined whether net energy available for external uses exists or not, once the needs of the process (heat demand and electricity demand) have been met, for example for supply in the residential sector. In this way, if net available energy exists, the process would be considered as energy efficient.

2.7. Environmental Suitability of the Proposed Process: Emissions Reduction by Using Generated Biogas as an Energy Source Rather Than a Conventional Fossil Fuel

The objective of this section is to estimate the reduction in emissions that would occur if it is decided to use the biogas generated in the proposed solution as an energy source, rather than obtaining such energy by conventional means in the combustion of a fossil fuel. To this end, the "Methodology for thermal energy projects aimed at reducing the consumption of fossil fuels in a new or existing facility" [34], developed by the Carbon Fund for a Sustainable Economy (FES-CO_2) of the Spanish Ministry of Environment, has been followed.

The methodology is based on calculating the emission reductions associated with the project as the difference between emissions from a base scenario (final energy is obtained through the burning of natural gas) and project emissions (the final energy is obtained by burning the generated methane).

3. Results and Discussion

3.1. Characterization of Samples, Substrates, Inoculum and Mixtures in the Digester

Studying the characterization of substrates, of the inoculum and from the mixture that exists in the digester, before and at the end of the BMP test, is essential to understand the development of the process and to be able to make decisions about the results obtained during the AD process. All tests have been conducted as described in Section 2.2, and the results are shown below.

Table 2 shows characterization results of the substrate (CrM), of the inoculum (S) and the mixture in the digester (CrM + S) before the BMP test, and after completion of the latter at 20 days. All results are indicated in the respective units, per gram of substrate, or per milliliter of sludge or reactor mix.

Table 2. Characterization results for BMP tests of residue CrM, at the start and after completion of the test.

	Substrate Material CrM	Inoculum Material S	Initial Reactor Mix CrM + S	Final Reactor Mix CrM + S
		Physical parameters		
Hum [% $_{wet\ weight}$]	62.60	94.30	85.37	78.14
TS [% $_{wet\ weight}$]	26.31	5.70	10.75	5.18
VS [% $_{wet\ weight}$]	25.44	4.92	11.05	4.96
		Macronutritional Analysis (LPCH Content)		
Lipids (L) [% $_{wet\ weight}$]	0.98	0.47	0.55	
Proteins (P) [% $_{wet\ weight}$]	7.37	0.53	2.25	
Carbohydrates (CH) [% $_{wet\ weight}$]	77.37	0.56	19.87	
		Organic Content Analysis (COD)		
COD_t [mg O_2/g—mL]	168.80	101.65	117.52	105.40
COD_f [mg O_2/g—mL]	87.77	37.08	49.70	22.06
Solubility [%]	58.81	36.48	42.30	20.93
		Nitrogen Content Analysis		
TKN [mg N/g—mL]	12.31	2.00	4.55	4.72
AN [mg N/g—mL]	0.53	1.15	0.96	1.24
ON [mg N/g—mL]	11.78	0.85	3.58	3.23
		pH and Alkalinity Analysis		
pH	5.22	7.46	6.85	6.97
TA [mg $CaCO_3$/g—mL]	5.81	8.88	8.06	11.06
PA [mg $CaCO_3$/g—mL]	-	5.22	5.35	5.15
IA [mg $CaCO_3$/g—mL]	5.81	3.65	2.71	4.47
		Elemental Analysis		
C [% $_{dry\ weight}$]	52.32	11.19	21.48	
H [% $_{dry\ weight}$]	4.43	9.00	7.75	
N [% $_{dry\ weight}$]	4.24	2.22	2.71	
S [% $_{dry\ weight}$]	0.07	0.18	0.15	
C/N Ratio	12.33	5.04	7.93	

CrM substrate is presented as a substrate with a high humidity level (62.60%) which makes its solubilization, in principle, easy and the process fast and profound. If the LPCH content is compared, it is observed that it is a carbohydrate-type substrate, especially rich in simple cellulose-type carbohydrates. The digestion of this type of substrate is stable and fast, but with the likelihood of releasing VFA during its digestion from the acid digestion of monosaccharides and other simple carbohydrates. However, because of the presence of a certain protein content, a small amount of ammoniacal nitrogen is likely to be released that can compensate by acting as a buffer for the slight acidification caused by VFA. In terms of COD it is a relatively rich substrate in carbonous matter and organic matter, with a particularly high solubility (of 58.81%), indicating that, of the entire COD, almost 60%

is directly accessible to microorganisms without the need to hydrolyze or release from encapsulation due to being a particulate substrate. For this easy accessibility, a rapid degradation is expected, even though the inoculum is a granular UASB sludge, difficult to solubilize. In terms of nitrogen content, it is not very high. It is distributed in ON that comes from the light protein content, and in AN, whose content is not excessive, in fact, is below the limit of accumulation studied (2 g/L), so it is expected that the AN will be released as a buffer for pH control and compensate for possible acidifications, without accumulating and ending up inhibiting the process by excess ammonia in the reactor. The CrM material has a C/N ratio of 12.33, which is close to the optimal C/N ratio (approx. 20) [33,35,36] to ensure stable digestion so it is expected that the process will have alterations, but without impact, such as a release of VFA due to the carbohydrate content, compensated with a slight release of ammoniacal nitrogen that will act as buffer dampening its effect.

Analyzing changes in the composition of the mixture in the reactor, before and after the BMP test, it can be observed that the humidity has been slightly reduced, by −8.50%, which is logical when it comes to anaerobic, closed, and wet digestion digesters, as is the case of a UASB digester specially designed for liquid substrates. The reduction of vs. and TS is very remarkable, of −51.82% and −55.11% respectively, which gives an idea that the process has developed correctly because the organic matter present in solids form has been digested. The reduction of COD has been very low, of −10.31%, indicating that, although the digestion process has been correct, the methanization has not been completely profound (later it will be analyzed with the content in methane, since the carbon contained in the COD is the one that is transformed into methane). However, the reduction in COD_f has been very noticeable, of −55.61%, indicating that, although only 10% of the COD has been reduced to be converted into methane, this has been practically a reduction in COD that is not encapsulated, and is therefore directly accessible to micro-strategies. This means that the methanization process has not been complete, by a failure of the disintegration + hydrolysis stage, which will provide a lower-than-expected methane content, below 60% which is considered the stable development limit. According to the nitrogen content, TKN is slightly increased by 3.89% by the release of some of the nitrogen encapsulated in the proteins, the ON. In fact, the ON is reduced by −9.75% precisely by the release of this nitrogen, by degrading the proteins. For its part, the AN is increased by 28.29% when the ON is released, until it reaches a value of 1.24 mg/mL, falling below the accumulation limit (2 g/L) in this way it is expected that the released ammoniacal nitrogen will act as a buffer, dampening any acidification, for example, that coming from the release of VFA when digesting carbohydrates. The initial and final pH values are very similar; however, they are expected to have varied during the 20 days of process. As for alkalinity, TA is increased by 43.92%, giving an idea of the stability of the process. Intermediate alkalinity is also increased by 64.94%, indicating that either no VFA has been released, or on the contrary these have been neutralized by the buffer effect of the released AN.

All these assumptions should be checked later with the analysis of the process that develops.

3.2. Biogas Production

To determine biogas production, the internal pressure generated inside the digesters during the development of BMP tests is measured, and then translated into the volume of gas generated, as described in Section 2.3.

The most convenient measure to express biogas generation is gross production, that is, biogas generated by a certain amount of waste. However, several authors [37,38] recommend expressing it in terms of specific production, that is, in biogas generation for each vs. containing the residue, in this way it can be compared with another substrate more quickly. Therefore, the results are commented based on gross production but are also displayed based on specific production.

As shown in Figure 1, biogas generation is completed on day 10, producing an average of 1604 (±19.980%) NmL of biogas measured under normal conditions, per 100 g of digested

CrM residue (Table 3). Although there are some failed assays, based on the results of the ANOVA analysis developed, all curves can be considered equal and assumed by the mean curve. This is because, after performing the ANOVA analysis, the significance level is, in all curves except those that the process has been failed, 1000, so that the null assumption of average equality can be accepted, and therefore all curves equal to the mean curve can be assumed. This convergence of results was to be expected since the substrate to be treated, as advanced in Section 3.1 is a substrate rich in simple carbohydrates, which gives it the property to generate stable digestions.

Table 3. Numerical results of the BMP tests when digestion residue CrM. Gross and specific production of biogas, methane and hydrogen, methane and hydrogen content of the produced biogas and descriptive statistics.

	Production [NmL/100 g of residue CrM]	Standard Deviation	Coefficient of Variation	Relative Error
Biogas	1604.22 NmL	525.39	0.32	19.98%
Methane	458.55 NmL	189.61	0.41	24.83%
Hydrogen	0.69 NmL + 0.33 NmL	0.60 + 0.15	0.87 + 0.48	105.72% + 134.45%
	Specific Production [NmL/g of vs. of residue CrM]			
Biogas	63.05 NmL	20.65	0.01	96.53%
Methane	18.02 NmL	7.45	0.01	96.80%
Hydrogen	0.02 NmL + 0.01 NmL	0.02 + 0.02	0.03 + 0.08	92.60% + 5.28%
	Content [% vol]			
Methane	27.48	7.22	0.26	16.20%
Hydrogen	0.28% + 0.03%	0.14 + 0.14	0.43 + 4.37	39.90% + 349.72%

Figure 1 shows the average curve of all the 21 curves obtained, which can be assumed as determined by the ANOVA analysis. It can be seen in them that the initial part of the curves, there are two clearly differentiated slope changes. During the first day there is a delay, and on days 1–4 there is a linear generation, which stops until day 8–9 in which generation resumes to stabilize on day 12. Although it will be demonstrated later in the analysis of the results of the following sections, it is an indicator that anaerobic digestion occurs in two phases. Given the characteristic particulate substrate, in the first place the organic matter is digested directly accessible, that is, the solubilized, and subsequently, after hydrolyzing the particulate matter, it begins to digest it. This assumption will be demonstrated later with the joint analysis of all the variables studied, especially with the evolution of hydrogen, as well as evolutions and compositional changes.

If the generation of biogas is compared with that obtained by the inoculum alone, the effect of adding the CrM substrate has been positive, by increasing the generation of gas by 311.33%, from the 390 mL of biogas produced by the inoculum to the 1604.22 NmL produced in joint digestion and by increasing the process speed, as you can see by comparing the slope of the start of the curves.

As for the specific production of biogas, the conclusions obtained are analogous to those of gross biogas production, as well as their curves, which are proportional. Specific production is nothing more than the production of biogas expressed, not by amount of gross waste, but for each gram of volatile solid contained in the residue, hence the curves are proportional. Specific production, as shown in Table 3, is estimated at 63,058 NmL of biogas per gram of vs. of CrM waste introduced into the reactor.

3.3. Methane Production

Figure 2 represents all the mean methane generation curves obtained during BMP tests. According to the ANOVA analysis performed, all curves can be assumed by the average curve, so this one will be used for the study, along with the error bars and the diagram of boxes and whiskers. The shape of the curves and the information obtained from them is similar to that obtained in biogas generation curves. Digestion occurs in the same way and clearly in two phases, and the stabilization of the process and therefore the generation of methane occurs around 10 days. In particular, the generation of methane obtained is 458.550 (±24.838%) Biogas NmL measured under normal conditions, per 100 g of digested CrM residue (Table 3).

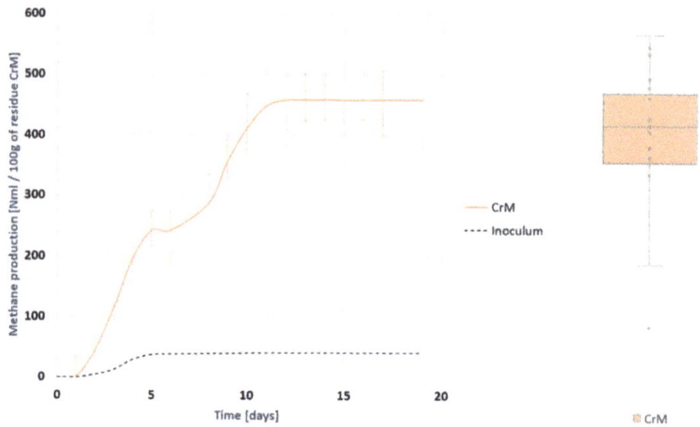

Figure 2. Gross methane production mean curve from the anaerobic digestion of 100 g of residue CrM and box diagram of its variation.

Although the dispersion between curves is similar to that obtained in the production of biogas, as inferred from the box and whisker diagram depicted in Figures 1 and 2 the descriptive statistics in Table 3 show that this variability between curves is not pronounced, in fact the dispersion between methane generation curves is less than the deviation between biodegradability curves or biogas production. It follows then that the process is quite stable, and the changes between them are not due to failures in the process, but to changes in the proportion of methane in biogas, which is a good indicator of the stability and development of the process and is therefore studied in a section of its own.

As for the specific production of methane, this is determined at 18.024 NmL of methanol per gram of vs. of CrM residue introduced into the reactor.

3.4. Methane Content of Biogas Generated

It can be seen in Figure 3 that the proportion of methane in biogas begins to detect something before the first day, begins to grow until day 5, slows slightly, resumes until moderately stabilized on day 10. The curves converge around 27% of methane. This double growth and observed slope change reconfirms the two-phase digestion phenomenon that occurs, a first phase in which directly accessible organic matter is digested, and secondly particulate or encapsulated organic matter, although this must be corroborated with a somewhat more thorough analysis such as the evolution of H_2 to be done in subsequent sections.

As noted, and as seen in the mean curve of Figure 4, the average methane content in the generated biogas is 27.485 (±16.201%) % of CH_4, resulting in an increase of 128.098% compared to the methane content of the biogas generated by the inoculum, so it is inferred again that the effect of adding CrM substrate has been positive.

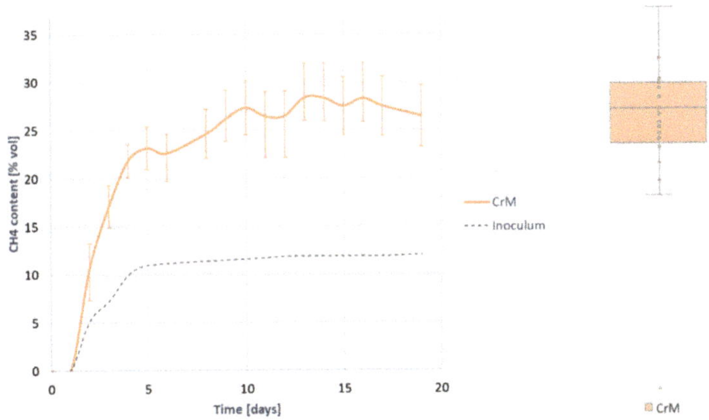

Figure 3. Methane content mean curve from the biogas produced in the anaerobic digestion of 100 g of residue CrM and box diagram of its variation.

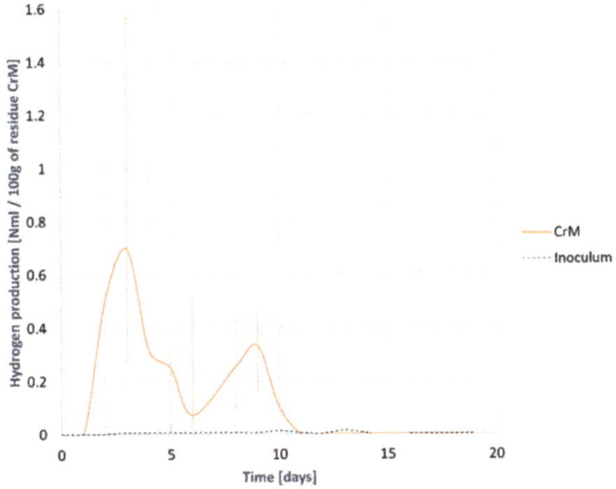

Figure 4. Gross hydrogen production mean curve from the anaerobic digestion of 100 g of residue CrM.

The methane content is below 60%, which is the limit considered acceptable for a stable and deeply developed process. In this case it is around 30%, so the process is inferred that it is not entirely complete, as indicated in previous paragraphs and is probably due to low degradation levels, which will be demonstrated later with the mathematical analyses and determinations including the level of biodegradation of the substrate. In this case it has already been indicated that there is first digestion, only of the organic matter directly accessible or solubilized, and subsequently the encapsulation, which will logically not be digested in depth or in full (and will be determined by the analysis of the level of degradation of the substrate). Although some of the results of previous research conducted by other authors yield similar values of methane content, this study explains by the nature of the inoculum, a granular UASB substrate, specially designed to digest solubilized organic matter, and that its characteristic of granule grouping makes the hydrolysis of particulate matter complicated, since microorganisms and hydrolytic enzymes, grouped in a granule have fewer freedoms and the ability to adhere to the walls of the particulate substrate and start hydrolyzing it, being easier directly to act with that COD that is solubilized and directly accessible (COD_f).

3.5. Hydrogen Production

Hydrogen generation occurs once hydrolysis has been completed and the acidic stages of acidogenesis and acetogenesis begin, which will consequently become CH_4 through acetoclastic methanogenesis. Hydrogen once generated begins to disappear when it is transformed into methane using hydrogenotrophic methanogenesis. It is therefore a good indicator of process development and hydrolysis speed, although its use is not widespread as it is difficult to detect and measure against other indicators such as pH. This research will use it for this purpose, providing added value and novelty to the study.

It is to be expected that hydrogen will form during the first few days and the larger its production and the faster the production peak is reached, the faster and deeper the hydrolysis process will be. As soon as hydrogen is generated, that is, as soon as hydrolysis is complete, the production of acidic elements by acidic elements through acidogenesis and acetogenesis begins, so it is accompanied by a decrease in pH. When the peak is reached it is understood that hydrolysis has completed, and when it begins to disappear hydrogen it is inferred that hydrogenotrophic methanogenesis begins. Methanogenesis should also be acetoclastic, that is, it must be formed from acetic acid and acidic elements. If methanogenesis occurs correctly, hydrogen reduction (by its transformation into methane) is accompanied by an increase of pH (by the transformation of acidic elements into methane). Any other development with hydrogen is identified with a process failure, inhibition or stop, for example if hydrogen reduction is slowed and also coincides with a pH that is maintained at low levels, it is understood that there has been a build-up of acidic elements that has led to inhibition of methanogenesis, and the process has been inhibited without methane production through acetoclastic methanogenesis.

Figure 4 represents the average of gross hydrogen generation curves obtained in each BMP test, along with the error bars. The ANOVA analysis demonstrates by the level of significance that in all cases curves can be resembling the middle curve and studying the process through it.

There is a first hydrogen peak on day 3. This means that hydrolysis ends on day 1, and as hydrogen grows the acidic phases begin. The hydrogen generation for this peak is 0.690 ($\pm$105.727%) NmL of hydrogen measured under normal conditions, per 100 g of digested CrM residue (Table 3). The variability of the data is very high, which is logical since production occurs in less than a day and many averages are performed after this period, on days 1–3. In any case, the presence of a maximum hydrogen on day 3 is clear in any test curve. From this day the amount of hydrogen begins to decrease, at a certain rate, the reduction stops between days 4 and 5, and resumes until the 6th day. This change in slope in the reduction of H_2 may be due to some kind of slowdown, which should be studied and determined with subsequent analyses, such as pH evolution, although it is likely to be due to an accumulation of VFA, since this substrate is prone to release them as stated in Section 3.1.

There is a second peak of H_2 generation on day 9 although lower value 0.330 ($\pm$134.450%) NmL for every 100 g of CrM residue introduced into the reactor). The appearance of this second peak may be due; well begins the digestion of particulate organic matter, which has released the components after hydrolyzing the outer membranes of the waste particles; or inhibition of methanogenesis, which subsequently resumes. Both options may be valid given the characteristics of the CrM residue studied in Section 3.1:

- CrM residue is susceptible to release VFA during digestion by having a significant carbohydrate content (although this does not seem likely due to changes occurring in AI, which does not look particularly altered and does not indicate excessive accumulation of VFA).
- CrM residue is strongly particulate, with resistant external membranes, especially for its content in lignocellulosic compounds. This will cause COD_f to be digested at

first, that is, The OM directly solubilized and accessible to microorganisms, and once hydrolysis has developed, the OM or COD that is encapsulated in the substrate begins to be digested.

In terms of specific production, the curves are proportional to those already described, there is a first generation peak on day 3 worth 0.027 (±39.903%) NmL per gram if vs. of CrM residue introduced, and a second peak on day 9 worth 0.033 (±349.718%) NmL for each gram of vs. of CrM waste introduced into the reactor.

3.6. Hydrogen Content of Biogas Produced

Like hydrogen generation, hydrogen content (Figure 5) can be used as a process indicator. Although methane is one of the final gases of AD and is therefore accumulated, H_2 is an intermediate gas that appears and subsequently transforms into a final compound. The hydrogen ratio of the biogas generated increases in the first few days when hydrolysis, acidogenesis and acetogenesis occur, until the peak is reached as soon as the latter ends. When methanogenesis begins, the ratio of H_2 should drop as hydrogen is transformed into methane by hydrogenotrophic route.

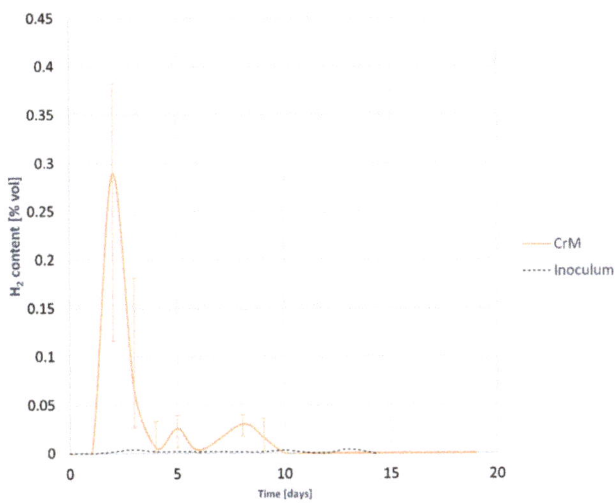

Figure 5. Hydrogen content mean curve from the biogas produced in the anaerobic digestion of 100 g of residue CrM.

As noted, there are the two peaks described above. The first and most pronounced occurs on day 3 with a hydrogen content of 0.289 (±9.903%) %, and the second occurs on day 9 reaching a hydrogen content of 0.033 (±349.718%) %. The occurrence of these two peaks reconfirms the possibility of two-phase digestion or a slowdown of the process that is subsequently recovered. To determine the reasons why this occurs, and to give a definitive explanation for the development of the process, we go on to study together all the generation curves together with the pH evolution curve, and to analyze them, taking into account the changes in the composition of the mixture at the end of the digestion process.

3.7. Assessment of the Evolution of the Anaerobic Process

Figure 6 represents the biogas, methane and hydrogen generation curves, along with the pH evolution recorded during the CrM residue AD process, and the process as a whole, can be evaluated. It is observed that:

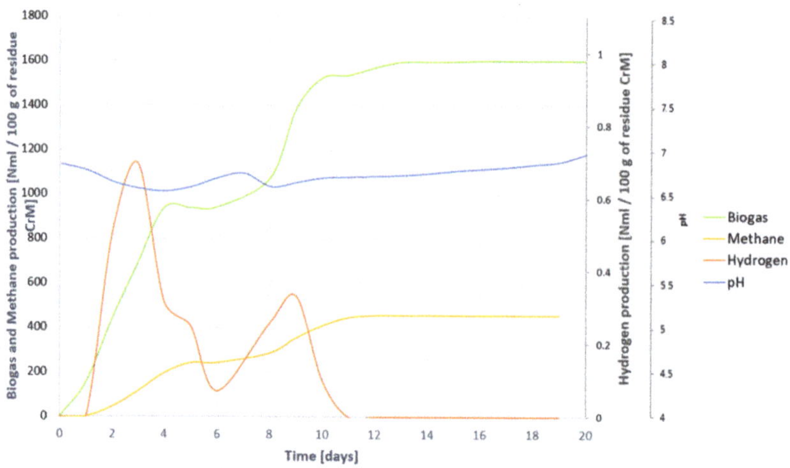

Figure 6. Evolution of the digestion process of residue CrM. Comparison of the generation of biogas, methane and hydrogen together with the evolution of pH.

During the first day (day 0–day 1), there is a delay in the generations without a trace of methane or hydrogen, so it is assumed that the phase of disintegration + hydrolysis occurs during the first day.

Between days 1 and 3 hydrogen generation begins until it reaches a peak on day 3. This means that during days 1 to 3 the phases of acidogenesis and acetogenesis occur. In fact, it is associated with a reduction in pH, indicating that acidic elements such as acetic acid and VFA have formed. At that point biogas and methane begin to appear. The growth of methane is slower (with a lower slope) as it is formed only by acetoclastic route.

During days 3–4 hydrogen disappears at a steady rate as methane and biogas continue to grow, indicating that methane is also generated by acetoclastic methanogenesis.

Between days 4 and 5 the reduction of hydrogen stops, and so do biogas and methane, so it is assumed that there is a slowdown of the process and not an inhibition, since it is subsequently resumed. In addition, being associated with a decrease in pH is assumed to be a slowdown by accumulation of acid elements, which is subsequently resumed when the pH is increased again during days 5 to 6. This is due to the small content in AN that is released and acts as a buffer, as expected as studied in Section 3.1.

From day 6 the second peak of generation and hydrogen appears, which is accompanied by a decrease in pH, so it is understood that the phases of acidogenesis and acetogenesis are resumed and the process of digestion resumes, In fact generations of biogas and methane are also resumed, which occurs mainly by acetoclastic route as the pH is increased by the transformation of the acid elements into methane.

From day 9 hydrogen reduction begins, so it is understood that hydrogenotrophic methanogenesis begins. This is confirmed by the generations of biogas and methane increasing its speed and slope, so methanogenesis is two-way, acetoclastic (pH increases) and hydrogenotrophic (hydrogen content is reduced).

On day 11 hydrogen is completely consumed, continuous methanogenesis by acetoclastic, pH is progressively increased to neutral values.

The process is stabilized on day 12–13 and ends without any inhibition, with correct pH parameters and with an evolution as expected and studied.

For all of the above, the process develops correctly, with a slight acidification by accumulation of acidic elements, but without impacts when it is dampened by the buffer effect of the AN released during the degradation of the small percentage of proteins. There are two phases of digestion, clearly identified by hydrogen peaks and the two pH changes.

It is therefore assumed that during the first phase the organic matter is digested directly accessible, and during the second stage part of the non-soluble organic matter begins to hydrolyze.

3.8. Determinations and Mathematical Adjustment of The Anaerobic Process

Depending on the COD values determined before the BMP test and once it has been completed, therefore the degraded COD values are known, along with the values obtained in terms of methane generation, the process can be mathematically adjusted (Table 4) as studied in Section 2.5, to extrapolate and compare it with other scenarios.

Table 4. Results obtained in mathematical processing of the parameters of waste CrM biodegradation.

		Standard Deviation	**Relative Error**
Theoretical methane generation	487.87 NmL/100 g of residue CrM	145.08	97.87%
Maximum methane generation (obtained)	458.82 NmL/100 g of residue CrM	143.19	106.02%
Disintegration constant	0.16 days^{-1}	0.05	0.03%
Substrate biodegradation [%]	15.28	2.83	1.91%

Based on the initial and final COD values in the reactor mixture, the theoretical production of methane is determined, that is, what is expected to be obtained if the process had been developed correctly and all that degraded COD would have been transformed into methane. This is estimated at 487.879 NmL of methane per 100 g of CrM residue introduced into the digester. The theoretical generation is only 6.396% higher than the average generation obtained, so it is understood that, within the degraded COD, the process has been complete since virtually all the COD consumed in methane has been transformed.

By adjusting the methane generation process to a kinetics of the first order, the maximum generation of methane obtained in all trials, quantified at 458.820 (±106.029%) is obtained. NmL of methane per 100 g of CrM residue, practically identical to the average generation of methane obtained. The disintegration constant, which provides information on the speed of the process and especially hydrolysis, can also be determined. In this case it is estimated at 0.164 days^{-1}, a fairly fast constant compared to other constants of disintegration of other elements, and that allows to compare the speed of the process with that of other substrates. This disintegration constant is an indication of the speed of hydrolysis, but not the depth of the process, for this it is necessary to go to another parameter.

It has been seen that of all the COD digested, almost everything has been transformed into methane, so the process has been robust, and has also been rapid according to the hydrolysis constant, but now arises the question of whether the process has been deep and complete, that is, whether much of the organic matter available in the substrate has been degraded in the substrate, or only a small part has been degraded. To do this, the degradation coefficient of the substrate that provides information on how much COD of the substrate has been biodegraded in the anaerobic process is analyzed. In this case it has been determined at 15.282%, that is, although almost all the degraded COD has been transformed into methane, with respect to the total COD available on the CrM substrate only 15.282%. This indicates that the COD_f which is directly accessible to microorganisms has probably been degraded, and subsequently, in the second stage of digestion, part of the COD that is encapsulated in the substrate is digested. It should be remembered that because of its high content of hemicellulose and lignin is a particulate substrate with strong outer membranes that makes it difficult to hydrolysate and release OM. In addition, it should be borne in mind that the nature of UASB sludge is granular, specially thought for wet digestion, combining two factors, that the substrate is strongly particulate by the high content of hemicellulose and lignin, and the granular nature of the inoculum, which complicates the adhesion of hydrolytic enzymes to the substrate. These low levels of degradation would also explain the methane content below 60%. These results open the door to process improvement through techniques such as pre-treated, which improve accessibility to the substrate, breaking the barriers created by lignocellulose and making the particulate and encapsulated COD solubilized and therefore directly accessible.

3.9. Comparison of Results with Previous Literature

This section compares the results obtained with some of previous research. This is intended to check whether the test conditions have been adequate, whether the use of the new process indicators can be reliable, and whether the methodology applied is valid. They also serve to leave the main innovations developed in this manuscript marked. Publications with a considerable time lapse and similar test characteristics have been selected, however they are not the same, since precisely one of the novelties introduced is the use of UASB sludge as an inoculum. The approach of most research is microbiological, not process development, so it is also another point in favor of innovation and original input.

Table 5 shows the results of other research, and in the variables that can be checked, the results are quite similar. It is then confirmed that the new methodology used can be assumed as correct, that the changes introduced are valid and that new information is provided about:

- Complete characterization of the substrate and in-depth study of the process development.
- UASB sludge can be used as a source of inoculum.
- Hydrogen is a reliable indicator of process development.
- The mathematical determinations developed are accurate and their results provide valuable information and complete the characterization of the process.

Table 5. Results obtained from bibliographic analysis, to be compared to the results of this research.

Compositional Studies							
	TS [%]	**VS** [%]	**COD** * [mg O_2/g—mL]	**TKN** * [mg N/g—mL]	**pH**	**TA** * [mg $CaCO_3$/g—mL]	**Reference**
Malt	17.35	16.76	204.40	9.18	6.40	-	[39,40]
Malt	25.4	25.0	168.80	-	5.4	5.50	
Sloe	32.3	31.3	260.9	-	6	7.1	[41]
Average	25.01	24.35	211.36	9.18	5.93	6.3	

Anaerobic digestion process in similar conditions					
	TS removal [%]	**Biogas yield** * [NmL/100 gr residue]	**Methane Content** [%]	k_{dis} [days^{-1}]	**Reference**
Malt (35%)	−41.1	838	25.2	-	[39]
Malt (35%)	−46.7	914	24.5	-	
Malt (55%)	−54.8	1370	36.6	-	
Malt (55%)	−64.4	876	23.7	-	[41]
Malt (1:3)	-	-	-	0.169	
Malt (1:3)	-	-	-	0.169	
Malt (1:3)	−69	-	-	0.188	
Average	−51.75	999.5	27.5	0.175	

* Some of the results may vary from the literature review as they were originally expressed in different units from the ones used in this manuscript. that is, NmL per gram of volatile soled degraded (unit used in other manuscript), instead of gram of volatile solid content in the residue (unit used in this manuscript).

3.10. Study of Full-Scale Application Potential: Energetic Feasibility and Environmental Suitability of the Solution. Releasing Its Potential in Spain

Once the laboratory tests have been developed, the potential of the solution is studied and whether it is applicable on a real scale. To do this, a study will be carried out at the Spanish level, estimating the amount of CrM waste that is generated in each Autonomous

Community. Subsequently, and with laboratory results, the amount of energy that is likely to be generated in each region will be estimated, and if once the needs of anaerobic reactors are met, energy is available for external use, for example for residential use, quantifying the number of homes that are estimated to be supplied. In addition, to check whether there is an environmental benefit, the reduction in emissions involved in the use of generated biogas rather than a conventional energy source such as natural gas is quantified.

3.10.1. Estimation of CrM Waste Generation

To estimate the amount of CrM waste generated, the statistics of the *Survey on Areas and Crop Yields (ESYRCE)* of the Ministry of Agriculture of Spain are thus determined the number of hectares cultivated of barley in each region, and their yield [42]. Taking into account the number of hectares of barley used for malt production, and the performance of this process, as specified in the "Guide to Technical Improvements Available in Spain of the Maltese processing sector" [43] obtains the amount of waste obtained in malting, which is identified with the CrM residue. These results are shown in Table 6, separating them by region, and next to a color scale map representing the distribution of crops in the Spanish geography.

Table 6. Estimation of CrM waste generation based on data from cultivated barley hectares from [43].

	Barley Cultivation [kha] [43]	**CrM Generation [kton]**
Andalusia	71.7	18.9
Asturias	0	0
Aragon	373.8	98.7
Balearic Islands	10	2.6
Canary Islands	57	0.1
Cantabria	57	0.1
Castile-La Mancha	584.8	154.4
Castile and Leon	278	73.4
Catalonia	168.1	44.4
Valencian Community	8.5	2.2
Extremadura	34.3	9.1
Galicia	0.2	0.1
La Rioja	17.1	4.5
Community of Madrid	24.6	6.4
Murcia	6.2	1.6
Navarre	100.8	26.6
Basque Country	18.1	4.7

3.10.2. Estimation of Biogas and Methane Generation in Each Region

Once estimated, CrM production is estimated at the amount of gas and methane that would be generated if this residue is digested in anaerobic digesters. For this purpose, the data obtained in the previous sections, summarized in Table 6, are used. The results of the estimation of biogas and methane for each region of the Spanish geography are shown in Table 7. Logically these results are proportional to those of the amount of CrM waste generated, and in those areas where barley production is much higher, more biogas is produced.

Table 7. Estimation of the biogas and methane generated with the anaerobic digestion of CrM.

	Biogas Generation [Nm^3/year]	Methane Generation [Nm^3/year]
Andalusia	303,724.9	83,481.8
Asturias	0	0
Aragon	1,583,362.6	435,203.1
Balearic Islands	42,352.1	11,640.9
Canary Islands	242	66.5
Cantabria	242	66.5
Castile-La Mancha	2,476,991.5	680,825.9
Castile and Leon	1,177,387.7	323,616.8
Catalonia	712,119.9	195,733.3
Valencian Community	36,301.7	9977.9
Extremadura	145,449.1	39,978.1
Galicia	1210.1	332.6
La Rioja	72,603.5	19,955.8
Community of Madrid	104,186.1	28,636.6
Murcia	26,621.3	7317.1
Navarre	427,150.9	117,406.7
Basque Country	76,959.7	21,153.1
Relative error	19.980%	21.421%

3.10.3. Estimation of the Energy Balance and Energy Available for External Uses in Each Region

The calculations in this section have been developed as explained in Section 2.6. It has been assumed that biogas is generated in anaerobic digesters of 4500 m^3 capacity, and thus determine the heat and electricity needs of the installation. Once covered, it is calculated whether there is excess energy or net available energy. Figure 7 shows the amount of energy that can be extracted from the biogas generated in each region, the needs required by the installation, and the excess energy that is available for external uses. Table 8 collects all this data, determines the power of a plant equivalent to that generation and to give more applicability to the results is estimated how many homes could be sufficient in one year with the available energy.

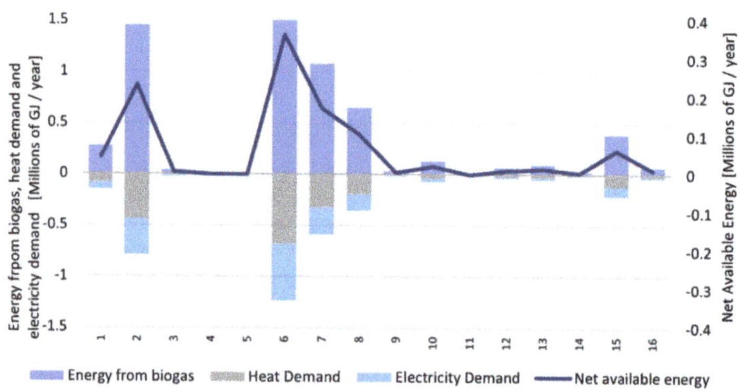

Figure 7. Energy balance of each region when treating CrM residue by anaerobic digestion.

Table 8. Estimation of the energy balance of CrM's anaerobic digestion.

	Energy from Biogas [GJ/year]	Energy Needs [GJ/year]		Net Available Energy [GJ/year]	Energy Efficiency of the AD Process [%]	Equivalent Power Plant[MW]	Supplied Homes
		Heat Demand	Electricity Demand				
Andalusia	278,829.3	83,987.9	68,158.3	43,703.2	15.6%	13.8	1228
Asturias	0	0	0	0	0	0	0
Aragon	1,453,578.2	434,587.7	355,319.4	231,659.6	15.9%	73.4	6507
Balearic Islands	38,880.5	12,376.1	9504.1	5312.1	13.6%	1.6	149
Canary Islands	222.1	838.6	54.3	−873.0	−392.9%	−0.2	
Cantabria	222.1	838.6	54.3	−873.0	−392.9%	−0.2	
Castile-La Mancha	2,273,958.4	679,427.4	555,857	362,918.0	15.9%	115.1	10,194
Castile and Leon	1,080,880.1	323,357.2	264,215.3	172,029.0	15.9%	54.5	4832
Catalonia	653,749.1	195,881.5	159,805.4	103,689.3	15.8%	32.8	2913
Valencian Community	33,326.2	10,718.4	8146.4	4423.4	13.2%	1.4	124
Extremadura	133,527.0	40,622.9	32,639.9	20,455.3	15.3%	6.4	575
Galicia	1110.8	1103.8	271.5	−730.8	−65.7%	−0.2	
La Rioja	66,652.4	20,664.5	16,292.8	9755.5	14.6%	3.0	274
Community of Madrid	95,646.2	29,317.6	23,380.2	14,394.5	15.0%	4.5	404
Murcia	24,439.2	8066.1	5974	3001.5	12.2%	0.9	84
Navarre	392,138.4	117,804.6	95,856.1	61,832.3	15.7%	19.6	1737
Basque Country	70,651.5	21,858.1	17,270.3	10,395.4	14.7%	3.2	292

As can be observed, the region that is the most energy capable of obtaining is Castille-La Mancha, which with this solution could become a generating power plant of 115 MW, followed by Aragon and Castile and Leon. The regions that would take the least advantage of this solution are Cantabria, Galicia and Canary Islands, which are precisely the regions that produce the least barley. Precisely for this reason the solution is not energy effective, when generating little amount of barley, the substrate level is low, little biogas is generated and is not enough to meet the needs of the reactors. The energy profitability limit can be set in the case of Murcia, that is, the solution is interesting from an energy point of view in regions that cultivate at least around 6000 hectares of barley and collect at least 1500 tons of CrM waste. As for the amount of homes that could be supplied, this solution is especially attractive for the regions with the highest available energy. For example, Castile-La Mancha could supply more than 10,000 homes in a year

This could be a solution to lack of energy supply or difficult access to energy in certain areas. The Energy Poverty report in Spain 2018 [44] of the Association of Environ-mental Sciences places the regions of Valencian Community, Murcia, Cantabria, Andalusia and Castile-La Mancha, in this order, as the areas where there is the most inequality in terms of household warming, so this solution could be of help in these regions, with the exception of Cantabria, which, as has been seen, is one of the regions in which this solution is not profitable. It is also interesting for isolated areas, such as the Balearic Islands, as it is an autonomous generation system that takes advantage of local resources.

With this proposal, renewable energy is obtained from a waste to be disposed of, thus meeting the Sustainable Development Goals (SDG) #1 (in terms of energy poverty) #7 on affordable and non-polluting energy, #10 reducing inequalities in access to energy, #12 in sustainable waste management, and #13 of climate action by providing a renewable energy source that reduces emissions into the atmosphere, as demonstrated in the following section.

3.10.4. Estimate of the Reduction in Emissions Involved in the Use of Biogas Generated Instead of a Conventional Source of Natural Gas

Precisely, for the above reason of the impact on the SDG #13, and since this solution fits perfectly into these goals, it is decided to study the environmental benefit that comes with, in terms of reducing CO_2 emissions equivalent. The calculations have been developed in accordance with Section 2.7 of the methodology, and the results highlight the importance of the use of biogas rather than conventional sources such as natural gas.

The results are shown in Table 9, and it is confirmed that there is a 55.4% emission reduction if biogas methane is used to generate energy, rather than natural gas. Specifically, it is a reduction of 38,060 tons of CO_2 equivalent at the Spanish level, being more notable in the areas where the most biogas has been generated, with Castile-La Mancha at first position.

It is important, in turn, to take into account that straws are considered biomass according to the Commission Regulation (EU) No 601/2012 of 21 June 2012 on the monitoring and reporting of greenhouse gas emissions [...] [45], so that if they are issued in the combustion process the emissions can be considered zero, which represents a reduction in the impact of global warming in terms of the use of this fuel of 100%, quantified in approximately 68,750 tons of CO_2 equivalent per year less emitted into the atmosphere.

Table 9. Estimation of the biogas and methane generated with the anaerobic digestion of CrM.

	Conventional Energy Source		Energy from the Generated Methane		Emissions Reduction [tons CO_2eq/year]	
	Natural Gas [Nm^3/year]	Emissions [tons CO_2eq/year]	Methane Gas [Nm^3/year]	Emissions [tons CO_2eq/year]	Considering Methane Emissions	Considering Zero Emissions from Methane (Biomass)
Andalusia	1,339,157.2	2879.2	1,434,186.9	1285.4	−1593.8	−2879.2
Asturias	0	0	0	0	0	0
Aragon	7,098,524.5	15,262.2	7,602,252.4	6813.5	−844.7	−15,262.2
Balearic Islands	162,775.8	349.9	174,326.7	156.2	−193.7	−349.9
Canary Islands	0	0	0	0	0	0
Cantabria	0	0	0	0	0	0
Castile-La Mancha	11,120,550.8	23,909.8	11,909,691.1	10674	−13,235.8	−23,909.9
Castile and Leon	5,271,321	11,333.6	5,645,386.3	5059.6	−6274	−11,333.6
Catalonia	3,177,253.1	6831.3	3,402,718.5	3049.6	−3781.6	−6831.3
Valencian Community	135,544.7	291.4	145,163.3	130.1	−161.3	−291.4
Extremadura	626,792.9	1347.6	671,271.6	601.6	−746	−1347.6
Galicia	0	0	0	0	0	0
La Rioja	298,931	642.7	320,143.9	286.9	−355.8	−642.7
Community of Madrid	441,077.1	948.3	472,377	423.3	−524.9	−948.3
Murcia	91,975	197.7	98,501.8	88.2	−109.4	−197.7
Navarre	1,894,670.6	4073.6	2,029,120.9	1818.6	−2255	−4073.6
Basque Country	318,537.4	684.8	341,141.5	305.7	−379.1	−684.8
Total reduction of emissions					−38,059.6	−68,752.8
Total reduction of emissions (Compared to the conventional energy source)					−55.3%	−100%

4. Conclusions

This study has determined that the composition of the CrM residue presents it as a residue formed, for the most part, by carbohydrates. It has good solubility and is relatively resistant to pH changes. Despite these positive aspects, its high content of hemicellulose and lignin may compromise the proper development of AD.

It has been determined that the process develops correctly, with a slight accumulation of VFA without impact when it is dampened by the buffer effect of the released AN. With the novelty introduced of the analysis of the evolution of hydrogen it has been possible to determine that the process occurs in two stages, digesting in the first stage the soluble COD, and in the second phase that of the COD that is particulate and not directly accessible. Methanization has been correct, transforming almost all the digested COD into methane. However, only 15.282% of all available COD is digested, due to the strong particulate characteristic of the substrate and the high content in lignin and hemicellulose, which makes it difficult for the for hydrolytic enzymes to adhere to particle membranes and solubilize the enclosed organic compounds; and is also due to the granular nature of the UASB inoculum. This explains why the methane content of the generated biogas is not very high.

Numerically, 100 g of CrM residue generates 1604.22 NmL of biogas, with a methane content of 27.485%. The disintegration constant is 0.164 days^{-1} and only 15.282% of the substrate is digested. This low level of degradation opens the door to process improvement through techniques such as pre-treatments, which improve accessibility to the substrate, breaking the barriers created by lignocellulose.

As for its applicability at the national scale, it is a particularly interesting solution from the point of view of energy. It has been determined that this solution is beginning to be energy-effective and therefore to produce enough energy available for external uses, in areas that have at least 6000 hectares of planted barley and collect 1500 tons of CrM waste. At best, it can be considered that this solution provides, in a given region, the equivalent of a 115 MW power plant, and could supply 10,000 households per year in that region. It is also considered an energy-efficient solution that complies with the SDGs #1, #7, #10, #12 and #13, and can guarantee access to energy in isolated areas or with supply problems. Not only is it an energy-efficient solution, but also, in the case of SDG #13, it has been estimated that its implementation would result in a 55.4% reduction in emissions if it were to replace a conventional natural gas energy source, reducing 38,060 tons of equivalent CO_2 released into the atmosphere at Spanish level (increased to 68,750 tons if zero emissions are considered from the burning of biogas, being a form of biomass).

Author Contributions: Conceptualization, C.M.-P. and M.d.M.C.-C.; methodology, C.M.-P.; validation, M.d.M.C.-C.; formal analysis, C.M.-P.; investigation, C.M.-P., M.d.M.C.-C., M.R.-A., K.H.-K.; resources, C.M.-P., M.d.M.C.-C.; data curation, C.M.-P.; Writing—Original draft preparation, C.M.-P.; Writing—Review and editing, C.M.-P., M.d.M.C.-C.; visualization, C.M.-P., M.d.M.C.-C.; supervision, C.M.-P., M.d.M.C.-C.; project administration, C.M.-P.; funding acquisition, M.d.M.C.-C. All authors have read and agreed to the published version of the manuscript.

Funding: This research received no external funding.

Acknowledgments: Materials and equipment used for experiments were provided by Department of Mechanical Engineering. ICAI School of Engineering. Comillas Pontifical University, Madrid, Spain.

Conflicts of Interest: The authors declare no conflict of interest.

Abbreviations

AD	Anaerobic Digestion
AN	Ammoniacal Nitrogen
ANOVA	Analysis of Variance
BD	Biodegradation
BMP	Biochemical Methane Potential
CHP	Combined Heat and Power
COD	Chemical Oxygen Demand
CrM	Crop Barley Residue (Substrate)
CV	Coefficient of variation
EU	European Union
FAOSTAT	Statistics from the Food and Agriculture Organization of the United Nations
FES-CO_2	Carbon Fund for a Sustainable Economy
FSC	Food Supply Chain
GC	Gas Chromatograph
Hum	Humidity
IA	Intermediate Alkalinity
k_{dis}	Disintegration constant
LCFA	Long Chain Fatty Acid
LPCH	Lipids, Proteins and Carbohydrates content
OM	Organic Matter
ON	Organic Nitrogen
PA	Partial Alkalinity
S	Sludge (Inoculum)
TA	Total Alkalinity
TCD	Thermal Conductivity Detector
TKN	Total Kjeldahl Nitrogen
TS	Total Solids
UASB	Upflow Anaerobic Sludge Blanket reactor
VFA	Volatile Fatty Acid
VS	Volatile Solids
WWTP	Wastewater Treatment Plant

References

1. EUROSTAT; FOODDRINK EUROPE. *Economic Bulletin Q1 2019*; EU: Brussels, Belgium, 2019.
2. EUROSTAT. *Extra-EU Trade in Agricultural Goods—Statistics Explained*; EU: Brussels, Belgium, 2019.
3. Database–Eurostat. Available online: https://ec.europa.eu/eurostat/data/database (accessed on 8 March 2020).
4. Chiu, S.L.H.; Lo, I.M.C. Reviewing the Anaerobic Digestion and Co-Digestion Process of Food Waste from the Perspectives on Biogas Production Performance and Environmental Impacts. *Environ. Sci. Pollut. Res.* **2016**, *23*, 24435–24450. [CrossRef] [PubMed]
5. Parfitt, J.; Barthel, M.; Macnaughton, S. Food Waste within Food Supply Chains: Quantification and Potential for Change to 2050. *Philos. Trans. R. Soc. B Biol. Sci.* **2010**, *365*, 3065–3081. [CrossRef]
6. Food and Agriculture Statisctics (FAOSTAT). Available online: http://www.fao.org/statistics/en/ (accessed on 13 April 2020).
7. EUROSTAT. *Agriculture, Forestry and Fishery Statistics*; EU: Brussels, Belgium, 2019; ISBN 978-92-79-94757-5.
8. Buyanovsky, G.; Wagner, G. Crop Residue Input to Soil Organic Matter on Sanborn Field. In *Soil Organic Matter in Temperate Agroecosystems—Long Term Experiments in North America*; CRC Press: Boca Raton, FL, USA, 2019; p. 73.
9. Kumar, K.; Goh, K. Crop residues and management practices: Effects on soil quality, soil nitrogen dynamics, crop yield, and nitrogen recovery. In *Advances in Agronomy*; Elsevier: Amsterdam, The Netherlands, 1999; Volume 68, pp. 197–319, ISBN 0065-2113.
10. Lal, R. The Role of Residues Management in Sustainable Agricultural Systems. *J. Sustain. Agric.* **1995**, *5*, 51–78. [CrossRef]
11. Mouratiadou, I.; Stella, T.; Gaiser, T.; Wicke, B.; Nendel, C.; Ewert, F.; van der Hilst, F. Sustainable Intensification of Crop Residue Exploitation for Bioenergy: Opportunities and Challenges. *GCB Bioenergy* **2020**, *12*, 71–89. [CrossRef]
12. De Molina, M.G.; Fernández, D.S.; Casado, G.G.; Infante-Amate, J.; Fernández, E.A.; Traver, J.V.; Ruiz, R.G. Environmental Impacts of Spanish Agriculture's Industrialization. In *The Social Metabolism of Spanish Agriculture, 1900–2008*; Springer: Berlin/Heidelberg, Germany, 2020; pp. 153–179.
13. Wych, R.; Rasmusson, D. Genetic Improvement in Malting Barley Cultivars Since 1920. *Crop Sci.* **1983**, *23*, 1037–1040. [CrossRef]
14. U.S. Energy Information. *International Energy Outlook*; U.S. Energy Information: Washington, DC, USA, 2017.
15. Aliane, A.; Abboudi, S.; Seladji, C.; Guendouz, B. An Illustrated Review on Solar Absorption Cooling Experimental Studies. *Renew. Sustain. Energy Rev.* **2016**, *65*, 443–458. [CrossRef]
16. International Energy Agency. *Global Energy & CO2 Status Report*; OECD/IEA: Paris, France, 2017.
17. Ashraf, M.T.; Fang, C.; Bochenski, T.; Cybulska, I.; Alassali, A.; Sowunmi, A.; Farzanah, R.; Brudecki, G.P.; Chaturvedi, T.; Haris, S.; et al. Estimation of Bioenergy Potential for Local Biomass in the United Arab Emirates. *Emir. J. Food Agric.* **2016**, 99–106. [CrossRef]
18. Jeguirim, M.; Limousy, L. Strategies for Bioenergy Production from Agriculture and Agrifood Processing Residues. *Biofuels* **2018**, *9*, 541–543. [CrossRef]
19. De Sanctis, M.; Chimienti, S.; Pastore, C.; Piergrossi, V.; Di Iaconi, C. Energy Efficiency Improvement of Thermal Hydrolysis and Anaerobic Digestion of Posidonia Oceanica Residues. *Appl. Energy* **2019**, *252*, 113457. [CrossRef]
20. Morales-Polo, C.; Cledera-Castro, M.D.M.; Moratilla Soria, B.Y. Reviewing the Anaerobic Digestion of Food Waste: From Waste Generation and Anaerobic Process to Its Perspectives. *Appl. Sci.* **2018**, *8*, 1804. [CrossRef]
21. Morales-Polo, C.; Cledera-Castro, M.d.M.; Moratilla Soria, B.Y. Biogas Production from Vegetable and Fruit Markets Waste—Compositional and Batch Characterizations. *Sustainability* **2019**, *11*, 6790. [CrossRef]
22. Gujer, W.; Zehnder, A.J.R. Conversion Processes in Anaerobic Digestion. *Water Sci. Technol.* **1983**, *15*, 127–167. [CrossRef]
23. Hawkes, F.R. The biochemistry of anaerobic digestion. In *Biomethane: Production and Uses*; Roger Bowskil Printing Ltd.: Exeter, UK, 1980; pp. 41–60.
24. Owens, J.M.; Chynoweth, D.P. Biochemical Methane Potential of Municipal Solid Waste (MSW) Components. *Water Sci. Technol.* **1993**, *27*, 1–14. [CrossRef]
25. VDI. *VDI 4630 Fermentation of Organic Materials. Characterisation of the Substrate, Sampling, Collection of Material Data, Fermentation Tests*; Verlag des Vereins Deutscher Ingenieure: Düsseldorf, Germany, 2016.
26. Elbeshbishy, E.; Nakhla, G.; Hafez, H. Biochemical Methane Potential (BMP) of Food Waste and Primary Sludge: Influence of Inoculum Pre-Incubation and Inoculum Source. *Bioresour. Technol.* **2012**, *110*, 18–25. [CrossRef] [PubMed]
27. Li, Y.; Chen, Y.; Wu, J. Enhancement of Methane Production in Anaerobic Digestion Process: A Review. *Appl. Energy* **2019**, *240*, 120–137. [CrossRef]
28. Owen, W.; Stuckey, D.; Healy, J.; Young, L.; Mccarty, P. Bioassay for Monitoring Biochemical Methane Potential and Anaerobic Toxicity. *Water Res.* **1979**, *13*, 485–492. [CrossRef]
29. Chynoweth, D.P.; Turick, C.E.; Owens, J.M.; Jerger, D.E.; Peck, M.W. Biochemical Methane Potential of Biomass and Waste Feedstocks. *Biomass Bioenergy* **1993**, *5*, 95–111. [CrossRef]
30. Granular Sludge Formation in Upflow Anaerobic Sludge Blanket (UASB) Reactors-Schmidt-1996-Biotechnology and Bioengineering-Wiley Online Library. Available online: https://onlinelibrary.wiley.com/doi/abs/10.1002/(SICI)1097-0290(19960205)49:3%3C229::AID-BIT1%3E3.0.CO;2-M (accessed on 8 March 2020).
31. AENOR. *UNE-EN ISO 11734:1999 Calidad Del Agua. Evaluación de La Biodegradabilidad Anaerobia "Final" de Los Compuestos Orgánicos Con Lodos En Digestión. Método Por Medida de La Producción de Biogas. (ISO 11734:1995)*; AENOR: Madrid, Spain, 1999.
32. Veeken, A.; Hamelers, B. Effect of Temperature on Hydrolysis Rates of Selected Biowaste Components. *Bioresour. Technol.* **1999**, *69*, 249–254. [CrossRef]

33. Morales-Polo, C.; del Mar Cledera-Castro, M.; Hueso-Kortekaas, K.; Revuelta-Aramburu, M. Anaerobic Digestion in Wastewater Reactors of Separated Organic Fractions from Wholesale Markets Waste. Compositional and Batch Characterization. Energy and Environmental Feasibility. *Sci. Total Environ.* **2020**, *726*, 138567. [CrossRef]
34. FES-CO2; Ministerio de Alimentación y Medio Ambiente. *Metodología Para Los Proyectos de Energía Térmica Destinados a La Reducción Del Consumo de Combustibles Fósiles En Una Instalación Nueva o Ya Existente*; Ministerio de Alimentación y Medio Ambiente: Madrid, Spain, 2014.
35. Beniche, I.; Hungría, J.; El Bari, H.; Siles, J.A.; Chica, A.F.; Martín, M.A. Effects of C/N Ratio on Anaerobic Co-Digestion of Cabbage, Cauliflower, and Restaurant Food Waste. *Biomass Conv. Bioref.* **2020**. [CrossRef]
36. El-Shinnawi, M.M.; El-Din, M.N.A.; El-Shimi, S.A.; Badawi, M.A. Biogas Production from Crop Residues and Aquatic Weeds. *Resour. Conserv. Recycl.* **1989**, *3*, 33–45. [CrossRef]
37. Neshat, S.A.; Mohammadi, M.; Najafpour, G.D.; Lahijani, P. Anaerobic Co-Digestion of Animal Manures and Lignocellulosic Residues as a Potent Approach for Sustainable Biogas Production. *Renew. Sustain. Energy Rev.* **2017**, *79*, 308–322. [CrossRef]
38. Schievano, A.; D'Imporzano, G.; Adani, F. Substituting Energy Crops with Organic Wastes and Agro-Industrial Residues for Biogas Production. *J. Environ. Manag.* **2009**, *90*, 2537–2541. [CrossRef]
39. Kang, H.; Weiland, P. Ultimate Anaerobic Biodegradability of Some Agro-Industrial Residues. *Bioresour. Technol.* **1993**, *43*, 107–111. [CrossRef]
40. Mohan, B.; Malleshi, N. Characteristics of Native and Enzymatically Hydrolyzed Common Wheat (Triticum Aestivum) and Dicoccum Wheat (Triticum Dicoccum) Starches. *Eur. Food Res. Technol.* **2006**, *223*, 355–361. [CrossRef]
41. De Diego-Díaz, B.; Fernández-Rodríguez, J.; Vitas, A.I.; Peñas, F.J. Biomethanization of Solid Wastes from the Alcoholic Beverage Industry: Malt and Sloe. Kinetic and Microbiological Analysis. *Chem. Eng. J.* **2018**, *334*, 650–656. [CrossRef]
42. Ministerio de Agricultura Pesca y Alimentación. *Encuesta Sobre Superficies y Rendimientos Cultivos (ESYRCE). Encuesta de Marco de Áreas de España*; Ministerio de Agricultura Pesca y Alimentación: Madrid, Spain, 2019.
43. Ministerio de Medio Ambiente y Medio Rural y Marino. *Guía de Mejores Técnicas Disponibles en España del Sector de Elaboración de Malta*; Ministerio de Medio Ambiente y Medio Rural y Marino, Centro de Publicaciones: Madrid, Spain, 2009.
44. Asociación de Ciencias Ambientales. *Pobreza Energética En España 2018*; Asociación de Ciencias Ambientales: Madrid, Spain, 2018.
45. The Commision of the European Communities. *Commission Regulation (EU) No 601/2012 of 21 June 2012 on the Monitoring and Reporting of Greenhouse Gas Emissions Pursuant to Directive 2003/87/EC of the European Parliament and of the Council Text with EEA Relevance*; EU: Brussels, Belgium, 2012.

agronomy

MDPI

Review

Consolidated Bioprocessing, an Innovative Strategy towards Sustainability for Biofuels Production from Crop Residues: An Overview

Edgar Olguin-Maciel [1,†], Anusuiya Singh [2,†], Rubi Chable-Villacis [1], Raul Tapia-Tussell [1,*] and Héctor A. Ruiz [2,*]

1 Renewable Energy Department, Yucatan Center for Scientific Research, Merida 97302, Mexico; eolmac.agro@gmail.com (E.O.-M.); rubi.chable@cicy.mx (R.C.-V.)

2 Biorefinery Group, Food Research Department, Faculty of Chemistry Sciences, Autonomous University of Coahuila, Saltillo 25280, Mexico; anusuiyasingh@uadec.edu.mx

* Correspondence: rtapia@cicy.mx (R.T.-T.); hector_ruiz_leza@uadec.edu.mx (H.A.R.); Tel.: +52-999-930-0760 (R.T.-T.); +52-844-416-1238 (H.A.R.)

† These authors contributed equally to this work.

Received: 12 November 2020; Accepted: 19 November 2020; Published: 22 November 2020

check for updates

Abstract: Increased energy demands in today's world have led to the exploitation of fossil resources as fuel. Fossil resources are not only on the verge of extinction but also causing environmental and economic issues. Due to these reasons, scientists have started focusing their interest on other eco-friendly processes to biofuel and recently, second-generation biorefinery is gaining much more attention. In second-generation biorefinery, the main objective is the valorization of lignocellulosic biomass cost-effectively. Therefore, many scientists started different bioprocessing techniques like Consolidated Bioprocessing (CBP) to produce ethanol by using a single or plethora of microorganisms to produce ethanol in a single process. In this review, in-depth study on CBP is assessed as well as biofuel's socio-economic value and a brief study of biorefineries. The study not only involves innovative approaches used in CBP but their effect on society and economic aspects.

Keywords: biomass; biorefinery; circular economy; enzyme; consolidated bioprocessing

1. Introduction

The finite nature and rapid depletion of fossil fuels due to growing global energy demands have negatively impacted the environment as their combustion entails have led to the search for alternative ways of producing fuels from sustainable, renewable, green, and economical energy sources [1–5]. One source of renewable energy production is biomass, which can be transformed into biofuels, which provide many advantages over fossil fuels in terms of mitigating residues generation and greenhouse gas (GHG), increasing energy independence, and improving the economy of agriculturalists [1,2,6,7].

The term biofuel refers to any liquid, gas, or solid fuel predominantly produced from a renewable biomass feedstock. Biofuel examples are bioethanol, biomethanol, biosynthetic gas (biosyngas), biodiesel, biogas (biomethane), biochar, bio-oil, biohydrogen, and Fischer_Tropsch produced liquids [1,8]. Biofuels serve as a bridge between the agricultural and energy markets as agricultural commodities are the significant feedstocks in biofuel production [9].

Recently, crop residues as a potential source of feedstock to produce bioenergy have been gaining importance as they are not in competition with food production for human consumption. They have high availability, wide distribution, and almost zero cost [1,10,11]. Crop residues are generally the waste generated in the harvesting of any agricultural crop. These are typically the parts of the products that are often of no value, including the stem, empty fruit bunches, leaves, and stalks. The amount of agricultural waste is variable and reaches 50% for specific crops [12]. In the production of commodities like sugar, rice, flour, starch, and oil, only a fraction of the main crop is recovered and used. The leftover fraction of the processed products are referred to as agro-industrial crop residue. Generally, these residues are either bagasse, molasses, the spent fibrous pulp left behind from stalks of crops, or seed coats, shells, and husks [1,12]. The optimal use of agricultural wastes has excellent economic and ecological advantages due to the possibility of recycling and producing materials with added value [2]. Wheat straw, rice straw, corn straw, and sugarcane bagasse are the agricultural residues available in the approximate amounts of 354.34, 731.30, 128.02, and 180.73 million tons, respectively. The primary composition of these residues is cellulose (20 to 60%), hemicellulose (20 to 40%), and lignin (10 to 25%) [1,13].

Production and use of biofuels made from crop residues could prevent the over-exploitation of fossil fuel and low waste management problems in the field. However, this potential can only be unlocked if they are cost-competitive to petroleum and starch and sucrose-based biofuels. To reduce capital and processing costs, simplifying the process scheme, and integrating as many unit operations as possible is necessary. Therefore, the consolidated bioprocessing CBP is a strategy where all four steps occur in a single reactor. A single microorganism or microbial consortium converts pretreated biomass to a commodity product such as ethanol without adding saccharolytic enzymes, which would represent an innovative breakthrough for low-cost biomass processing. In principle, a CBP strategy can be applied to produce a broad range of chemicals from natural biomass. It requires degrading recalcitrant biomass substrates into solubilized sugars and metabolic intervention to direct metabolic flux toward desired products at high yield and titer [14–16].

The increase in population brought a greater demand for food and energy; this requires an increase in agricultural production and an increase in agricultural waste, which presents enormous challenges for its proper management. Although crop residues represent a potential source for the generation of biofuels, the environmental impacts that their use could cause must also be critically evaluated [16]. Therefore, this work aims to briefly describe the current situation of biofuels, available raw materials, and their production strategies, especially the CBP approach, a strategy that is aimed at the best option for the sustainable production of biofuels.

2. Biofuels

2.1. Worldwide Scenario of Biofuel Production

We are strongly dependent on fossil fuels due to the extensive use and utilization of petroleum derivatives limiting the use of petroleum resources, resulting in environmental and political issues [17]. Questions regarding ecological stagnation and unpredictability dependent on the future reservoir and increasing oil prices have been inspiring methods relying on alternative energy sources [18]. At a national, regional, and global level, there is acceptance of plants' raw material (i.e., biomass) at an increased level. It can reinstate a large portion of fossil resources as feedstock for the industrial generation of both the energy and non-energy (i.e., chemicals and materials) area [17]. It has been reported to be approximately 2.7% of global transportation (~2.9 TW) that is dominated by first-generation biofuels. Ethanol derived from sugarcane and cornstarch (28 billion gallons in 2018) and biodiesel from rapeseed or soybean oils (8 billion gallons in 2018). These raw materials are also a segment of the food chain supply and do not serve a long-term, high-scale solution; therefore, first-generation biofuels will not be a considerable section of the transportation fuel supply in 2050. Therefore, it is essential to replace the 1G biofuels gradually with lignocellulosic biofuels [18].

2.2. Challenges to Biofuel Production

2.2.1. Policy Initiatives

Several problems are faced in the production of biofuels, among which the production costs above the fluctuating price of fossil fuels stand out, and the ecological challenges during biofuels production. The challenges faced are explained in this section.

Various industry considerations (climate change, pricing ambiguity, and geopolitical uncertainty) are taken into account to explore the utility of substituting biofuels for traditional petroleum-based fuels as a clean energy alternative. Therefore, the U.S. renewable fuel standard (RFS_2) was mainly designed to motivate biofuel production, relying on reducing the greenhouse gas emission related to petroleum fuels [19].

Different types of biofuels produced at a global level have overcome 35 billion gallons during transportation in 2015, and by 2023 will reach 62 billion gallons. The International Energy Agency (IEA) evaluate that biofuels have the probability of supplying continuously from their ongoing share of 3% to about 27% of total transportation by liquid [20].

At this moment, our world´s main objective is to maintain economic development without increasing environmental degradation. Bio-based industries can dramatically alleviate pollution and ecological damage after the industrial revolution [21]. Twenty years ago, Rio's acknowledgment of environment and development took the first step to solve this unorganized scheme for the future. In this session, several countries reviewed the matter of social, scientific, and technological development. The use of agricultural residues was one of the activities affecting all the fields [22]. This valorization consolidates the utilization of second-generation biorefineries.

2.2.2. Market Challenges

One of the dominant objectives in manufacturing renewable fuels is encompassing the target price of 0.79 $\$L^{-1}$ [23], agreed by the Bioenergy Technologies Office (BETO). Scientists can enhance renewable fuel generation's industrial growth by utilizing non-profitable biomass such as forest residues as a feedstock and producing yields from selling co-products [23]. The shortage of fossil fuel has led to increasing concern regarding greenhouse gas emissions and air pollution. This has resulted in rising interest in bioethanol production from natural resources like algal strains, notably from lignocellulosic biomass via enzymatic route. The estimation for biofuel growth by 2020 will be 4% of road transport applications, and according to USDA, global ethanol production will grow to 40% by 2022 [24]. The main aim of the biorefinery is to produce broad-spectrum merchandise in a cost-effective and eco-friendly way. Therefore, second-generation biofuel production is the most feasible process to generate bioethanol in competition with food resources and fertile land. Hence, in the long run, bioethanol production from lignocellulosic material through enzymatic hydrolysis is a valuable process, along with a probability of huge output [25]. It has been reported that 44% cost of biofuel production in 2G ethanol production is from enzymes. The researchers recommended that on-site or near-site enzyme generation encourages access to the remarkable decrease in enzymes' value up to 30–70% carrying its interpreted purification and logistics [25,26]. According to another author, the cost of enzymes at 15% and 35% solid pretreated loading during enzymatic hydrolysis varies from 34.63% to 36.38%, respectively [25,27].

2.2.3. Social and Socio-Economic Issues

The problem that is being faced regarding the depletion of fossil fuels is the involvement in energy security, particularly in the countries that are energy-dependent because of constant clashes in the oil-exporting countries and climate change as a result of the burning of fossil fuels. All these problems are arising due to fossil fuels: society realized biofuels to solve these problems. Specifically, the thriving biofuel sector could administer the opportunity to develop rural areas and generate job opportunities for the local residents, working towards the development of healthy, efficient communities and reducing the emission of greenhouse gases. Recently, some issues regarding biofuel generation and its use have appeared, such as biofuels' effect on energy and food markets, working conditions and workers' rights regarding occupational health, the disparity of biofuel policies, the land change, etc. should also be recognized [28].

2.2.4. Sustainability

For biofuel sustainability, it is necessary to deal with the various complex and conflicting estimates at stake. Hence, the biofuel capability to grant to one particular value cannot guide any definite conclusion regarding biofuel's overall sustainability. The range of the sustainability opinion may differ depending on individuals' inclination, the time scale, and the geographical region. The five pillars of the sustainability theory encompasses social, economic, environmental, legal, and cultural considerations. In a recent study, research proposed estimating various biofuel sustainability objectives for France by 2030, along with a stakeholder-driven perspective; the stakeholders recognize 22 different sustainability standards for biofuels. Therefore, they had a shallow level of agreement between the various professions (feedstock producers, biofuel producers, refining industry, fuel distributors, car manufacturers, end-users, government, and non-governmental organization (NGOs)). The sustainability evaluation initiation, a set of indicators, has been recognized by stakeholders authorized to quantify biofuel's scope to fulfill each of their criteria. There were seventeen biofuel objectives evaluated regarding economic, social, environmental, cultural, and legal considerations, allowing the identification of each biofuel's strengths and weaknesses. Biofuel sustainability evaluation needs a review of a wide variety of different objectives to endure with their multidimensional impacts. Therefore, the biofuel capability to accomplish one particular goal cannot reach any infinite conclusion regarding overall biofuel sustainability, policy efficiency, and if it has a positive or negative impact [29].

Many studies are done on the use of various organic wastes to produce biofuel. These processes have been reported to be successful in response to global challenges like fossil fuel dependency, production cost optimization, and waste management. In addition, the matter of emission control and after so many investigations, everyone concluded that biofuel production from organic waste has an excellent possibility for sustainable and economic development while confirming minimal environmental influence and overall production cost [30].

Despite the advantages that lignocellulosic residues represent for the production of biofuels e.g., value added to crops, reducing GHG emissions and not competing with food, some issues must be critically evaluated [1,16]. Currently, crop residues are used in animal feed or left in the field to avoid erosion, or incorporated by plowing for the recycling of nutrients. Therefore, its use in a massive way could break the balance in the carbon-nitrogen relationship, decrease the amount of nutrients, and affect the soil, causing a severe problem in agricultural land quality [16]. Therefore, it must be established for residues, and the extraction limit of biomass in the corn crop case must also be established. Some data mention that 40% of the residues can be collected in an intensive cultivation system and up to 70% of the residues in a no-tillage system to keep the low risk of material organic and soil structure loss [31].

2.2.5. Biorefineries

Plant-based raw materials (i.e., biomass) are progressively recognized as a possible future resource to supersede large fractions of fossil resources as feedstocks. Therefore, to meet both the energy and non-energy demands for industrial purposes (i.e., chemicals and materials) sectors, fossil fuel exploitation at an outside scale occurs [32]. Bioenergy, biofuels, and biochemicals are the three primary national, regional, and global biorefineries. Building policies to focus on one driver can have destructive effects on others like local coal reserve exploitation, which increased greenhouse gases, resulting in global warming. There are other alternative renewable resources (wind, sun, water, biomass, etc.). However, biomass alone is often a feasible substitute for fossil resources to transport fuels and chemicals, because it is rich in carbon materials present on earth and in fossil fuels [32]. The perception of biorefinery influences a vast scope of technologies competent of separating biomass resources. i.e., wood, grass, corn, etc. into their constituents, i.e., carbohydrates, proteins, triglycerides, etc., resulting in being reconstructed into superior products, which are biofuels and chemicals [32]. Current industrial science trying to approach compounds from a single agriculture residue is among the new investigation exercises worldwide. Biorefineries are eco-friendly and very similar to oil refineries [33]. The objective of oil refineries is to procure comprehensive products from petroleum through a diverse system. The same theory is practiced on biorefineries. The first basic concepts of biorefinery were administered in 1990. Before 1997, images of continuity, environmental awareness, and biorefinery merged with the notion of green technologies [34].

2.3. *Generations of Biofuels: A Brief Review*

Fuels produced by the biological process are known as biofuel. Primary biofuels are mostly fuelwood, wood chips, pellets, and organic materials, generally used for heating generation, cooking, or electricity purposes in a crude state. Secondary biofuels come from energy crops, agricultural and forestry wastes and by-products, manure, or microbial biomass used extensively in transportation and industrial purposes [35,36]. Innovative research on biomass for secondary biofuel production has been continuous for the last five decades of biofuels production. A wide variety of potential feedstock from all around the world is utilized for biofuels production. Based on the feedstock, biofuels are classified into four groups, namely: first, second, third, and fourth-generation biofuels [37–39].

- First-generation. 1G biofuels are produced from specific parts (usually edible) of oil-based plants and starch and sugar crops. Initially, 1G biofuels showed a promising capability to reduce fossil fuel combustion and lower atmospheric levels of CO_2, which is consumed by crops as they grow [39,40]. Ethanol represents the most common biofuel produced to date. Nowadays, 78% of biofuel's total production contributed by bioethanol produced around 28 billion gallons per year from central corn in the U.S.A. and sugarcane in Brazil [38,41]. However, this generation's biofuels increased production by raising questions due to its production-generating competition food production vs. fuels, arable lands, and biodiversity loss, in addition to being responsible for ecological degradation [37,42]. Studies have shown that biofuels obtained in this manner frequently do not contribute to greenhouse gas reduction, and they require a large amount of energy for their production [43].
- Second generation. 2G biofuels, known as "advanced biofuels," overcome the problem of competition "food versus fuels" by using inedible raw materials, in addition the net carbon (emitted–consumed) from combusting second-generation biofuels is neutral or even negative [42]. Innovative processes producing 2G biofuels from non-food feedstocks sources of biomass included: residues produced by agricultural and food processing systems (discarded biomass), manure, used cooking oil, wood and sawdust to garbage, food waste, and energy crops [35,36,38,42]. Among these sources, the lignocellulosic residues of crops would more likely to be the primary candidates for its abundance, wide availability throughout the world and at times of the year, and its low cost [1,41]. However, the structural heterogeneity in these residues' composition requires more complex production processes, making 2G biofuels not industrially profitable [39,41,42]. Today, the estimated production of 2G biofuels is around 0.4 billion liters/year, i.e., <0.4% of the overall ethanol production. IEA for advanced biofuels estimates a cellulosic ethanol production increase to 0.8 billion liters in 2023 [44].
- Third generation. According to the IEA definition, third-generation biofuels are bio-based fuels produced from aquatic feedstock (usually algae) [36]. Algae are a promising alternative feedstock due to their high lipid and carbohydrate contents, increased carbon dioxide absorption, and the possibility of cultivating wastewater and seawater. Unproductive drylands and marginal farmlands do not compete with food crops on arable land or in freshwater environments [36,38]. Another characteristic that makes algae interesting for biofuels is the low level of lignin. The growth rate is high [38,42] and the possible production of biodiesel, butanol, and methane ethanol. Some green algal species can photolyze mediated biohydrogen (H_2) production [35,45].

However, this type of biomass has disadvantages such as its high initial investment for its production. The biofuel produced from algae is less stable than that produced from other sources, mainly because the oil generated by the algae is highly unsaturated, which means it is more volatile at high temperatures, so it is more likely to degrade. Furthermore, the high water quantity is also a problem when lipids have to be extracted from the algal biomass, requiring dewatering via either centrifugation or filtration before extracting lipids [46].

- Fourth generation. These biofuels, which are still in the developmental stage, use bioengineered microorganisms as microalgae, yeast, fungi, and cyanobacteria. Genetically altered crops used to consume more CO_2 from the environment than they emit. These microorganisms are used to produce different fuels, including ethanol, butanol, hydrogen, methane, vegetable oil, biodiesel, isoprene, gasoline, and jet fuel [39]. Fourth-generation biofuel research started in 2006, and significant results have not been published yet in peer-review journals [35]. Figure 1 shows the schematic representation of the integrated biorefinery for 1G to 4G.

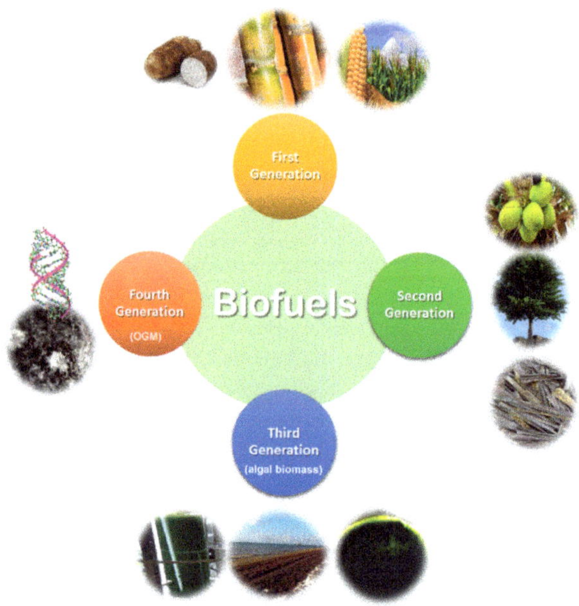

Figure 1. Schematic representation of integrated biorefinery for first, second, third, and fourth generation of biomass.

2.4. Scale-Up

Bioprocessing Techniques

The dominant predicament in cellulase manufacturing is its low enzyme titer. Numerous scholars have endeavored to crack distinctive representations to amplify cellulase enhancement in fermentation technology. It encompasses superior bioprocess performance, employing low-cost material or rudimentary raw materials as a substrate to produce enzymes and genetically obtained microorganisms etc. [47]. There are commonly two kinds of fermentation techniques that focus on enhanced cellulase production: solid-state fermentation (SSF) and submerged fermentation (SMF). These fermentation techniques use microorganisms such as fungus and bacteria. Recently, a new bioprocessing practice developed successively by using bioprocessing techniques in a single reactor. All the bioprocessing practices are discussed in detail in this section, including advancements in new bioprocessing techniques, i.e., CBP and other methods for better enzyme generation.

Currently, scientists mainly focus on the scale-up process of both 2G and 3G biofuel. However, both the sectors are facing problems that are causing interferences in the scaling-up process.

In the 2G biorefinery, in-house enzyme production and enzymatic hydrolysis play a crucial role in reducing the cost and scaling up 2G-ethanol production. From the two types of fermentation, SSF has many advantages but its use is confined at mostly lab scale because of some obstacles in controlling specific parameters and operating variables, which hugely impact microbial growth and metabolite production [25,29]. Moreover, some other drawbacks corresponding to SSF are heat and mass transfer, and high equipment costs [25].

As stated by techno-economic research, the potential to operate biomass at high solid loading during enzymatic hydrolysis will be essential in 2G biorefineries, principally. The equipment cost will come down and can lead to minimizing the requirement process energy and wastewater treatment. However, the enzymatic hydrolysis at high solid loading is somehow not an easy occupation to handle as pretreated solid biomass viscosity increases enormously and starts showing firmly non-Newtonian flow properties. The complications with both mass and heat transfer in the material and mechanical problems are pumping and adequate mixing of the biomass slurry. In SSF, researchers observed ethanol production at high solid loading during simultaneous saccharification and fermentation (SSnF) requires approximately 60% of the energy content while mixing in SSF. All processes are studied at lab scale. Nevertheless, there are several parameters like flow behavior (or even the flow regime) and particular power inputs that change with the reactor scale [48]. According to [49], their study focused on evaluating a prospective fungal strain *Penicillium oxalicum* IODBF5 in the production of cellulase under submerged fermentation. The first shake flask experiment was performed with this strain at 28 °C, 180 RPM and incubated for 8 days and initial cellulase activity reported was 0.7 FPU/mL, which later increased by 1.7 fold i.e., 1.2 FPU/mL in 8 days. After optimizing specific parameters, the same experiment was performed at 7 L reactor level in which the temperature maintained was at 28 °C and pH at 5.0 by adding 1 M HCl or 1 M NaOH. Fungal spores inoculated around 10^9 spores/L, and initial airflow kept was of 1vvm which after 72 h increased to 1.5 vvm and the incubation time was reduced from eight to four days with the same reproducibility. For effective enzymatic hydrolysis with the crude enzyme, the temperature and pH were maintained at 50 °C and 5.0, respectively. At 50 °C, the natural enzyme preserved their 50% and 26% of their activity at 48 h and 72 h, respectively. This property of Truffle was reported by [50] in their study where the capability of *Tuber maculatum* to secrete an extracellular cellulase during SMF was explained for the first time in their research. In SMF basal salt media was used with sodium carboxymethyl cellulose (0.5% *w*/*v*) as a carbon source at pH 7 and the cellulase activity reported was 1.70 U/mL after seven days of incubation. The stability test was performed, and it was observed that the enzyme was stable at 50 °C and pH 5.0. The enzyme was thermostable as well as maintained its 99% activity at 50 °C temperature. Tray Bioreactor, Packed bed Bioreactor (PBB), Rotary Drum Bioreactor (RDB), Fluidized Bed Bioreactor (FBB) and Instrumented Labscale Bioreactor (ILB) are usually used as bioreactors for SSF systems [51]. According to the study performed by [52], cellulase was produced by microorganism *Trichoderma reessei* RUT C-30 under SSF. The substrate used in SSF was wheat bran and cellulose, where cellulose was used as an inducer for further enhanced cellulase production. The combination of wheat bran and cellulose was added in a 250 mL Erlenmeyer flask. The substrate moistened with mineral salt medium and pH was maintained at 4.8 by adding 1 N HCl and 1 N NaOH. This experiment was optimized by involving a two-stage statistical design of experiments, resulting in an increase of CMCase by 3.2 fold i.e., 959.53 IU/gDS. This method was repeated at pilot scale and the reactors used were tray fermenters, and the same lab conditions were used, resulting in 457 IU/gDS yield. Cellulase produced in this experiment was used for enzymatic hydrolysis of alkali pretreated sorghum stover with or without BGL and the BGL used in this experiment was extracted from *Aspergillus niger*. The result from hydrolysis with BGL was reported to be 174% efficient. The hydrolysate produced from hydrolysis of sorghum stover was used in the fermentation process to generate ethanol, which resulted in being around 80% effective. In another study, crude olive pomace and exhausted olive pomace was used in SSF as substrate. They both were pretreated by ultrasound pretreatment, and their comparative study was also performed. In SSF, both wastes were checked and used as a substrate in SSF to yield cellulase and xylanase by fungi. The use of exhausted olive pomace as substrate by *Aspergillus* resulted in higher titers in the screening. They were therefore used in the experiment of ultrasound pretreatment. The outcome of sonication showed a 3-fold increase of xylanase activity and harmed cellulase activity. Furthermore, the liquid part acquired from ultrasound pretreatment was used to maintain substantial moisture, resulting in a positive effect on enzyme activity, which caused a 3.6-fold increase and 1.2-fold increase, respectively [53].

To date, large-scale algal cultivation still presents various mechanical threats that interfere with the process of commercialization as a promising aspect of algal biomass as a renewable feedstock for biofuels production. These threats or challenges are related to upstream or downstream processing.

In upstream processing, these are the algae and algal cultivation process selection process, the energy input for handling closed-photobioreactors, nutrient sources, water reusability, and footprint and susceptibility of algae towards the surrounding.

In downstream processing, this is attributed to harvesting and drying technology for algal cells, efficient algal lipid extraction methods, biodiesel conversion technologies from algae and biodiesel, and probability to transform biofuels production from the algal residue after extracting the lipid. After the technological hurdles, merchandising algal biofuels production's profitable viability is still dubious because algal cultivation and correlated biofuels production technologies are in shortage. Enhance algal biofuels' commercial potential: it is crucial to understand and recognize technical and economically related problems [54].

3. Consolidated Bioprocessing

3.1. *Strategies for Development Ideal Consolidated Bioprocessing-Enabling Microorganisms*

In searching for CBP functional organism, two different strategies have been applied to endow the microorganism with the capabilities required, such as high hydrolysis rate, tolerance to a compound derived from pretreatment, and proper fermentation.

3.1.1. Native Strategy

The native strategy focuses on studying organisms with the natural capacity to produce different enzymes and use different substrates to improve biofuels' performance. Some approaches to enhance bioprocessing capabilities include adaptive evolution and isolation of new strains for CBP [55].

The native strains proposed for use in CBP are mainly wild-type strains that are generally poorly characterized. To date, genetic manipulation tools are established for only some of them, and in a few cases, their metabolisms are investigated in-depth. Concerning cellulolytic fungi, most genetic engineering efforts are focused on increasing cellulase production. However, there is a growing interest in biofuel production using these organisms. Candidates for the native strategy can be classified into three groups: fungi, bacteria that produce and excrete enzymes, and cellulosome-forming bacteria [56–58].

However, to date, the organisms discovered with the capacity to carry out CBP are well below efficient alcohol production expectations. Therefore, the co-cultivation of two or more different organisms (consortium) taking advantage of their capacities-specific metabolism is a promising method to improve substrate conversion and ethanol yield [56]. A microbial consortium is an association of two or more organisms acting together as a community in a complex system, where all benefits from others' activities can be enjoyed [59]. Consortia can be classified into natural, artificial, and synthetic. Natural consortia are symbiotic due to co-evolution; in contrast, artificial consortia and synthetic consortia are defined as mixed crop systems, differing from each other. In synthetic consortia, genetic modifications are performed to achieve specific interactions between mixed strains [60].

3.1.2. Recombinant Strategy

The recombinant strategy aims to provide the fermenting organisms with the hydrolytic capacity. The hydrolytic organisms to provide them with the fermentation capacity. Therefore, before designing microorganisms for biomass conversion, it is crucial to select host organisms with the desired characteristics, emphasizing strains that can use low-cost substrates, resistance to environmental stress, and a high yield of the desired product [61,62].

The main challenges in the recombinant strategy include the adverse effects of the co-expression of multiple unwanted genes, the modulation of the expression of different genes at the appropriate levels, and the improper folding of proteins, which can prevent their secretion, in addition to inadequate fermentation pathway [61,63].

Recent advances in the development of modified microorganisms through evolutionary, metabolic and genetic engineering approaches have paved the way for using lignocellulosic biomass as a substrate for biofuel production [63].

3.2. Aspects of Consolidated Bioprocessing

3.2.1. Economic Viability of Consolidated Bioprocessing

CBP is a promising strategy for effective biofuel production due to the combination of three processes (enzyme production, saccharification, and fermentation) in CBP. This strategy can reduce the reactor and the enzyme cost, which are the major impediments to low-cost biomass processing [61]. In the conventional ethanol production process, the cost by CBP production can be reduced by 25%. When considering capital, raw materials, utilities, and yield loss expenditures, a comparative cost analysis conducted on ethanol production resulted in a projection of \$0.04 gal^{-1} for CBP. At the same time, saccharification and co-fermentation was projected at \$0.19 gal^{-1} [64].

Despite the potential advantages that CBP represents, some limitations have not allowed bioethanol's industrial production to use this strategy. When using native organisms, the main problem is low yield and productivity despite long fermentation process periods of between 3 and 12 days [55]. In the case of using engineered modified organisms, a significant hurdle for industrial CBP organism development is achieving high levels of enzyme production without compromising ethanol productivity because there are generally problems associated with the co-expression of multiple genes other than those of interest, tolerance to the culture medium, in addition to the high cost inherent in the production of this type of organism [39,65] (Table 1).

Table 1. Advantages and disadvantages of CBP [63].

Advantages of CBP	Disadvantages of CBP
• Simplification of total operation process • Fermentation and saccharification vessels are reduced • Enzymes addition are totally eliminated • Risk of contamination is minimal by reducing carbohydrates and producing ethanol • Co-fermentation of pentose and hexose • Capital investment is highly reduced to the lowest minimum • Less energy is required during the process	• To date, only engineered microbial strains are known to perform optimally in CBP. • The use of recombinant organism is highly restricted in some countries, and there is growing public health concern and environmental risk associated with this. • Wild-type bacterial/fungal species tested to date on CBP produce very low concentrations of ethanol. • CBP typically requires long periods of fermentation

Some pioneering companies that use CBP to produce lignocellulosic ethanol are the American companies Qteros and Mascoma, who used the microorganism *C. phytofermentans*, registered as Q Microbe® (Marlborough, MA, USA). In 2014 Mascoma was acquired by the Canadian company Lallemand and all its assets, and currently continues to work in the production of advanced biofuels [55].

3.2.2. Long Term Economics of Consolidated Bioprocessing

The increasing demand for oil and cleaner energy research has led to an escalation in biofuels production (mainly bioethanol). The worldwide production from 2008 to 2017 presents an annual growth of 5.07% of bioethanol. The ethanol consumption is projected to reach 164 billion liters by 2030, 1.98 times the consumption of 2018. Increased demand for bioethanol is motivated by increased concern about climate change, the annual increase of 3.8% ± 0.5 in the global car fleet, and the fact that ethanol's production and consumption blended with gasoline has been promoted through subsidies, mandates, and financing for research in 65 countries [44,66].

Even though the large-scale techno-economic analysis of consolidated bioprocessing [67] has hardly been reported in the literature, the strategy is presented as a promising technology for biofuels. Different chemical compounds with high value-added compounds, from different types of residual biomass under the circular economy's perspective, are the resource of other biochemical processes. Hence, more beneficial impact is generated long-term. Although it is still a challenge, it is a way to take advantage of waste and convert it into a source to build capital rather than waste it [68]. Any technology that involves the transformation of biomass is aligned with a sustainable model.

3.2.3. Consolidated Bioprocessing (CBP) and Some Case Studies

In recent years, the consolidated bioprocess CBP, initially known as direct microbial conversion (D.M.C.), has been investigated. CBP's critical difference with other biomass bioprocessing strategies lies in the use of a single organism or their consortium for enzyme production, hydrolysis, and fermentation. These are generally carried out at room temperature, which results in the reduction of production costs [64,65,69,70]. Furthermore, the hydrolytic and fermentative processes' compatibility means that a single reactor is required, simplifying its operation. Although CBP has great advantages compared to other production processes, there are issues such as long fermentation periods and low biofuel yields due to the formation of various by-products (organic acids), the sensitivity of microorganisms to alcoholic solvents, and growth limited in the supernatant of hydrolysis, which requires the continuous search for more efficient strategies [64,69].

However, the industrial-scale application of CBP remains challenging so far because of the low efficiency. Derived from CBP: Consolidated Bio-Saccharification (CBS) strategy proposed fermentation separated from the integrated process. CBS produces fermentable sugars as the target product rather than end metabolites such as ethanol. Consequently, fermentation would not be limited by hydrolysis condition. The cellulolytic capability is maximized as well. Thus, CBS provides new insight into lignocellulose bioconversion [11,71]. Figure 2 shows the schematic representation of the consolidated bioprocessing process in terms of biorefinery. The advantages of SSF are linked to the utilization of agro-industrial residues as solid substrate substituting for carbon and energy sources, however enzyme production for industrial purposes still surfaces into certain industrial restrictions [72]. On the contrary, SMF is a universally acknowledged fermentation process for industrial enzyme production as it is more convenient to control all the variables such as pH, temperature, and operational techniques. The CBP strategy proposed in 1996 and increasing evidence supports that CBP may be feasible. CBP's research has focused on developing new and even more effective CBP microorganisms, a critical challenge [73,74]. Several studies were performed for sustainable ethanol production using advanced consolidated bioprocessing. In one research study, the subsequent pairing used in CBP was *Clostridium thermocellum* ATCG 27,405 with mesophilic microaerobe, *Pichia stiptis* NCIM-3498. The biogenic municipal solid waste was pretreated with 0.5% NaOH for reducing the recalcitrance property. It was observed that subsequent CBP (23.99 g/L) was better than CBP alone (18.10 g/L).

In subsequent CBP, exogenous xylanase was added to enhance xylan hydrolysis. Subsequent CBP II biosystem was observed to give maximum ethanol production (36.90 g/L) at pH 5 in a single reactor [75]. In another investigation, the author used yeast engineered with five functional cellulase genes (BGL, XYNII, EGII, CBHI, and CBHII) where ionic pretreated bagasse and laubholz unbleached kraft pulp was used as targeted biomass. The screening was performed to observe that it is essential to breed CBP yeast having the optimized cellulase expressing the ratio for the proposed biomass. In CBP, yeast development is considered a promising and cheap way in consolidated bioprocessing for ethanol production [76]. In one such study, the author used a new approach that is partially consolidated bioprocessing (PCBP). In this process, a mixture of lignocellulosic material was used (mentioned in Table 2). PCBP was used to prepare mixed lignocellulosic substrate by utilizing non-isothermal simultaneous pretreatment. Saccharification methods were used for the hydrolysis of the pretreated substrate. In the saccharification process (*Trichoderma Reesei* RUT C 30) was exploited the combination of laccase (*Pleurotus djamor*) and holocellulase. Then, it was later pursued by the co-fermentation process in a single reactor. The artificial neural network (Feedforward ANN) model was used to optimize all parameters in PCBP that resulted in increased ethanol productivity [77]. In one of the studies, the authors blend two characteristics, i.e., the saccharolytic and fermentation integrated into one microorganism. It diverted a huge interest towards this process for production of ethanol in terms of CBP causing lignocellulosic biorefineries to decrease environmental pollution and economic cost. Therefore, in this study, industrial *S. cerevisiae* strains are used, expressing strong characteristics such as thermotolerance and enhanced resistance to inhibitors. The strain was estimated to be great as it expresses hemicellulolytic enzyme activity on its cell surface. It resulted in increased ethanol productivity with the addition of commercial cellulase [78]. In one of the investigations, a thermophilic anaerobic bacterium was isolated from Himalayan hot spring *Clostridium* sp. DBT-IOC-C19. This strain was considered suitable for consolidated bioprocessing. It expressed a wide range of substrate conversion into ethanol, acetate, and lactate in a single step, where ethanol is considered the main product. DBT-IOC-C19 displayed 94.6% degradation at 5 g/L and 82.74% degradation at 10 g/L of avicel concentration or loading in 96 h of incubation time during fermentation. Rice straw was used as a lignocellulosic substrate for comparative analysis with different strains but total product yield and ethanol production increased in *Clostridium* DBT-IOC-C19 [79]. The same author studied different strains for CBP, e.g., *Clostridium thermocellum* ATCC31924, where crystalline cellulose was used as the sole carbon source. It was reported that cellulosome extracted from the bacterial strain when purified resulted in concentrated cellulase and xylanase enzyme [80]. Ionic liquid pretreated pine needle biomass was used as a substrate in CBP. First, saccharification was performed by using cellulase and xylanase in a single pot. Then, fermentation of hydrolysate was performed by using yeast to yield ethanol [81]. CBP cannot only be used for the production of ethanol, but there are other yields as primary productivity. In one study, the author worked with CBP along with co-cultivation. The author designed a microbial consortium composed of hemicellulose producing bacteria *Thermoanerobacterium, Thermosacchrolyticum* M5, and succinic acid production specialist *Actinobacillus succinogens.* This co-cultivation in CBP resulted in increased production of succinic acid under optimized conditions [82]. The three lignocellulolytic bacteria (*Clostridium thermocellum, C. stercorarium,* and *Thermoanaerobacter thermohydrosulfuricus*) were used as collegial co-culture strains in consolidated bioprocessing. This was performed in terms of a comparative analysis with the monoculture by employing a wide range of lignocellulosic material. The pretreated and unpretreated substrate was used to study the productivity (yield) produced from monoculture and triculture, and co-culture [83]. In another investigation, CBP was used to produce H_2. The cellulose-degrading microorganism was retrieved from the bovine ruminal fluid (BRF). *Clostridium acetobutylium* had great potential in increasing H_2 yield [84]. Figure 3 shows the schematic representation of the consolidated bioprocessing process using strategy hydrothermal pretreatment processing in the production of biofuels.

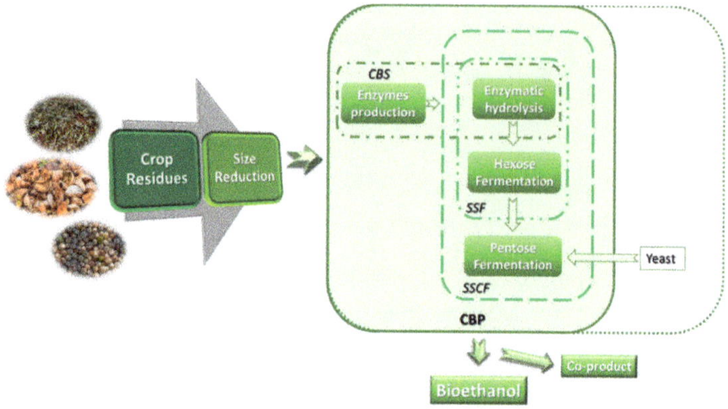

Figure 2. Schematic representation of consolidated bioprocessing processing in terms of biorefinery from lignocellulosic biomass.

Table 2. Consolidated bioprocessing (CBP) for a mixture of lignocellulosic material in the enzyme production.

Microorganism	Substrate	Enzyme	Operational Conditions	Reference
S. cerevisiae MT8-1	Laubholz unbleached kraft pulp and ionic liquid pre-treated bagasse	Cellulolytic enzymes	Incubation time 72–96 h	[76]
Saccharomyces cerevisiae	*Ricinus communis* and *Saccharum spontaneum* and top portions of *Saccharum officinarum*	Laccase, Holo-cellulase	Substrate loading-10% to 30% (*v*/*v*), incubation time (12–24 h), temperature 30–37 °C.	[77]
Thermophilic bacterium Clostridium sp. DBT-IOC-C19	Rice straw, crystalline cellulose	Cellulosomes	Temperature 55 to 65 °C, pH-7 to 8, time-96 h	[79]
C. thermocellum strain ATCC 31,924,	Cellulose	Cellulosomes	pH-8, temperature-55 °C, inoculum size 4% (*v*/*v*) & 0.5% (*w*/*v*) substrate concentration	[80]
B. subtilis G2, *S. cerevisiae* and *P. stipitis*.	Pine needle biomass	Cellulase, Xylanase	pH-5.6, inoculum size-1.5% (*v*/*v*), temperature-30–37 °C, incubation time 24–72 h	[81]
Thermoanaerobacterium thermosaccharolyticum M5 & *Actinobacillus succinogenes* 130Z	Corn cob	Xylanase & β-Xylosidase	Temp-55 °C & 37 °C, pH-6 to 7, incubation time-24 h to 48 h	[82]
CBP thermophile, *Clostridiumthermocellum* ATCC-27405 with mesophilic microaerobe, *Pichia stipitis* NCIM-3498	Biogenic municipal solid waste (BMSW)	Xylanase	Temp (45–70°C), initial pH (6–8), inoculum volume (4–12% *v*/*v*), pretreated BMSW Loading (10–80 g/L cellulose equivalent)	[75]

Table 2. *Cont.*

Microorganism	Substrate	Enzyme	Operational Conditions	Reference
Clostridiumthermocellum, C. stercorarium, and *Thermoanaerobacter thermohydrosulfuricus*	cattail (*Typha* spp.) and wheat straw	Cellulase & Hemicellulase	10% (*v*/*v*, monoculture), 5/5% (*v*/*v*, dual cultures), or 3.3/3.3/3.3% (*v*/*v*, tri-culture), temp-45 °C, incubation-4 days	[83]
Clostridium acetobutylicum	Agave biomass	Enzymatic commercial complex Cellic CTec3® Novozymes (Bagsvaerd, Denmark)	10% solid loading, pH-5.5, temp-35 °C, incubation-264	[84]
Industrial *S. cerevisiae* strains	Corn cob	Hemicellulolytic enzymes	pH-5, temp-35 °C, incubation time-48 h	[78]

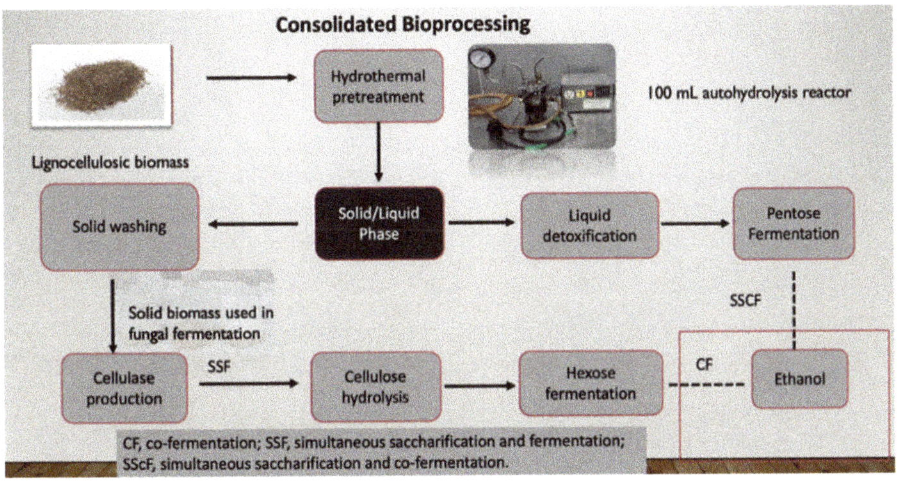

Figure 3. Schematic representation of consolidated bioprocessing processing using as hydrothermal strategy pretreatment. Adapted and modified from [15].

3.2.4. Physiochemical Conversion Routes in CBP with Pretreatment Process

The scientific lead that summons the continuous supply of biomass as the natural way for biomass saccharification with high sugar productivity is still a challenge. The problem is not yet presented in research programs dealing with biomass processing. However, many scientists aspire to work on biomass optimization for several decades. The merged yield from biofuels and chemicals still requires boosting from both technical and economic perspectives. None of the current pretreatment processes has improved, since their results rely on the type of feedstock, downstream process configuration, and many other factors. Consolidated bioprocessing (CBP) associates an individual or consortium of microbes to deteriorate biomass. Physicochemical pretreatment and biological methods are considered a better approach to reduce the recalcitrant property, leading to high sugar yield.

The hydrothermal pretreatment, i.e., liquid hot water pretreatment, does not require fast decompression and does not need any chemicals or catalysts. The temperature lifted between 160–240 °C, which caused an increase in pressure that required preserving the water in the liquid state and activating the lignocellulose composition [85]. The main objective of the liquid hot water pretreatment is to dissociate and solubilize hemicellulose. It makes easy cellulose availability to cellulase and limits inhibitory compounds during the reaction. Nonetheless, for avoiding the generation of inhibitors, the response must be performed at a low pH, i.e., between 4 and 7, reducing monomers' production [85,86]. Scientists recently established that heating lignin and raising the temperature above glass transition temperature could split the bond's existence between carbohydrate lignin. It emerged into the migration of hydrophobic lignin present in liquid media from the cell walls and getting re-precipitated into small spherical droplets assembled on both sides (inside and outside) of fibers [87]. Almost all hemicellulose contents got detached, leaving cellulose fiber bundles behind with free lignin, because of the lignin's realignment on both fibers' sides. After hydrolysis, fast disintegration was reported by researchers, along with a higher degradation temperature [87]. In another study, material balance and multiscale characterization techniques utilized the systematic analysis of the effects of severity factor and pH on the lignin-carbohydrate complex (LCC). Researchers discovered that the severity factor affects xylan removal under a diverse range of temperature profiles, which means a high severity factor causes high xylan removal. The temperature does not affect xylan removal. The liquid phase (hydrolysate) causes pH to drop due to hemicellulose depolymerization and degradation. It results in the accumulation and production of the acetic acid, causing hydrothermal pretreatment, and hemicellulose sugar's high yielding capacity causes increased furfural productivity and xylose loss. Rare, increased hemicellulose sugar degradation causes reduced furfural production. In contrast, approximately 80% of the original xylan was lost [88].

3.2.5. Integrated Technologies Based on Hydrolysis for Biofuel (Ethanol) Production

The quest to achieve a higher yield of biofuel per unit of biomass has led to the integration of different phases of the process; it reduces the cost of capital and makes biofuels more efficient and economically viable. Given that bioethanol is the most produced secondary biofuel today, the search for more efficient configurations for its production has been sought.

Simultaneous Saccharification and Fermentation (SSnF)

Conventionally, ethanol production is carried out in separate phases, first the hydrolysis and later the fermentation. Due to this separation, each one is carried out under optimal conditions [89]. However, the excessive accumulation of sugars during the hydrolysis inhibits the enzymatic activity, reducing the process [63,69,90]. This problem led to the development of simultaneous saccharification and fermentation (SSnF). In this process, hydrolysis and fermentation occur together in the same reactor, reducing enzymatic inhibition due to sugars' presence since the monosaccharides produced are immediately consumed by fermenting microorganisms [65,70,90]. Compared to the conventional process, SSnF achieved a higher hydrolysis rate, and it conducted higher ethanol concentration. This strategy requires less equipment and a more straightforward operation, and the presence of ethanol in the wort makes it less susceptible to contamination by unwanted microorganisms [61,70,90].

However, this method's disadvantage is that the optimal operating conditions for hydrolysis and fermentation are different, implying difficulty for its control and optimization. Enzymatic hydrolysis is performed optimally at temperatures above 50 °C. For most fermenting microorganisms, the optimum temperature for their performance is between 28 and 37 °C; likewise, the optimum pH for the hydrolysis and fermentation stages is different [65,70,90].

Work performed on the search and selection of suitable enzymes and microorganisms for this strategy, and a promising option to overcome this difficulty, is the use of thermotolerant yeasts such as *Kluyveromyces marxianus*, which is a promising species since some of its strains grow at temperatures between 45 and 52 °C [91].

Simultaneous Saccharification and Co-Fermentation (SSCF)

This configuration aims to complete all the sugars released during the pretreatment and hydrolysis of the biomass through mixed yeast cultures that can assimilate both the pentoses and the hexoses produced in the same reactor. This strategy offers this advantage. By continuously removing the final hydrolysis products, causing cellulase and glucosidase activity inhibition can lead to the high productivity of ethanol, giving greater yield compared to the SFS process. One drawback of this method is that the organisms that use hexoses grow faster than those that use pentoses, leading to inhibition by the high ethanol concentration. Genetically modified yeast or bacteria can be used to achieve efficient ethanol production. To carry out the fermentation of both, pentoses and hexoses are required [65,70,92,93].

4. Concluding Remarks and Prospects

When scientists started working on the valorization of lignocellulosic biomass, different bioprocessing technique such SSF and SMF, CBP emerged to play a critical role in the environment and economy. Scientists are using the genetic engineering method on microorganisms to play all the functions with or without pretreatment. CBP is timesaving and cost-effective in biorefineries as it is a one-process technique involving less instruments or one single reactor. In the future, CBP will play a vital role in converting lignocellulosic biomass to ethanol and other essential products in a single process or reactor with or without pretreatment because of some microorganisms capable of doing all at once, making it more economical and timesaving.

Author Contributions: Conceptualization, H.A.R. and R.T.-T.; writing—original draft, E.O.-M. and A.S.; writing—review and editing, H.A.R., R.T-T., E.O.-M., A.S. and R.C.-V.; Project administration, H.A.R. and R.T.-T.; funding acquisition, H.A.R. All authors have read and agreed to the published version of the manuscript.

Funding: This research was funded by Secretary of Public Education of Mexico–Mexican Science and Technology Council (SEP-CONACYT, Mexico) for the Basic Science Project -2015–01, grant number 254808. Anusuiya Singh also thanks the National Council for Science and Technology (CONACYT, Mexico) for her Ph.D. Fellowship support. Publication funded by "Programa de Fortalecimiento a la Excelencia Educativa" (PROFEXCE-2020).

Conflicts of Interest: The authors declare no conflict of interest.

References

1. Ali, M.; Saleem, M.; Khan, Z.; Watson, I.A. The use of crop residues for biofuel production. In *Biomass, Biopolymer-Based Materials, and Bioenergy*; Verma, D., Fortunati, E., Jain, S., Zhang, X., Eds.; Woodhead Publishing: Oxford, UK, 2019; pp. 369–395.
2. Taghizadeh-Alisaraei, A.; Motevali, A.; Ghobadian, B. Ethanol production from date wastes: Adapted technologies, challenges, and global potential. *Renew. Energy* **2019**, *143*, 1094–1110. [CrossRef]
3. Nargotra, P.; Sharma, V.; Bajaj, B.K. Consolidated bioprocessing of surfactant-assisted ionic liquid-pretreated Parthenium hysterophorus L. biomass for bioethanol production. *Bioresour. Technol.* **2019**, *289*, 121611. [CrossRef]
4. Patel, A.; Hrůzová, K.; Rova, U.; Christakopoulos, P.; Matsakas, L. Sustainable biorefinery concept for biofuel production through holistic volarization of food waste. *Bioresour. Technol.* **2019**, *294*, 122247. [CrossRef] [PubMed]
5. Popp, J.; Lakner, Z.; Harangi-Rákos, M.; Fari, M. The effect of bioenergy expansion: Food, energy, and environment. *Renew. Sustain. Energy Rev.* **2014**, *32*, 559–578. [CrossRef]
6. Brinkman, M.L.; Wicke, B.; Faaij, A.P.; Van Der Hilst, F. Projecting socio-economic impacts of bioenergy: Current status and limitations of ex-ante quantification methods. *Renew. Sustain. Energy Rev.* **2019**, *115*, 109352. [CrossRef]
7. Rajarathinam, R.; Rajarathinam, R.; Kanagaraj, L.P.; Ranganathan, R.V.; Dhanasekaran, K.; Manickam, N.K. Evaluation and characterization of novel sources of sustainable lignocellulosic residues for bioethanol production using ultrasound-assisted alkaline pre-treatment. *Waste Manag.* **2019**, *87*, 368–374.
8. Carrillo-Nieves, D.; Alanís, M.J.R.; Quiroz, R.D.L.C.; Ruiz, H.A.; Iqbal, H.M.N.; Parra, R. Current status and future trends of bioethanol production from agro-industrial wastes in Mexico. *Renew. Sustain. Energy Rev.* **2019**, *102*, 63–74. [CrossRef]
9. Debnath, D.; Giner, C. Interaction between biofuels and agricultural markets. In *Biofuels, Bioenergy and Food Security*; Debnath, D., Babu, S., Eds.; Elsevier: Amsterdam, The Netherlands, 2019; pp. 61–76.

10. Antonopoulou, G.; Kampranis, A.; Ntaikou, I.; Lyberatos, G. Enhancement of liquid and gaseous biofuels production from agro-industrial residues after thermochemical and enzymatic pretreatment. *Front. Sustain. Food Syst.* **2019**, *3*, 92. [CrossRef]
11. Liu, Y.J.; Li, B.; Feng, Y.; Cui, Q. Consolidated bio-saccharification: Leading lignocellulose bioconversion into the real world. *Biotechnol. Adv.* **2020**, *40*, 107535. [CrossRef]
12. Go, A.W.; Conag, A.T.; Igdon, R.M.B.; Toledo, A.S.; Malila, J.S. Potentials of agricultural and agro-industrial crop residues for the displacement of fossil fuels: A Philippine context. *Energy Strat. Rev.* **2019**, *23*, 100–113. [CrossRef]
13. Bernal-Lugo, I.; Jacinto-Hernandez, C.; Gimeno, M.; Montiel, C.C.; Rivero-Cruz, F.; and Velasco, O. Highly efficient single-step pretreatment to remove lignin and hemicellulose from softwood. *BioResources* **2019**, *14*, 3567–3577.
14. Yan, Q.; Fong, S.S. Challenges and advances for genetic engineering of non-model bacteria and uses in consolidated bioprocessing. *Front. Microbiol.* **2017**, *8*, 2060. [CrossRef] [PubMed]
15. Brethauer, S.; Studer, M.H. Consolidated bioprocessing of lignocellulose by a microbial consortium. *Energy Environ. Sci.* **2014**, *7*, 1446. [CrossRef]
16. Prasad, S.; Singh, A.; Korres, N.E.; Rathore, D.; Sevda, S.; Pant, D. Sustainable utilization of crop residues for energy generation: A life cycle assessment (LCA) perspective. *Bioresour. Technol.* **2020**, *303*, 122964. [CrossRef]
17. Cherubini, F. The biorefinery concept: Using biomass instead of oil for producing energy and chemicals. *Energy Convers. Manag.* **2010**, *51*, 1412–1421. [CrossRef]
18. Caspeta, L.; Caro-Bermúdez, M.A.; Ponce-Noyola, T.; Martinez, A. Enzymatic hydrolysis at high-solids loadings for the conversion of agave bagasse to fuel ethanol. *Appl. Energy* **2014**, *113*, 277–286. [CrossRef]
19. Collier, Z.A.; Connelly, E.B.; Polmateer, T.L.; Lambert, J.H. Value chain for next-generation biofuels: Resilience and sustainability of the product life cycle. *Environ. Syst. Decis.* **2016**, *37*, 22–33. [CrossRef]
20. Birur, D.; Chapagain, A.; Devadoss, S.; Krishna, P. Assessing sustainability of biofuels production in china. In Proceedings of the 19th Annual Conference on Global Economic Analysis, Washington, DC, USA, 15–17 June 2016.
21. Lara-Flores, A.A.; Araújo, R.G.; Rodríguez-Jasso, R.M.; Aguedo, M.; Aguilar, C.N.; Trajano, H.L.; Ruiz, H.A. Bioeconomy and biorefinery: Valorization of hemicellulose from lignocellulosic biomass and potential use of avocado residues as a promising resource of bioproducts. In *Waste to Wealth*; Singhania, R., Agarwal, R., Kumar, R., Sukumaran, R., Eds.; Springer: Singapore, 2018; pp. 141–170.
22. Morais, A.R.; Bogel-Lukasik, R. Green chemistry and the biorefinery concept. *Sustain. Chem. Process.* **2013**, *1*, 18. [CrossRef]
23. Dalvand, K.; Rubin, J.; Gunukula, S.; Wheeler, M.C.; Hunt, G. Economics of biofuels: Market potential of furfural and its derivatives. *Biomass Bioenergy* **2018**, *115*, 56–63. [CrossRef]
24. Singh, A.; Adsul, M.; Vaishnav, N.; Mathur, A.; Singhania, R.R. Improved cellulase production by *Penicillium janthinellum* mutant. *Indian J. Exp. Biol.* **2017**, *55*, 436–440.
25. Singh, A.; Jasso, R.M.R.; Gonzalez-Gloria, K.D.; Rosales, M.; Cerda, R.B.; Aguilar, C.N.; Singhania, R.R.; Ruiz, H.A. The enzyme biorefinery platform for advanced biofuels production. *Bioresour. Technol. Rep.* **2019**, *7*, 100257. [CrossRef]
26. Khajeeram, S.; Unrean, P. Techno-economic assessment of high-solid simultaneous saccharification and fermentation and economic impacts of yeast consortium and on-site enzyme production technologies. *Energy* **2017**, *122*, 194–203. [CrossRef]
27. Solarte-Toro, J.C.; Romero-García, J.M.; Martínez-Patiño, J.C.; Ruiz-Ramos, E.; Castro-Galiano, E.; Cardona, C.A. Acid pretreatment of lignocellulosic biomass for energy vectors production: A review focused on operational conditions and techno-economic assessment for bioethanol production. *Renew. Sustain. Energy Rev.* **2019**, *107*, 587–601. [CrossRef]
28. Avramović, J.; Veličković, A.; Veljković, V. Challenges in biodiesel industry: Socio-economic, occupational health, and policy issues. *Saf. Eng.* **2018**, *8*, 79–83. [CrossRef]
29. Baudry, G.; Delrue, F.; Legrand, J.; Pruvost, J.; Vallée, T. The challenge of measuring biofuel sustainability: A stakeholder-driven approach applied to the French case. *Renew. Sustain. Energy Rev.* **2017**, *69*, 933–947. [CrossRef]

30. Stephen, J.L.; Periyasamy, B. Innovative developments in biofuels production from organic waste materials: A review. *Fuel* **2018**, *214*, 623–633. [CrossRef]
31. Lal, R. World crop residues production and implications of its use as a biofuel. *Environ. Int.* **2005**, *31*, 575–584. [CrossRef]
32. Cherubini, F.; Mandl, M.; Philips, C.; Wellisch, M.; Jørgensen, H.; Skiadas IV, B.L.; Dohy, M.; Pouet, J.C.; Wilke, T.; Walsh, P.; et al. IEA bioenergy Task 42 on biorefineries: Co-production of fuels, chemicals, power and materials from biomass. In *IEA Bioenergy Task*; IEA Bioenergy: Copenhagen, Denmark, 2007; pp. 1–37.
33. Aguilar-Reynosa, A.; Romaní, A.; Rodríguez-Jasso, R.M.; Aguilar, C.N.; Garrote, G.; Ruiz, H.A. Comparison of microwave and conduction-convection heating autohydrolysis pretreatment for bioethanol production. *Bioresour. Technol.* **2017**, *243*, 273–283. [CrossRef]
34. Maity, S.K. Opportunities, recent trends and challenges of integrated biorefinery: Part I. *Renew. Sustain. Energy Rev.* **2015**, *43*, 1427–1445. [CrossRef]
35. Datta, A.; Hossain, A.; Roy, S. An overview on biofuels and their advantages and disadvantages. *Asian J. Chem.* **2019**, *31*, 1851–1858. [CrossRef]
36. Saladini, F.; Patrizi, N.; Pulselli, F.M.; Marchettini, N.; Bastianoni, S. Guidelines for emergy evaluation of first, second and third generation biofuels. *Renew. Sustain. Energy Rev.* **2016**, *66*, 221–227. [CrossRef]
37. Abdullah, B.; Muhammad, S.A.F.S.; Shokravi, Z.; Ismail, S.; Kassim, K.A.; Mahmood, A.N.; Aziz, M.A. Fourth generation biofuel: A review on risks and mitigation strategies. *Renew. Sustain. Energy Rev.* **2019**, *107*, 37–50. [CrossRef]
38. Jambo, S.A.; Abdulla, R.; Azhar, S.H.M.; Marbawi, H.; Gansau, J.A.; Ravindra, P. A review on third generation bioethanol feedstock. *Renew. Sustain. Energy Rev.* **2016**, *65*, 756–769. [CrossRef]
39. Alalwan, H.A.; Alminshid, A.H.; Aljaafari, H.A. Promising evolution of biofuel generations. Subject review. *Renew. Energy Focus* **2019**, *28*, 127–139. [CrossRef]
40. Zentou, H.; Rosli, N.S.; Wen, C.H.; Abdul Azeez, K.; Gomes, C. The viability of biofuels in developing countries: Successes, failures, and challenges. *Iran. J. Chem. Chem. Eng.* **2019**, *38*, 173–182.
41. Boboescu, I.Z.; Chemarin, F.; Beigbeder, J.B.; De Vasconcelos, B.R.; Munirathinam, R.; Ghislain, T.; Lavoie, J.M. Making next-generation biofuels and biocommodities a feasible reality. *Curr. Opin. Green Sustain. Chem.* **2019**, *20*, 25–32. [CrossRef]
42. Chowdhury, H.; Loganathan, B. Third-generation biofuels from microalgae: A review. *Curr. Opin. Green Sustain. Chem.* **2019**, *20*, 39–44. [CrossRef]
43. Stolarski, M.J.; Krzyżaniak, M.; Łuczyński, M.; Załuski, D.; Szczukowski, S.; Tworkowski, J.; Gołaszewski, J. Lignocellulosic biomass from short rotation woody crops as a feedstock for second-generation bioethanol production. *Ind. Crop. Prod.* **2015**, *75*, 66–75. [CrossRef]
44. Sydney, E.B.; Letti, L.A.J.; Karp, S.G.; Sydney, A.C.N.; Vandenberghe, L.P.D.S.; De Carvalho, J.C.; Woiciechowski, A.L.; Medeiros, A.B.P.; Soccol, V.T.; Soccol, C.R. Current analysis and future perspective of reduction in worldwide greenhouse gases emissions by using first and second generation bioethanol in the transportation sector. *Bioresour. Technol. Rep.* **2019**, *7*, 100234. [CrossRef]
45. Khan, S.; Fu, P. Biotechnological perspectives on algae: a viable option for next generation biofuels. *Curr. Opin. Biotechnol.* **2020**, *62*, 146–152. [CrossRef]
46. Lee, R.A.; Lavoie, J.M. From first to third-generation biofuels: Challenges of producing a commodity from a biomass of increasing complexity. *Anim. Front.* **2013**, *3*, 6–11. [CrossRef]
47. Singh, P. Microbial enzymes with special characteristics for biotechnological applications. *Biomolecules* **2013**, *3*, 597–611.
48. Palmqvist, B.; Kadić, A.; Hägglund, K.; Petersson, A.; Lidén, G. Scale-up of high-solid enzymatic hydrolysis of steam-pretreated softwood: the effects of reactor flow conditions. *Biomass Convers. Biorefinery* **2015**, *6*, 173–180. [CrossRef]
49. Saini, R.; Saini, J.K.; Adsul, M.; Patel, A.K.; Mathur, A.; Tuli, D.; Singhania, R.R. Enhanced cellulase production by *Penicillium* oxalicum for bio-ethanol application. *Bioresour. Technol.* **2015**, *188*, 240–246. [CrossRef] [PubMed]
50. Bedade, D.K.; Singhal, R.S.; Turunen, O.; Deska, J.; Shamekh, S. Biochemical characterization of extracellular cellulase from tuber maculatum mycelium produced under submerged fermentation. *Appl. Biochem. Biotechnol.* **2016**, *181*, 772–783. [CrossRef]

51. Behera, S.S.; Ray, R.C. Solid state fermentation for production of microbial cellulases: Recent advances and improvement strategies. *Int. J. Biol. Macromol.* **2016**, *86*, 656–669. [CrossRef]
52. Singhania, R.R.; Sukumaran, R.K.; Patel, A.K.; Larroche, C.; Pandey, A. Advancement and comparative profiles in the production technologies using solid-state and submerged fermentation for microbial cellulases. *Enzym. Microb. Technol.* **2010**, *46*, 541–549. [CrossRef]
53. Leite, P.; Salgado, J.M.; Venâncio, A.; Domínguez, J.M.; Belo, I. Ultrasounds pretreatment of olive pomace to improve xylanase and cellulase production by solid-state fermentation. *Bioresour. Technol.* **2016**, *214*, 737–746. [CrossRef]
54. Lam, M.K.; Khoo, C.G.; Lee, K.T. Scale-up and commercialization of algal cultivation and biofuels production. In *Biofuels from Algae*; Elsevier: Amsterdam, The Netherlands, 2019; pp. 475–506.
55. Jouzani, G.S.; Taherzadeh, M.J. Advances in consolidated bioprocessing systems for bioethanol and butanol production from biomass: A comprehensive review. *Biofuel Res. J.* **2015**, *2*, 152–195. [CrossRef]
56. Olson, D.G.; E McBride, J.; Shaw, A.J.; Lynd, L.R. Recent progress in consolidated bioprocessing. *Curr. Opin. Biotechnol.* **2012**, *23*, 396–405. [CrossRef]
57. Zuroff, T.R.; Xiques, S.B.; Curtis, W.R. Consortia-mediated bioprocessing of cellulose to ethanol with a symbiotic Clostridium phytofermentans/yeast co-culture. *Biotechnol. Biofuels* **2013**, *6*, 59. [CrossRef] [PubMed]
58. Nagarajan, D.; Lee, D.-J.; Chang, J.-S. Recent insights into consolidated bioprocessing for lignocellulosic biohydrogen production. *Int. J. Hydrog. Energy* **2019**, *44*, 14362–14379. [CrossRef]
59. Liu, J.; Shi, P.; Ahmad, S.; Yin, C.; Liu, X.; Liu, Y.; Zhang, H.; Xu, Q.; Yan, H.; Li, Q.X. Co-culture of Bacillus coagulans and Candida utilis efficiently treats Lactobacillus fermentation wastewater. *AMB Express* **2019**, *9*, 15. [CrossRef] [PubMed]
60. Bernstein, H.C.; Paulson, S.D.; Carlson, R.P. Synthetic Escherichia coli consortia engineered for syntrophy demonstrate enhanced biomass productivity. *J. Biotechnol.* **2012**, *157*, 159–166. [CrossRef] [PubMed]
61. Fan, Z. Consolidated bioprocessing for ethanol production. In *Biorefineries*; Qureshi, N., Hodge, D.B., Vertès, A.A., Eds.; Elsevier: Amsterdam, The Netherlands, 2014; pp. 141–160.
62. Ábrego, U.; Chen, Z.; Wan, C. Consolidated bioprocessing systems for cellulosic biofuel production. In *Advances in Bioenergy*; Li, Y., Ge, X., Eds.; Elsevier: Amsterdam, The Netherlands, 2017; Volume 2, pp. 143–182.
63. Hasunuma, T.; Kondo, A. Consolidated bioprocessing and simultaneous saccharification and fermentation of lignocellulose to ethanol with thermotolerant yeast strains. *Process. Biochem.* **2012**, *47*, 1287–1294. [CrossRef]
64. Chinn, M.S.; Mbaneme, V. Consolidated bioprocessing for biofuel production: recent advances. *Energy Emiss. Control Technol.* **2015**, *3*, 23–24. [CrossRef]
65. Devarapalli, M.; Atiyeh, H.K. A review of conversion processes for bioethanol production with a focus on syngas fermentation. *Biofuel Res. J.* **2015**, *2*, 268–280. [CrossRef]
66. Olguin-Maciel, E.; Larqué-Saavedra, A.; Lappe-Oliveras, P.E.; Barahona-Pérez, L.F.; Alzate-Gaviria, L.; Chablé-Villacis, R.; Maldonado, J.A.D.; Pacheco-Catalán, D.; Ruiz, H.A.; Tapia-Tussell, R. Consolidated bioprocess for bioethanol production from raw flour of brosimum alicastrum seeds using the native strain of trametes hirsuta Bm-2. *Microorganisms* **2019**, *7*, 483. [CrossRef]
67. Raftery, J.P.; Karim, M.N. Economic viability of consolidated bioprocessing utilizing multiple biomass substrates for commercial-scale cellulosic bioethanol production. *Biomass Bioenergy* **2017**, *103*, 35–46. [CrossRef]
68. Saldarriaga-Hernandez, S.; Hernandez-Vargas, G.; Iqbal, H.M.; Barceló, D.; Parra-Saldívar, R. Bioremediation potential of *Sargassum* sp. biomass to tackle pollution in coastal ecosystems: Circular economy approach. *Sci. Total Environ.* **2020**, *715*, 136978. [CrossRef]
69. Parisutham, V.; Kim, T.H.; Lee, S.K. Feasibilities of consolidated bioprocessing microbes: From pretreatment to biofuel production. *Bioresour. Technol.* **2014**, *161*, 431–440. [CrossRef] [PubMed]
70. Vohra, M.; Manwar, J.; Manmode, R.; Padgilwar, S.; Patil, S. Bioethanol production: Feedstock and current technologies. *J. Environ. Chem. Eng.* **2014**, *2*, 573–584. [CrossRef]
71. Liu, S.; Liu, Y.J.; Feng, Y.; Li, B.; Cui, Q. Construction of consolidated bio-saccharification biocatalyst and process optimization for highly efficient lignocellulose solubilization. *Biotechnol. Biofuels* **2019**, *12*, 35. [CrossRef] [PubMed]

72. Farinas, C.S.; Vitcosque, G.L.; Fonseca, R.F.; Neto, V.B.; Couri, S. Modeling the effects of solid state fermentation operating conditions on endoglucanase production using an instrumented bioreactor. *Ind. Crops Prod.* **2011**, *34*, 1186–1192. [CrossRef]
73. Lynd, L.R.; Van Zyl, W.H.; E McBride, J.; Laser, M. Consolidated bioprocessing of cellulosic biomass: An update. *Curr. Opin. Biotechnol.* **2005**, *16*, 577–583. [CrossRef]
74. Okamoto, K.; Uchii, A.; Kanawaku, R.; Yanase, H. Bioconversion of xylose, hexoses and biomass to ethanol by a new isolate of the white rot basidiomycete Trametes versicolor. *SpringerPlus* **2014**, *3*, 1–9. [CrossRef]
75. Althuri, A.; Mohan, S.V. Sequential and consolidated bioprocessing of biogenic municipal solid waste: A strategic pairing of thermophilic anaerobe and mesophilic microaerobe for ethanol production. *Bioresour. Technol.* **2020**, *308*, 123260. [CrossRef]
76. Amoah, J.; Ishizue, N.; Ishizaki, M.; Yasuda, M.; Takahashi, K.; Ninomiya, K.; Yamada, R.; Kondo, A.; Ogino, C. Development and evaluation of consolidated bioprocessing yeast for ethanol production from ionic liquid-pretreated bagasse. *Bioresour. Technol.* **2017**, *245*, 1413–1420. [CrossRef]
77. Althuri, A.; Gujjala, L.K.S.; Banerjee, R. Partially consolidated bioprocessing of mixed lignocellulosic feedstocks for ethanol production. *Bioresour. Technol.* **2017**, *245*, 530–539. [CrossRef]
78. Cunha, J.T.; Romaní, A.; Inokuma, K.; Johansson, B.; Hasunuma, T.; Kondo, A.; Domingues, L. Consolidated bioprocessing of corn cob-derived hemicellulose: engineered industrial *Saccharomyces cerevisiae* as efficient whole cell biocatalysts. *Biotechnol. Biofuels* **2020**, *13*, 1–15. [CrossRef]
79. Singh, N.; Mathur, A.S.; Tuli, D.K.; Gupta, R.P.; Barrow, C.J.; Puri, M. Cellulosic ethanol production via consolidated bioprocessing by a novel thermophilic anaerobic bacterium isolated from a Himalayan hot spring. *Biotechnol. Biofuels* **2017**, *10*, 1–18. [CrossRef] [PubMed]
80. Singh, N.; Mathur, A.S.; Gupta, R.P.; Barrow, C.J.; Tuli, D.; Puri, M. Enhanced cellulosic ethanol production via consolidated bioprocessing by Clostridium thermocellum ATCC 31924☆. *Bioresour. Technol.* **2018**, *250*, 860–867. [CrossRef] [PubMed]
81. Vaid, S.; Nargotra, P.; Bajaj, B.K. Consolidated bioprocessing for biofuel-ethanol production from pine needle biomass. *Environ. Prog. Sustain. Energy* **2017**, *37*, 546–552. [CrossRef]
82. Lu, J.; Lv, Y.; Jiang, Y.; Wu, M.; Xu, B.; Zhang, W.; Zhou, J.; Dong, W.; Xin, F.; Jiang, M. Consolidated bioprocessing of hemicellulose enriched lignocellulose to succinic acid through a microbial co-cultivation system. *Authorea* **2019**, *8*, 9035–9045.
83. Froese, A.G.; Nguyen, T.N.; Ayele, B.T.; Sparling, R. Digestibility of wheat and cattail biomass using a co-culture of thermophilic anaerobes for consolidated bioprocessing. *BioEnergy Res.* **2020**, *13*, 325–333. [CrossRef]
84. Morales-Martínez, T.K.; Medina-Morales, M.A.; Ortíz-Cruz, A.L.; La Garza, J.A.R.-D.; Moreno-Dávila, I.; López-Badillo, C.M.; Rios-González, L.J. Consolidated bioprocessing of hydrogen production from agave biomass by *Clostridium acetobutylicum* and bovine ruminal fluid. *Int. J. Hydrogen Energy* **2020**, *45*, 13707–13716. [CrossRef]
85. Ruiz, H.A.; Conrad, M.; Sun, S.L.; Sanchez, A.; Rocha, G.J.; Romaní, A.; Castro, E.; Torres, A.; Rodríguez-Jasso, R.M.; Andrade, L.P.; et al. Engineering aspects of hydrothermal pretreatment: From batch to continuous operation, scale-up and pilot reactor under biorefinery concept. *Bioresour. Technol.* **2020**, *299*, 122685. [CrossRef]
86. Ruiz, H.A.; Rodríguez-Jasso, R.M.; Fernandes, B.D.; Vicente, A.A.; Teixeira, J.A. Hydrothermal processing, as an alternative for upgrading agriculture residues and marine biomass according to the biorefinery concept: A review. *Renew. Sustain. Energy Rev.* **2013**, *21*, 35–51. [CrossRef]
87. Ruiz, H.A.; Thomsen, M.H.; Trajano, H.L. *Hydrothermal Processing in Biorefineries*, 1st ed.; Springer: Cham, Switzerland, 2017.
88. Yao, K.; Wu, Q.; An, R.; Meng, W.; Ding, M.; Li, B.-Z.; Yuan, Y. Hydrothermal pretreatment for deconstruction of plant cell wall: Part I. Effect on lignin-carbohydrate complex. *AIChE J.* **2018**, *64*, 1938–1953. [CrossRef]
89. Canilha, L.; Chandel, A.K.; Milessi, T.S.D.S.; Antunes, F.A.F.; Freitas, W.L.D.C.; Felipe, M.D.G.A.; Da Silva, S.S. Bioconversion of sugarcane biomass into ethanol: An overview about composition, pretreatment methods, detoxification of hydrolysates, enzymatic saccharification, and ethanol fermentation. *J. Biomed. Biotechnol.* **2012**, *2012*, 1–15. [CrossRef]

90. Dahnum, D.; Tasum, S.O.; Triwahyuni, E.; Nurdin, M.; Abimanyu, H. Comparison of SHF and SSF processes using enzyme and dry yeast for optimization of bioethanol production from empty fruit bunch. *Energy Procedia* **2015**, *68*, 107–116. [CrossRef]
91. Kurtzman, C.; Fell, J.W.; Boekhout, T. *The Yeasts. A Taxonomic Study*, 5th ed.; Elsevier: Amsterdam, The Netherlands, 2011.
92. Bhatia, L.; Johri, S. Optimization of simultaneous saccharification and fermentation parameters for sustainable ethanol production from sugarcane bagasse by pachysolen tannophilus MTCC 1077. *Sugar Tech.* **2015**, *18*, 457–467. [CrossRef]
93. Koppram, R.; Nielsen, F.; Albers, E.; Lambert, A.; Wännström, S.; Welin, L.; Zacchi, G.; Olsson, L. Simultaneous saccharification and co-fermentation for bioethanol production using corncobs at lab, PDU and demo scales. *Biotechnol. Biofuels* **2013**, *6*, 2. [CrossRef] [PubMed]

Publisher's Note: MDPI stays neutral with regard to jurisdictional claims in published maps and institutional affiliations.

© 2020 by the authors. Licensee MDPI, Basel, Switzerland. This article is an open access article distributed under the terms and conditions of the Creative Commons Attribution (CC BY) license (http://creativecommons.org/licenses/by/4.0/).

MDPI
St. Alban-Anlage 66
4052 Basel
Switzerland
Tel. +41 61 683 77 34
Fax +41 61 302 89 18
www.mdpi.com

Agronomy Editorial Office
E-mail: agronomy@mdpi.com
www.mdpi.com/journal/agronomy

www.ingramcontent.com/pod-product-compliance
Lightning Source LLC
Chambersburg PA
CBHW061848210926
51850CB00006B/1009

9783036514109